Lecture Notes in Physics

For information about Vols. 1–131, please contact your bookseller or Springer-Verlag.

Vol. 132: Systems Far from Equilibrium. Proceedings, 1980. Edited by L. Garrido. XV, 403 pages. 1980.

Vol. 133: Narrow Gap Semiconductors Physics and Applications. Proceedings, 1979. Edited by W. Zawadzki. X, 572 pages. 1980.

Vol. 134: γγ Collisions. Proceedings, 1980. Edited by G. Cochard and P. Kessler. XIII, 400 pages. 1980.

Vol. 135: Group Theoretical Methods in Physics. Proceedings, 1980. Edited by K. B. Wolf. XXVI, 629 pages. 1980.

Vol. 136: The Role of Coherent Structures in Modelling Turbulence and Mixing. Proceedings 1980. Edited by J. Jimenez. XIII, 393 pages. 1981.

Vol. 137: From Collective States to Quarks in Nuclei. Edited by H. Arenhövel and A. M. Saruis. VII, 414 pages. 1981.

Vol. 138: The Many-Body Problem. Proceedings 1980. Edited by R. Guardiola and J. Ros. V, 374 pages. 1981.

Vol. 139: H. D. Doebner, Differential Geometric Methods in Mathematical Physics. Proceedings 1981. VII, 329 pages. 1981.

Vol. 140: P. Kramer, M. Saraceno, Geometry of the Time-Dependent Variational Principle in Quantum Mechanics. IV, 98 pages. 1981.

Vol. 141: Seventh International Conference on Numerical Methods in Fluid Dynamics. Proceedings. Edited by W. C. Reynolds and R. W. MacCormack. VIII, 485 pages. 1981.

Vol. 142: Recent Progress in Many-Body Theories. Proceedings. Edited by J. G. Zabolitzky, M. de Llano, M. Fortes and J. W. Clark. VIII, 479 pages. 1981.

Vol. 143: Present Status and Aims of Quantum Electrodynamics. Proceedings, 1980. Edited by G. Gräff, E. Klempt and G. Werth. VI, 302 pages. 1981.

Vol. 144: Topics in Nuclear Physics I. A Comprehensive Review of Recent Developments. Edited by T.T.S. Kuo and S.S.M. Wong. XX, 567 pages. 1981.

Vol. 145: Topics in Nuclear Physics II. A Comprehensive Review of Recent Developments. Proceedings 1980/81. Edited by T. T. S. Kuo and S. S. M. Wong. VIII, 571-1.082 pages. 1981.

Vol. 146: B. J. West, On the Simpler Aspects of Nonlinear Fluctuating. Deep Gravity Waves. VI, 341 pages. 1981.

Vol. 147: J. Messer, Temperature Dependent Thomas-Fermi Theory. IX, 131 pages. 1981.

Vol. 148: Advances in Fluid Mechanics. Proceedings, 1980. Edited by E. Krause. VII, 361 pages. 1981.

Vol. 149: Disordered Systems and Localization. Proceedings, 1981. Edited by C. Castellani, C. Castro, and L. Peliti. XII, 308 pages. 1981.

Vol. 150: N. Straumann, Allgemeine Relativitätstheorie und relativistische Astrophysik. VII, 418 Seiten. 1981.

Vol. 151: Integrable Quantum Field Theory. Proceedings, 1981. Edited by J. Hietarinta and C. Montonen. V, 251 pages. 1982.

Vol. 152: Physics of Narrow Gap Semiconductors. Proceedings, 1981. Edited by E. Gornik, H. Heinrich and L. Palmetshofer. XIII, 485 pages. 1982.

Vol. 153: Mathematical Problems in Theoretical Physics. Proceedings, 1981. Edited by R. Schrader, R. Seiler, and D.A. Uhlenbrock. XII, 429 pages. 1982.

Vol. 154: Macroscopic Properties of Disordered Media. Proceedings, 1981. Edited by R. Burridge, S. Childress, and G. Papanicolaou. VII, 307 pages. 1982.

Vol. 155: Quantum Optics. Proceedings, 1981. Edited by C.A. Engelbrecht. VIII, 329 pages. 1982.

Vol. 156: Resonances in Heavy Ion Reactions. Proceedings, 1981. Edited by K.A. Eberhard. XII, 448 pages. 1982.

Vol. 157: P. Niyogi, Integral Equation Method in Transonic Flow. XI, 189 pages. 1982.

Vol. 158: Dynamics of Nuclear Fission and Related Collective Phenomena. Proceedings, 1981. Edited by P. David, T. Mayer-Kuckuk, and A. van der Woude. X, 462 pages. 1982.

Vol. 159: E. Seiler, Gauge Theories as a Problem of Constructive Quantum Field Theory and Statistical Mechanics. V, 192 pages. 1982.

Vol. 160: Unified Theories of Elementary Particles. Critical Assessment and Prospects. Proceedings, 1981. Edited by P. Breitenlohner and H.P. Dürr. VI, 217 pages. 1982.

Vol. 161: Interacting Bosons in Nuclei. Proceedings, 1981. Edited by J.S. Dehesa, J.M.G. Gomez, and J. Ros. V, 209 pages. 1982.

Vol. 162: Relativistic Action at a Distance: Classical and Quantum Aspects. Proceedings, 1981. Edited by J. Llosa. X, 263 pages. 1982.

Vol. 163: J. S. Darrozes, C. Francois, Mécanique des Fluides Incompressibles. XIX, 459 pages. 1982.

Vol. 164: Stability of Thermodynamic Systems. Proceedings, 1981. Edited by J. Casas-Vázquez and G. Lebon. VII, 321 pages. 1982.

Vol. 165: N. Mukunda, H. van Dam, L.C. Biedenharn, Relativistic Models of Extended Hadrons Obeying a Mass-Spin Trajectory Constraint. Edited by A. Böhm and J.D. Dollard. VI, 163 pages. 1982.

Vol. 166: Computer Simulation of Solids. Edited by C.R.A. Catlow and W.C. Mackrodt. XII, 320 pages. 1982.

Vol. 167: G. Fieck, Symmetry of Polycentric Systems. VI, 137 pages, 1982.

Vol. 168: Heavy-Ion Collisions. Proceedings, 1982. Edited by G. Madurga and M. Lozano. VI, 429 pages. 1982.

Vol. 169: K. Sundermeyer, Constrained Dynamics. IV, 318 pages. 1982.

Vol. 170: Eighth International Conference on Numerical Methods in Fluid Dynamics. Proceedings, 1982. Edited by E. Krause. X, 569 pages. 1982.

Vol. 171: Time-Dependent Hartree-Fock and Beyond. Proceedings, 1982. Edited by K. Goeke and P.-G. Reinhard. VIII, 426 pages. 1982.

Vol. 172: Ionic Liquids, Molten Salts and Polyelectrolytes. Proceedings, 1982. Edited by K.-H. Bennemann, F. Brouers, and D. Quitmann. VII, 253 pages. 1982.

Lecture Notes in Physics

211

Resonances – Models and Phenomena

Proceedings of a Workshop
Held at the Centre for Interdisciplinary Research
Bielefeld University, Bielefeld, Germany, April 9–14, 1984

Edited by S. Albeverio, L. S. Ferreira and L. Streit

Springer-Verlag
Berlin Heidelberg GmbH 1984

Editors

S. Albeverio
Mathematisches Institut, Ruhr-Universität
D-4630 Bochum 1

L. S. Ferreira
Departamento de Física, Universidade de Coimbra
P-30100 Coimbra, Portugal

L. Streit
Fakultät für Physik, Universität Bielefeld
D-4800 Bielefeld 1

ISBN 978-3-540-13880-8 ISBN 978-3-540-39077-0 (eBook)
DOI 10.1007/978-3-540-39077-0

Originally published by Springer-Verlag Berlin Heidelberg New York Tokyo in 1984

2153/3140-543210

PREFACE

"Resonances - Models and Phenomena" originated during the Research Year "Project No. 2 - Mathematics + Physics" at the Centre for Interdisciplinary Research (ZiF) of Bielefeld University. Project No. 2 has brought more than a hundred mathematicians and physicists from 25 nations to Bielefeld for extended periods of residence with the purpose of studying problems, concepts and methods in an interdisciplinary setting.

At an early stage of the project the nuclear physicists and mathematicians at ZiF started discussing a mathematically consistent inclusion of resonances in models of nuclear structure. The presence of a large and diverse group of scientists made it tempting to try and base these discussions on a broad overview of the subject of resonances which - as a phenomenon - occur in so many fields not only in physics, and for which so many methods and models have been and are still being developed.

Preparations for the Symposium have profited from the advice of Professors E. Balslev, A. Grossmann, J. Hinze, and H. Weidenmüller. Thanks also go to the staff of ZiF, in particular to Ms. M. Hoffmann, who expertly handled the organization of the meeting, and to Ms. L. Jegerlehner for the patience with which she prepared the manuscript for publication.

Finally, the support of ZiF and of the Volkswagen Foundation are gratefully acknowledged, support that was particularly valuable because it was granted quickly and unbureaucratically.

S. Albeverio
L.S. Ferreira
L. Streit

Bielefeld, July 1984

CONTENTS

RESONANCE : ITS DESCRIPTION, CRITERIA AND SIGNIFICANCE

R.H. Dalitz
Department of Theoretical Physics
Oxford University, England.

1. INTRODUCTION

In this lecture, I shall discuss resonance as it arises in quantum-mechanical problems. Of course, this necessarily also includes classical situations although these may look rather complicated in this framework. I shall take my formalism and examples from the field of elementary particle physics, since my own work has been concerned mostly with this field, but similar phenomena and the same principles also hold for the fields of nuclear physics and atomic physics.

The notion of an isolated resonance state is an ideal, of course, but the case of the Λ hyperon comes rather close to this ideal, I think. This is a strange particle which has mass $\approx$ 1116 MeV and can decay to a π^- meson (mass $\approx$ 140 Mev) and a proton (mass $\approx$ 938 MeV) with a lifetime τ_Λ close to 2.6×10^{-10} sec.[1] It should therefore appear as a resonance in π^-+P elastic scattering at c.m. kinetic energy $\approx$ 38 MeV. It undoubtedly does so, but it has not been detected in π^-P scattering, and probably never will be, since the width of this resonance is only $\hbar/\tau_\Lambda \approx 2.5\times10^{-6}$ eV. This example also illustrates well the fact that the forces giving rise to a resonance state may be quite different from those between the particles which excite the resonance. Today, we know that the Λ hyperon is composed of three quarks, one of them being a strange quark, whereas the pion and proton coming together to form this resonance in π^-P scattering are composed of non-strange quarks, to a sufficient approximation.

In Sec.2, we shall set up the S-matrix formalism appropriate for the description of scattering in quantum-mechanical systems and discuss the phenomenological description of resonances and the criteria adopted for their identification in various contexts. In Sec.3, we shall discuss, as case studies, a series of systems where resonances have been sought, in order to illustrate a range of possibilities, including situations where the occurrence of resonance states is unclear and controversial. In Sec.4, we conclude with a brief discussion as to whether the criteria which are generally adopted are really necessary, illustrated by situations where there are striking phenomena closely allied with resonance, but where resonance in the usual sense may not occur.

2. THE S-MATRIX AND ISOLATED RESONANCE POLES

We consider here a system with n channels, each involving two particles. We denote the c.m. momentum in channel α by p_α. The wavefunction for the system has n components, denoted by ψ_α , for α = 1,2.. n, or by the column matrix $[\psi]$. Scattering states are characterised by the asymptotic form of the wavefunction, which we may write in matrix form as

$$[r\psi] \xrightarrow[r\to\infty]{} [Ae^{-ikr}] + [S.Ae^{ikr}], \tag{2.1}$$

in terms of ingoing and outgoing waves at infinity, assuming tacitly that the interactions effective are of short range. A is an arbitrary column matrix, and the nxn matrix S is known as the S-matrix. This is a function of the total c.m. energy E, and the channel momenta k_α are given in terms of E by the relation

$$\sqrt{(m_\alpha^2 + k_\alpha^2)} + (M_\alpha^2 + k_\alpha^2) = E. \tag{2.2}$$

For a state with outgoing waves only, the appropriate boundary condition may be written

$$\{\frac{d}{dr}(r\psi_\alpha) - i\, k_\alpha(r\psi_\alpha)\}_{r\to\infty} = 0. \tag{2.3}$$

This may be regarded as an eigenvalue problem to determine the energy $E = E_R$ such that the second term of (2.1) dominates infinitely in the asymptotic form (2.1). To achieve this, S(E) must be singular at E_R, the simplest and most natural possibility being the occurrence of an isolated pole in S(E) at $E = E_R$.

In general, the outflow of probability in channel α is given by the expression

$$\underline{J}_\alpha = -\frac{i}{2m_\alpha}(\psi_\alpha^* \underline{\nabla}\psi_\alpha - (\underline{\nabla}\psi_\alpha)^*\psi_\alpha) \tag{2.4}$$

For the eigenstate defined by (2.3), the radial component of $\underline{J}_\alpha$ takes the form

$$(r^2 J_\alpha)_r = -\frac{i}{2m_\alpha}\{(r\psi_\alpha^*)\frac{d}{dr}(r\psi_\alpha) - \frac{d}{dr}(r\psi_\alpha^*)(r\psi_\alpha)\}_{r\to\infty} \tag{2.5a}$$

$$= \{(k_{R\alpha}^* + k_{R\alpha})/2m_\alpha\}|(r\psi_\alpha)|^2_{r\to\infty} \tag{2.5b}$$

If E_R is real, then each $k_{R\alpha}$ is either real or pure imaginary, according as E_R lies above or below the threshold energy E_α. When $k_{R\alpha}$ is real, it is necessarily positive and expression (2.5b) is positive; when $k_{R\alpha}$ is pure imaginary, expression (2.5b) vanishes. Unless E_R lies below all the thresholds E_α, there is necessarily a net outflow of probability, a situation incompatible with the conservation of probability. It follows that the eigenvalue E_R cannot generally be real, and we shall write it

$$E_R = E_r - \frac{i}{2}\Gamma. \tag{2.6}$$

The term Γ is necessarily positive, in order that the time development of the eigenstate,

$$\exp(-iE_Rt) = \exp(-iE_rt).\exp(-\Gamma t/2), \tag{2.7}$$

be damped, as is required by the outflow of probability. The only exception to this conclusion is that where E_R lies below all the thresholds E_α, which is the case of a bound state. In this case, E_R is real but the momenta k_α are all pure imaginary, with $k_\alpha = +i|k_\alpha|$; there is no outflow of probability and indeed the eigenstate ψ_R then has the asymptotic form $\exp(-|k_\alpha|r)$ in all channels and so vanishes exponentially at infinity.

The simplest singularity at E_R is an isolated pole, the S-matrix elements having the form

$$S_{\beta\alpha} \sim i \frac{C_{\beta\alpha}}{E_R - E} + B_{\beta\alpha}(E), \tag{2.8}$$

where $B_{\beta\alpha}(E_R)$ is finite, in the vicinity of E_R. If time-reversal invariance is assumed to hold for the strong interactions, the S-matrix is necessarily symmetric[2], and the same must therefore be true for the matrices C and B, thus,

$$\text{(a)} \quad C_{\beta\alpha} = C_{\alpha\beta}, \qquad \text{(b)} \quad B_{\beta\alpha} = B_{\alpha\beta}. \tag{2.9}$$

The eigenvalue problem (2.3) may now be restated in terms of the S-matrix. We now seek the energy E_R at which S(E) has a pole, and this is given by the solution of the equation

$$\det[S^{-1}(E)] = 0. \tag{2.10}$$

In general, the resonance eigenstate will be non-degenerate. This requires that the eigenvalue equation should have only a simple root at $E = E_R$. In terms of the form (2.8), the condition for this is that $C_{\beta\alpha}$ be factorizable, i.e. have the product form $C_{\beta\alpha} = c_\beta c_\alpha$. In the vicinity of E_R, the S-matrix then has the form

$$S_{\beta\alpha} = i \frac{c_\beta c_\alpha}{E_R - E} + B_{\beta\alpha} \tag{2.11}$$

and the resonance eigenstate is given by the column matrix $[c_\alpha]$ in channel space. We note explicitly that the coefficients c_α are complex numbers, in general. In matrix notation, we now have

$$S = i \frac{c\tilde{c}}{(E_R - E)} + B. \tag{2.12}$$

We term the matrix B in (2.12) the background amplitude. For a sufficiently narrow resonance, B will be slowly varying across the resonance, and we shall assume that this is generally the case, in our remarks here, where we shall neglect the energy-dependence of B. Since unitarity holds generally for S,

$$S^\dagger S = SS^\dagger = I, \tag{2.13}$$

the same must hold for B since the first term of (2.12) quickly becomes negligible away from the resonance energy; thus

$$B^{\dagger}B = BB^{\dagger} = I \qquad (2.14)$$

Since B is unitary, it can be diagonalized. Its eigenvalues B_j have modulus unity and can be written in the form $\exp(2i\beta_j)$, where the phase angles β_j are real. Since B is also symmetric, it follows that the matrix L which diagonalizes B is real and orthogonal, i.e. satisfies the equation $\tilde{L}L = I$. With L, the matrix B may be written in terms of the diagonal matrix of eigenphases β, in the following form:

$$B = L.e^{2i\beta}.\tilde{L} \qquad (2.15)$$

The matrix L may now be used to transform S to the form S', where

$$S' = e^{-i\beta}.\tilde{L}.S.L.e^{-i\beta}. \qquad (2.16)$$

Although this transformation is not unitary, the matrix S' obtained as a result of it is unitary and symmetric provided that S is unitary and symmetric. The same holds for the inverse transformation

$$S = L.e^{i\beta}.S'.e^{i\beta}.\tilde{L} \qquad (2.17)$$

Using the form (2.12) in (2.16) leads us to the transformed expression

$$S' = i\,\frac{(e^{-i\beta}\tilde{L}c)(\tilde{c}Le^{-i\beta})}{(E_R-E)} + I \qquad (2.8)$$

Writing χ for the column matrix $[e^{-i\beta}\tilde{L}c]$, we have

$$S' = I + i\frac{\chi\tilde{\chi}}{E_R-E}\,. \qquad (2.19)$$

With this form for S', it is immediately apparent that χ is an eigenvector of the matrix S', since we have

$$S'\chi = (I + i\,\frac{\chi\tilde{\chi}}{E_R-E})\chi \equiv \chi(1 + i\,\frac{\tilde{\chi}\chi}{E_R-E}), \qquad (2.20)$$

where the last bracket is a pure number, the corresponding eigenvalue of S'. Since S' is both unitary and symmetric, the eigenvector can necessarily be made real, by including appropriate phase factors in the definition of the channel base states. Further, since S' is unitary, this eigenvalue must have modulus unity. Rewriting it in the form

$$\left(\frac{E_R-E + i\tilde{\chi}\chi}{E_R-E}\right) = e^{2i\delta_R}, \qquad (2.21)$$

where δ_R is a real phase angle, we see that its numerator must be identical with (E_R^*-E). In turn, this requires that

$$\chi\tilde{\chi} = \Gamma, \qquad (2.22)$$

a real quantity which is positive according to the arguments following Eq. (2.7).

The phase angle δ_R defined by (2.21) is then given by

$$\tan\delta_R = \frac{\Gamma}{2(E_r - E)} \tag{2.23}$$

and is known as the resonance phase, taking the value $\pi/2$ at $E = E_r$, the resonance energy. If we denote the eigenvectors of S' by ϕ_k for k=1,...n, where $\phi_1 = \chi$, we note that, for k≠1, we have

$$S'\phi_k = (I - i\frac{\chi\tilde{\chi}}{E_R - E})\,\phi_k = \phi_k, \tag{2.24}$$

since $[\tilde{\chi}\phi_k] = 0$ in view of the orthogonality between any two eigenvectors of S'. In other words, all other eigenvalues of S' are unity and their corresponding eigenphases are zero.

We can now use (2.17) to return to the form (2.12) for the S-matrix, with

$$c = L.e^{i\beta}.\chi \tag{2.25}$$

The elements of c are given in channel space by

$$\begin{aligned} c_\alpha &= (\alpha|L.e^{i\beta}.\chi) \\ &= \Sigma_j L_{\alpha j} e^{i\beta_j} \chi_j, \end{aligned} \tag{2.26}$$

where j refers to the eigenstates of B and χ_i denotes the i-th component of χ in that space. This element is generally written in the form

$$c_\alpha = \Gamma_\alpha^{1/2}.e^{i\phi_\alpha}, \tag{2.27}$$

where $\Gamma_\alpha = |c_\alpha|^2$ is referred to as the <u>partial width</u> of the resonance for channel α. With (2.25) and (2.26), we have

$$\begin{aligned} \Sigma_\alpha\Gamma_\alpha &= \Sigma_\alpha|c_\alpha|^2 = (\chi^\dagger e^{-i\beta} L^\dagger L e^{+i\beta}\chi) \\ &= \tilde{\chi}\chi = \Gamma, \end{aligned} \tag{2.28}$$

since L is real and orthogonal and χ is real, an equation which justifies the name "partial width" used for the Γ_α, since (2.28) then states that the total width of the resonance is equal to the sum of the partial widths. The phase ϕ_α depends generally on both the background scattering, through L and β, and on the structure of the resonant state, through χ. With (2.27), the S-matrix element (2.11) takes the form

$$S_{\beta\alpha} = B_{\beta\alpha} + i\,\frac{\sqrt{\Gamma_\alpha}.\sqrt{\Gamma_\beta}}{(E_r - E - i\frac{\Gamma}{2})}\,.\,e^{i(\phi_\alpha+\phi_\beta)} \tag{2.29}$$

The unitarity relations (2.13) also add some further constraints for all S-matrix elements. In terms of indivudal S-matrix elements, the unitarity relations may be written

$$\Sigma_\beta (S^\dagger)_{\gamma\beta} S_{\beta\alpha} = \delta_{\gamma\alpha} \tag{2.30}$$

Since $(S^\dagger)_{\gamma\beta} = S^*_{\beta\gamma}$, the case $\gamma = \alpha$ gives us

$$\Sigma_\beta |S_{\beta\alpha}|^2 = 1 \tag{2.31}$$

for each α. In particular, it follows from (2.31) that

$$|S_{\beta\alpha}| \leq 1, \tag{2.32}$$

for all S-matrix elements.

The case of elastic scattering, where $\beta = \alpha$, is of particular interest. The scattering amplitude $T_{\alpha\alpha}$ is related with the S-matrix element $S_{\alpha\alpha}$ by the equation

$$T_{\alpha\alpha} = \frac{S_{\alpha\alpha}-1}{2i}, \tag{2.33}$$

and expressed in terms of the resonance phase δ_R, as follows

$$T_{\alpha\alpha} = \frac{B_{\alpha\alpha}-1}{2i} + \left(\frac{\Gamma_\alpha}{\Gamma}\right) e^{2i\phi_\alpha} e^{i\delta_R} \sin\delta_R. \tag{2.34}$$

It is convenient to write the first term as $\eta_{B\alpha}\sin\delta_{B\alpha}\exp(i\delta_{B\alpha})$. Since the matrix element $B_{\alpha\alpha}$ must separately satisfy the inequality (2.32), we have that

$$\begin{aligned} |B_{\alpha\alpha}|^2 &= |1 + 2i\eta_{B\alpha}\sin\delta_{B\alpha}\exp(i\delta_{B\alpha})|^2 \\ &= 1 - 4\eta_{B\alpha}(1-\eta_{B\alpha})\sin^2\delta_{B\alpha} \leq 1, \end{aligned} \tag{2.35}$$

from which we deduce that $0 \leq \eta_{B\alpha} \leq 1$. The parameter $\eta_{B\alpha}$ is a measure of the inelasticity of the background scattering in channel α, $\eta_{B\alpha} = 1$ being the case of purely elastic scattering. With the notation $\eta_{R\alpha} = (\Gamma_\alpha/\Gamma)$ for the elasticity of the resonance scattering in channel α, expression (2.34) takes the form

$$T_{\alpha\alpha} = \eta_{B\alpha}\sin\delta_{B\alpha} e^{i\delta_{B\alpha}} + e^{2i\phi_\alpha}\eta_{R\alpha}\sin\delta_R\, e^{i\delta_R}. \tag{2.36}$$

The behaviour of the elastic scattering amplitude $T_{\alpha\alpha}$ is illustrated by Fig. 1, which shows $T_{\alpha\alpha}$ on an Argand plot, depicted by the vector $\overrightarrow{OP}$ from O. With the relation (2.33)

$$T_{\alpha\alpha} = +\frac{i}{2} + \frac{S_{\alpha\alpha}}{2i}, \tag{2.37}$$

and the elastic unitarity condition $|S_{\alpha\alpha}| \leq 1$ from (2.32), it follows that the amplitude $T_{\alpha\alpha}$ is limited to points less than distance 1/2 from a centre located at $+ i/2$, i.e the points lying inside the large circle on Fig. 1, known as "the unitarity circle". Purely elastic scattering corresponds to points P lying on the boundary of this circle; with $|S_{\alpha\alpha}| = 1$, the equality (2.31) requires $|S_{\beta\alpha}| = 0$ for all $\beta\neq\alpha$.

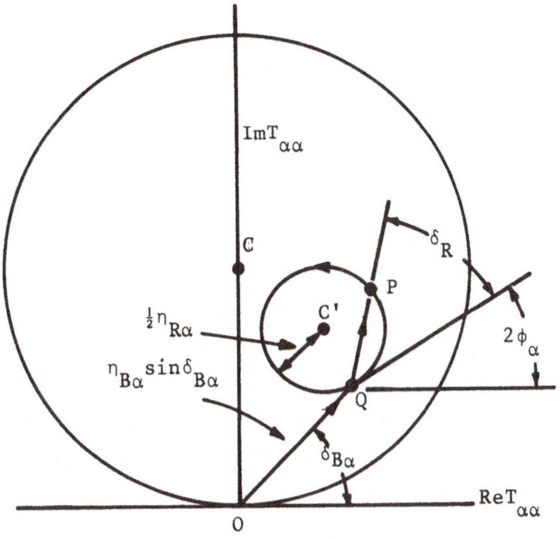

Fig. 1. Argand plot for the elastic scattering amplitude $T_{\alpha\alpha}$.

The elastic amplitude (2.36) consists of two terms:

(i) the background scattering amplitude $\eta_{B\alpha}\sin\delta_{B\alpha}\exp(i\delta_{B\alpha})$. On Fig. 1, this is depicted by the vector $\overrightarrow{OQ}$, making an angle $\delta_{B\alpha}$ with the real axis. The greatest length possible for OQ is $\sin\delta_{B\alpha}$, reached when Q is on the boundary of the unitarity circle. Hence the factor $\eta_{B\alpha}$ is limited to positive values $\eta_{B\alpha} \leq 1$, and its value gives a measure of the inelasticity of the background scattering. If $\eta_{B\alpha}$ were held fixed, and $\delta_{B\alpha}$ increased from 0 to π, the point Q would describe a circle of diameter $\eta_{B\alpha}$ inscribed within the unitarity circle and touching it at O.

(ii) the resonance scattering term $\eta_{R\alpha}\sin\delta_R\exp(i\delta_R)\exp(2i\phi_\alpha)$. On Fig. 1 above, this is depicted by the vector $\overrightarrow{QP}$. It has the same structure as the first term, except for the additional phase $2\phi_\alpha$, but the phase angle δ_R now increases from 0 to π as the energy traverses the resonant value E_r, from below to above. As we have just seen, this means that P will describe a circle of diameter $\eta_{R\alpha}$ passing through Q, but tangent to the line passing through Q at an angle of $2\phi_\alpha$ to the real axis. It will be apparent that, given the phase angles $\delta_{B\alpha}$ and ϕ_α, with any $\eta_{B\alpha} < 1$, there is an upper limit on the value possible for $\eta_{R\alpha}$, consistent with unitarity, i.e. an upper limit on the radius of the resonance circle such that it lies entirely within the unitarity circle.

We note explicitly that

(i) the point P corresponding to $T_{\alpha\alpha}$ traces out the resonance circle in the left-handed sense,

(ii) at the resonance energy E_r, P is diametrically opposite Q on the resonance circle,

(iii) $|T_{\alpha\alpha}|$ has its maximum value when P lies on the line through O and C', whereas Im $T_{\alpha\alpha}$ has its maximum value when PC' is parallel to the ordinate axis. We note without demonstration that $\sigma_{elastic}(\alpha) = 4\pi|T_{\alpha\alpha}|^2/k^2$, and $\sigma_{total}(\alpha) = 4\pi(\mathrm{Im}T_{\alpha\alpha})/k^2$, with our notation for T and neglecting spin factors.

(iv) The angular velocity with which P traverses the resonance circle is

$$\frac{d\delta_R}{dE} = \frac{\Gamma/2}{(E-E_r)^2+\Gamma^2/4} \tag{2.38}$$

This angular velocity rises as E increases towards E_r, has its maximum value at E_r and then slows down as E increases beyond E_r. This energy dependence must be regarded as an essential criterion for resonance. Other mechanisms can sometimes lead to an approximately circular path for the scattering or reaction amplitude on the Argand plot, at least for part of a circle, but they will not generally have the behaviour (2.38) for their rate of traversing this circular arc.

(v) in practice, the resonances observed in elementary particle collisions generally have a substantial width, typically of order 100 MeV. In such cases, the background amplitude may vary appreciably across the resonance, so distorting the resonance circle by movement of its centre C', and both the partial widths Γ_α and the full width may vary appreciably across the resonance, especially the inelasticity (Γ_α/Γ) when the resonance energy E_R lies not far above the threshold for channel α, in consequence of the energy-dependence of the centrifugal barrier penetration factor. We shall see examples of these behaviours below.

For resonance identification, it is always required that the data be well fitted by the Breit-Wigner form (2.29). If this is a good fit for the elastic channel, further checks on this identification are provided by requiring that the reaction amplitudes $T_{\beta\alpha}$ are also well fitted by this form (2.29) with the same values for E_r and Γ. Usually there is only one entrance channel for the resonance, available for experiment (albeit with several charge arrangements, like K^-p and $\bar{K}^0n$ for the $I_3 = 0$ component of a Σ^* resonance (this has isospin I = 1), or for different isospin components, like π^+p and π^-p, which excite the $I_3 = +3/2$ and $I_3 = -1/2$ components of a Δ resonance (this has I = 3/2)). When the resonant state is produced in the elastic channel, this is termed formation of the resonance. It sometimes happens that

(a) the resonant state has its largest partial width in an inelastic channel β. It may then be difficult to establish its existence from the study of elastic scattering, while it appears quite clearly in at least one reaction process $\alpha \to \beta$. However, the resonance circle for $T_{\beta\alpha}$ then has radius $\surd(\Gamma_\alpha\Gamma_\beta/4\Gamma^2)$ and detection of the resonance from a formation experiment may be difficult if (Γ_α/Γ) is too small.

(b) the resonant state lies below the threshold for the entrance channels available for experiment. A well established example is the $\Sigma^*(1382)$ state, the strange counterpart to the Δ(1232) state, both having spin-parity $3/2^+$ and belonging to the same SU(3) decuplet. $\Sigma^*(1382)$ lies below the K^-p threshold at 1432 MeV and decays dominantly to πΛ, both π and Λ being short-lived particles. In such cases, the resonance state R may become available for study as a result of production processes. A baryon resonance B^* may be produced directly in a meson-nucleon collision, for example, and then decay, as for

$$m + B_1 \to m' + B^*$$
$$\quad\quad\quad\quad\quad\quad \lfloor\!\!\longrightarrow m'' + B_2. \qquad (2.39)$$

It is a matter of experience that, for most resonances B^*, the state $m'B^*$ is generally well approximated as a two-body channel (especially if allowance is made for the finite width of B^*). An outstanding example of this kind is given by the well-known reaction sequences

$$K^- + p \begin{cases} \pi^- + \Sigma^{*+} \rightarrow \pi^- + (\Lambda+\pi^+) \\ \pi^+ + \Sigma^{*-} \rightarrow \pi^+ + (\Lambda+\pi^-) \end{cases} = \Lambda + \pi^+ + \pi^-, \qquad \begin{matrix} (2.40a) \\ (2.40b) \end{matrix}$$

where the same final state is reached by two parallel production and decay sequences. The interference between the amplitudes for these two sequences provides additional information and a check on this "intermediate resonance" interpretation for multi-particle production processes of the type (2.39). Another possibility of the same general kind is that where the resonance of interest is not directly produced but is the decay product from a directly produced resonance. A rather clear example of this kind is provided by the sequence

$$\left.\begin{matrix} K^- + p \rightarrow \pi^- + \Sigma^*(1660)^+ \\ \quad \hookrightarrow \pi^+ + \Lambda(1405) \\ \qquad\quad \hookrightarrow \Sigma^+ + \pi^- \end{matrix}\right\} = \pi^- + \pi^+ + \pi^- + \Sigma^+, \qquad (2.41)$$

which has given us our most detailed knowledge of $\Lambda(1405)$, another resonant state lying below the K^-p threshold. In this case, $\Sigma^*(1660)$ is a resonance which can be formed in K^-+p interactions, but it appears to have a rather small elasticity Γ_α/Γ, so that the low yield of $\Sigma^*(1660)$ in formation experiments has not allowed its use for the study of $\Lambda(1405)$ there. The $K^- \rightarrow \pi^-$ transition in (2.41) proceeds by transferring a virtual $\bar{K}^{*0}$ vector meson from the K^- to the target proton, and it appears that the formation of this $\Sigma^*(1660)$ state in the virtual $\bar{K}^{*0}p$ interactions occurring for $p_{K\ lab} = 4.2$ GeV/c is much stronger than is the case for K^-p interactions at 1660 MeV. c.m. energy. Spin-parity assignments are generally possible for baryonic resonances observed in production processes, from a study of the angular and polarisation angular distributions for their break up. A check on the resonance interpretation for a given final state system is provided by the observation of a resonance with the same mass, width, isospin, spin and parity in a variety of different production processes.

Finally, in some cases where cross sections are small, or for other reasons, it has not been possible to measure all of the distributions needed for such partial-wave or spin-parity analyses. Only bumps are seen in mass distributions, and the hypothesis of a resonance interpretation gains plausibility after a bump with the same width and the same mean mass is observed in a number of different processes.

However, bumps, and even bumps for the same final subsystem, with the same mass value being observed in different production processes, can result from another mechanism. Because hadronic interactions are strong, so that reaction cross sections

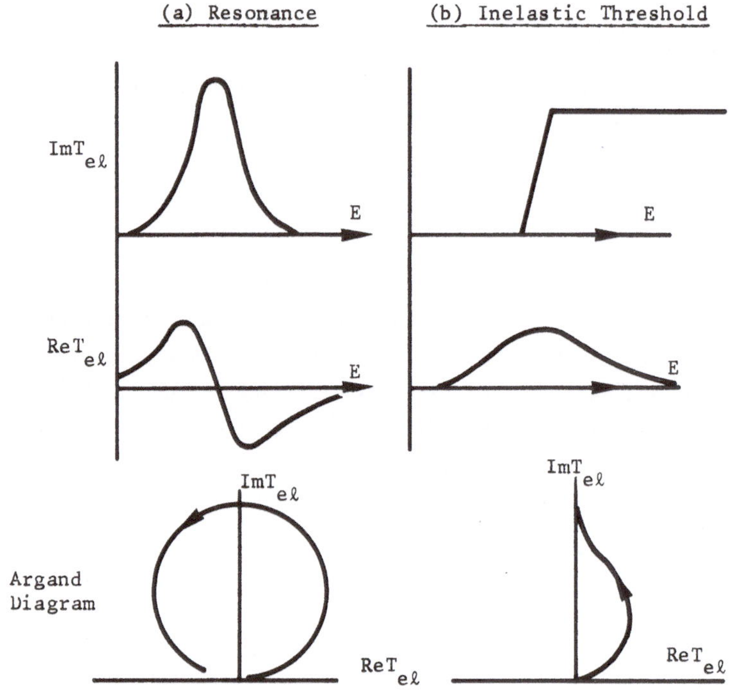

Fig. 2. Threshold inelasticity and resonance compared on an Argand plot.

are often comparable with the unitarity limit, the opening of new channels can often have a strong effect on the energy-dependence of the cross section. The simplest situation is that where the new threshold state is formed by the entrance channel (i.e. by the two initial particles). We illustrate this in Fig. 2. In the neighbourhood of the threshold energy E_t, the elastic amplitude $T_{\alpha\alpha}(E)$ will satisfy a dispersion relation of the form

$$\mathrm{Re}T_{\alpha\alpha}(E) = \text{Const.} + \frac{1}{\pi}\int_{E_t}^{\infty} \frac{\mathrm{Im}T_{\alpha\alpha}(E')dE'}{(E'-E)} \tag{2.42}$$

$\mathrm{Im}T_{\alpha\alpha}$ is, as always, directly linked with the total cross section $\sigma(\alpha)$ and hence with the inelasticity of the interaction. For the case of resonance, as shown on Fig. 2(a), $\mathrm{Im}T_{\alpha\alpha}$ rises and falls, as E crosses E_r. When there is such rapid energy dependence, (2.42) is given approximately by

$$\mathrm{Re}T_{\alpha\alpha}(E) \approx C' + \frac{d}{dE}(\mathrm{Im}T_{\alpha\alpha}(E)), \tag{2.43}$$

the curve of which is shown on Fig. 2(a). This is the behaviour we have already seen above, where the amplitude moves on a resonance circle in the Argand plane. On the other hand, for the case of a threshold, the cross section for the new channel can rise quite abruptly almost to its unitarity limit, and expression (2.43) then gives a function which rises and falls, as shown on Fig. 2(b). This behaviour corresponds to the Argand plot shown. Owing to the increasing inelasticity, the the amplitude moves in towards the inner region of the unitarity circle, and this part of its path may sometimes be well fitted by part of a circle. However, as the energy E increases, the circle is not completed and the path generally deviates towards the centre of the unitarity circle. The behaviour on the "part circle", the rate of movement along the curve with respect to E, does not follow the form (2.38) required for a resonance, and it is often possible to rule out a resonance interpretation, for a bump near a strongly excited threshold, on this ground.

3. SOME CASE STUDIES IN HADRONIC PHYSICS[1]

(a) πN and $\bar{K}N$ Interactions. The occurrence of resonances is common in meson-baryon systems which are coupled with three-quark configurations. For different strangeness values, they are linked by SU(3) symmetry, existing in unitary multiplets of dimension 1, 8 or 10. In the lowest configuration, the octets and decuplets differ through spin-flip. Higher baryonic resonances are the result of the excitation of rotations, of internal orbital motions, or of radial oscillations. All baryonic resonances established fit these descriptions, and all baryonic multiplets predicted by this three-quark model up to 2000 MeV have been observed. However, it is not our purpose to discuss here the origin of these baryonic resonances. These remarks are illustrated by Fig. 3, where the total cross sections observed are plotted vs. meson laboratory momentum, showing many overlapping peaks and bumps. Many further resonance peaks appear when the partial waves (with specific spin-parity values) are examined after analysis of the angular and polarization differential cross sections as function of energy. Since the total cross section is a sum of all partial wave cross sections, only the most striking resonances remain visible on Fig. 3.

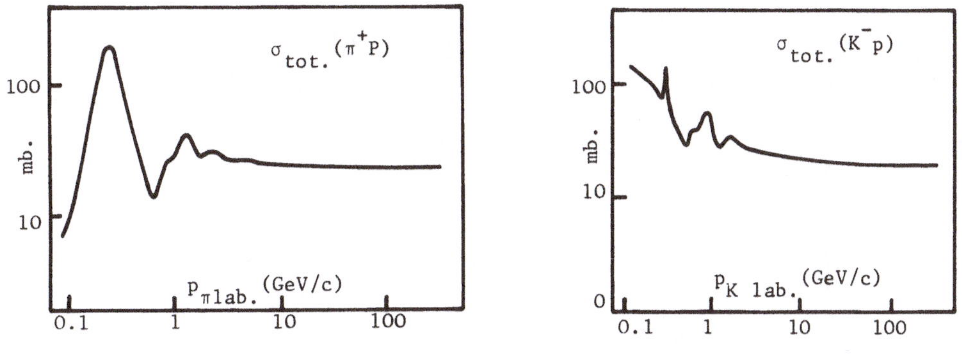

Fig. 3. Total cross sections for π^+p and K^-p collisions, as function of lab. momentum (log. scales).

Argand plots are given on Fig. 4 showing the elastic amplitudes for a few chosen πN partial waves. We now comment on each in turn:

NS11. The nucleon N has I = 1/2, so this resonance is seen in π^-p interactions, but not for π^+p. The scattering is elastic nearly up to the (ηN) threshold at 1490 MeV, above which the amplitude turns to the left, moving toward the centre of the unitarity circle. However, in this case, the behaviour observed does require the assumption of a resonance N(1535). For higher energy, another large circular motion is observed, due to the resonance N(1650).

ΔP33. The best known case, excited in π^+p collisions at 1232 MeV. This resonance is essentially elastic, being the least massive resonant state excited by pion-nucleon collisions. As the Argand path starts to enter the unitarity circle, it begins another circular path (Δ(1600)); before this circle is complete, there is the onset of a third resonance Δ(1920). We note that each of these three resonance circles has a smaller radius than those of lower energy, this being the result of the rise of the inelasticity with increasing mass value.

ΔS31. In this case, the scattering is repulsive at low energies. The first resonance, Δ(1620), is inelastic; the path moves inside the unitarity circle and traces

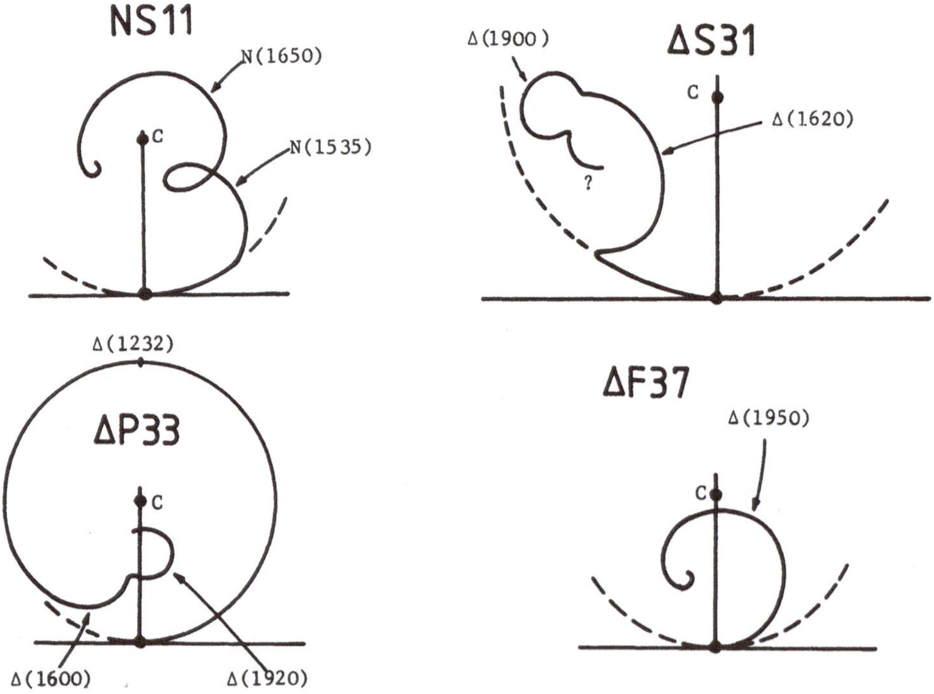

Fig. 4. Argand plots of some πN elastic scattering amplitudes as function of c.m. energy.

a left-hand path approximating to a circular path. Before its completion, the amplitude moves to follow a second circle and almost completes it. There is possibly an indication for the beginning of a third circle, but this is not established. We include this case in order to show that resonance can occur even when the long-range forces are repulsive, contrary to assumptions common in the early days of hadronic resonance physics.

ΔF37. With spin-parity $7/2^+$, this resonance is a rotational excitation of ΔP33(1232) and so lies high in mass. The amplitude becomes inelastic relatively late and then follows a spiralling path, the resonance circle being modified by gradually increasing inelasticity.

We note that, as the energy increases, more partial waves become effective, while the inelasticity continually increases, making the resonance circles smaller and smaller in radius. It is clear that the indentification of particular resonances becomes increasingly difficult, with increasing mass value, and that this situation settles in quite quickly.

The resonances excited by $\bar{K}N$ interactions belong to SU(3) multiplets of dimension 10, 8 and 1. The decuplet states are Σ^* states corresponding to the Δ^* states. The octet states are Λ^* and Σ^* states corresponding to the N^* states. The singlet states have no non-strange counterpart. Of these, the $\Lambda(1520)$ state has spin-parity $3/2^-$ and is coupled dominantly with the $\bar{K}N$ and $\pi\Sigma$ channels, being well above both thresholds. It is unusually narrow, with $\Gamma \approx 16$ MeV, since its mass is quite low relative to these thresholds, whereas its spin-parity requires both $\bar{K}N$ and $\pi\Sigma$ channels to be D-wave. It provides a text-book example of a Breit-Wigner resonance. The other singlet state is $\Lambda(1405)$, with spin-parity $1/2^-$. Its only open channel is S-wave $\pi\Sigma$ and its width is $\Gamma \approx 40$ MeV. We shall discuss it briefly below.

(b) K^+N and NN Interactions. The I=1 and I=0 KN states belong to SU(3) representations of dimension 27 and $\overline{10}$, respectively, which are not possible for three quarks. Any resonances in these states would therefore have quite different structure from the N^*, Δ^*, Σ^* and Λ^* states already discussed.

The K^+p total cross section is shown on Fig. 5 (a). The rise of cross section occurs just above the thresholds for Δ excitation,

$$K^+ + p \rightarrow \begin{cases} K^0 + \Delta^{++} , & (3.1a) \\ K^+ + \Delta^+ , & (3.1b) \end{cases}$$

and is now believed to be entirely due to this strong inelastic excitation, and not to any resonance states, as the result of a long series of measurements of the elastic angular and polarization differential cross sections and of these inelastic processes (3.1), and of their partial wave analysis.

The K^+n total cross sections shown on Fig. 5(b) were obtained from the measure-

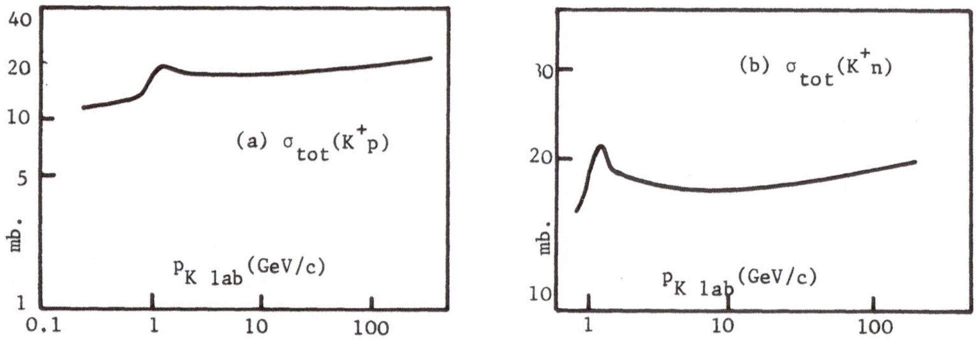

Fig. 5. The total cross sections for K^+ incident on (a) a target proton and (b) a target neutron.

ment of K^+-deuteron interactions, taken together with the data on K^+p interactions. The peak observed for $p_K \gtrsim 1$ GeV/c appeared to be a promising resonance state. Since there is no corresponding sharp peak in the K^+p cross sections, this peak can only be due to the I=0 KN system. However, the inelastic reactions KN → KΔ which would be analogous to the reactions (3.1) are actually forbidden by isospin conservation for the I=0. In the end, after a long series of investigations, it is generally agreed that the only possibility for resonance in these data is in the $P_{1/2}$ amplitude. This is shown on the Argand plot of Fig. 6. We see that the scattering becomes inelastic quite rapidly just above 1 GeV/c, but not strongly so, the amplitude following a circular curve with a left-hand sense up to about 1.2 GeV/c. Does this correspond to a resonant state? The main phenomenological argument against this conclusion is that the rate of movement along this path does not correspond to the Breit-Wigner requirement (2.38). For any who might wish to identify this state with some specific quark structure more complicated than three quarks, say to a

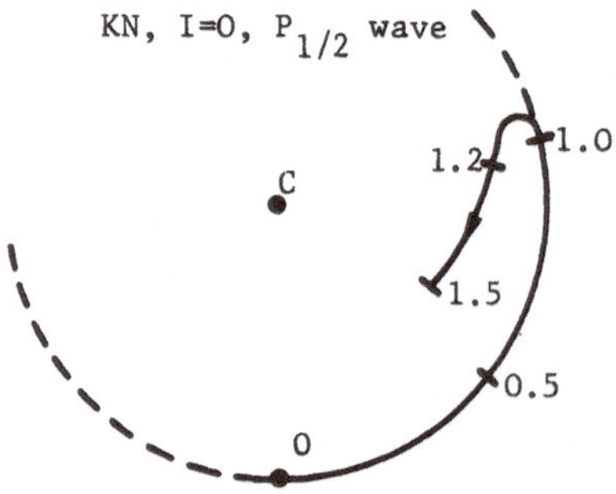

Fig. 6. Argand plot for the KN elastic scattering amplitude P01. The numbers give $p_{K\ lab}$ in GeV/c.

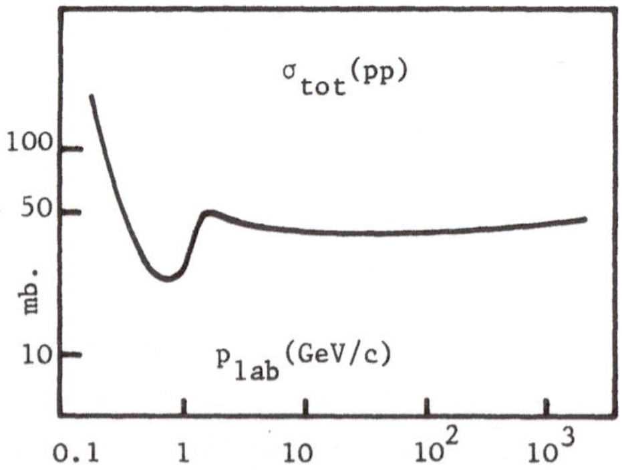

Fig. 7. Total proton-proton cross section as function of incident lab. momentum.

structure mode of four quarks and one antiquark, the absence of evidence for any other KN resonance states is a chilling circumstance. There is certainly a sharp peak here; what might its origin be? The most likely possibility appears to be that the I=0 KN interaction has a relatively long range (due to the exchange of a $\pi\pi$ pair between K and N) and is attractive to the $P_{1/2}$ state. Looking apart from the inelasticity, the phase shift simply rises to a maximum of about 45° and then falls back towards zero as is usual for a potential interaction. The P-wave centrifugal barrier sharpens the peak on the low-energy side; the inelasticity is only moderate but causes the amplitude to turn to the left as the $P_{1/2}$ cross section falls at higher energies. It is perhaps a coincidence that the onset of inelasticity should occur near the energy for which the (real) phase shift reaches its maximum value, but we conclude that there is no case for a P01 resonance on the basis of these data.

Nucleon-nucleon interactions are more complicated, in that both particles in the interaction have spin. This is one reason why partial wave analyses for proton-proton elastic scattering could first be made in full detail only rather recently, after experiments became possible using polarized beams incident on polarized targets. These experiments have shown up many striking spin-correlation phenomena and some of them have been interpreted as reflecting the existence of NN resonances. This is a large subject, and we shall confine our brief remarks here to several partial waves for the proton-proton system.

The pp total cross section is shown on Fig. 7. The fall of the cross section from high values at low energies is due both to the strength of the pp potential in the $^1S_0$ state - this is not far below the strength which would give rise to a pp bound state - and to its relatively long range character, since one-pion-exchange is possible. The cross section rises sharply at the onset of strong inelasticity due

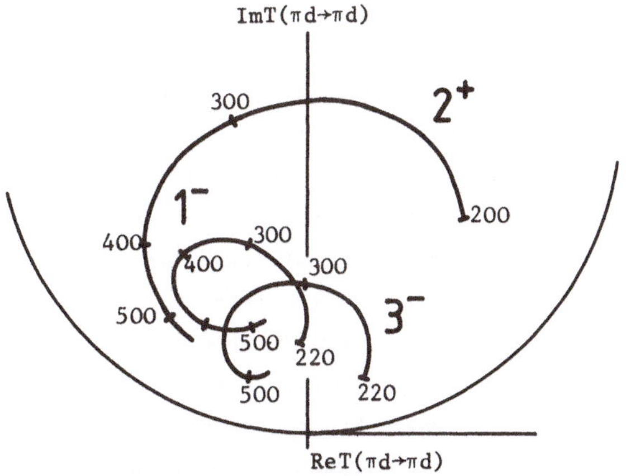

Fig. 8. The partial-wave amplitudes T($\pi d\to\pi d$) calculated for (2^+),(1^-) and (3^-) by Kaina et al. (see text). The numbers give p_π(MeV/c) for a stationary deuteron target.

to Δ excitation,

$$p + p \rightarrow \begin{cases} n + \Delta^{++}, & (3.2a) \\ p + \Delta^{+}, & (3.2b) \end{cases}$$

a situation quite analogous to that for the K^+p system discussed above. This inelasticity occurs first in the 2^+ state, the pp state which corresponds to the production of an S-wave Δ in the reactions (3.2). This is one of the two states for which Hoshizaki[3] and others who have made partial wave analyses of all the data on pp interactions have reported the existence of an NN resonance state, the other being the 3^- state, which corresponds to the NΔ P-wave state with the highest total spin. The question naturally raised, e.g. by Bugg , is to what extent these "resonant NN phenomena" may be accounted for as effects resulting from the strong inelastic processes (3.2). To illustrate this concern, we show on Fig. 8 some partial-wave amplitudes calculated by Kanai et al. for π^+-deuteron elastic scattering,[4] a channel strongly coupled to the channels pp and NΔ, e.g. through the reactions $\pi^+d \rightarrow pp$. These calculations take into account $\pi N \rightarrow \Delta$, i.e. Δ excitation and decay, with secondary scattering. The πd interaction is strongly inelastic, of course, in view of the ready break-up of the deuteron. What we wish to note here is the resonance-like character of these partial wave amplitudes, all corresponding to resonance at essentially the same energy. The dominant amplitude is that for spin-parity 2^+, the state in which the incident pion is P-wave and the NΔ intermediate system has L=0, the same orbital angular momentum as holds in the deuteron. The states with spin-parity 1^-, 2^- and 3^- are those where the NΔ intermediate system is P-wave; the last two require a D-wave πd interaction for their excitation, but this is readily achieved owing to the large diameter of the deuteron. If the empirical πd partial wave amplitudes had the form shown on Fig. 8, it would be difficult to consider them as evidence for dibaryon resonance states, since they are essentially the result of π^+ scattering off one nucleon in the deuteron, the other being essentially a bystander, complicated somewhat by the inclusion of some multiple scattering.

However, some elaborate K-matrix calculations on the NN system were carried out a few years ago by Edwards and Thomas[5] for the 2^+ state, and by Edwards[6] for the 3^- state, which are worth mentioning here. These calculations were carried out primarily for the case of two channels, pp and $n\Delta^{++}$. The energy-dependence of the K-matrix elements was constrained to be at most quadratic, and the T-matrix elements were calculated from them. The K-matrix parameters were then determined by fitting the resulting expressions to all the data available on the pp phase shifts for the spin-parity state considered; the T-matrix elements which resulted from this fit were then examined to determine whether or not they had a resonance pole. For the 2^+ case, several dozen solutions were obtained. For illustration, their T-matrix solutions 1 and 2 are displayed on Fig. 9(a) and compared with the phenomenological partial wave analyses reported in the literature. Their conclusion is that the fit

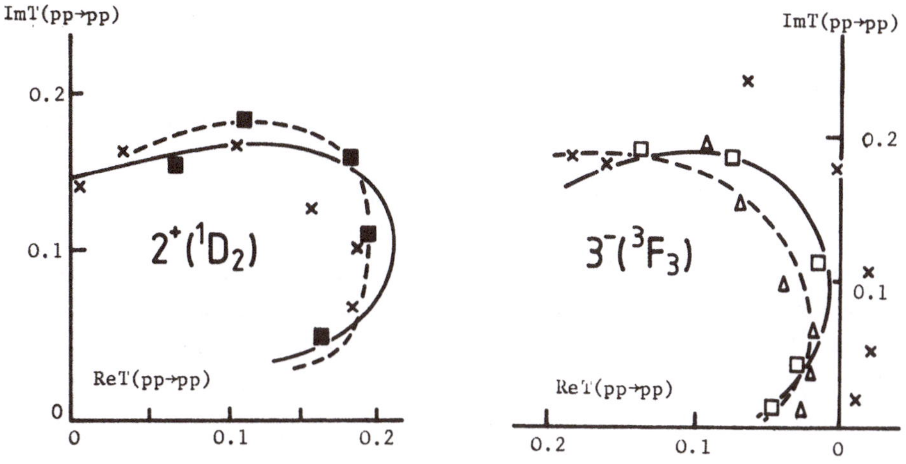

Fig. 9. Argand plot for T(pp→pp) for solutions 1 (solid line) and 2 (dashed line) obtained (a) by Edwards & Thomas for 2^+, and (b) by Edwards for 3^-, and compared with energy independent (triangles) and energy-dependent (squares) partial wave analyses by Arndt and the energy-independent partial wave analysis (crosses) by Hoshizaki.

to the data does require a pole in the T-matrix, i.e. a dibaryon resonance state, although its character depends on the details of the solution considered. For a large class of solutions, the dibaryon state appears to be primarily a feature of the NΔ system whose effects appear in the pp channel only through unitarity. For the 3^- case, quite similar conclusions were reached. The solutions 1 and 2 obtained are depicted on Fig. 9(b) and compared with corresponding partial wave analyses. Again, all the solutions found implied a resonance pole in the T-matrix. The physical origin of these T-matrix poles is not completely clear, and their relationship with the physical measurements is not simple. It is still possible that their occurrence may be only an artifact of the calculational procedure, but the degree of agreement between the T-matrix reached through these involved calculations and the T-matrix obtained by direct partial wave analysis is at least reassuring. This is a rather uncertain conclusion to be reaching after almost ten years of experimentation.

(c) Unstable Bound States. In many circumstances, especially where unitarity is important, it proves more convenient to use the K-matrix, defined by the boundary conditions for standing waves, rather than the T-matrix, which is defined in terms of outgoing waves. In place of (1.1), we then have

$$[r\psi] \xrightarrow[r \to \infty]{} [A.\frac{\sin kr}{k}] + [K.A.\cos kr] , \qquad (3.3)$$

to define the K-matrix elements. The K-matrix is similar is spirit to the reaction-matrix of the Wigner-Eisenbud formalism in nuclear reaction theory, the main difference being that the latter satisfy a boundary condition at $r = R_n$, the radius of the nucleus, rather than at infinity, as for the K-matrix. They both have the advantage that they do not have branch-cut singularities at two-particle thresholds. At

$k = 0$, both cos kr and (sin kr)/k are analytic functions of k^2, and hence of the energy E, and this observation holds also for K since the Schrodinger equations to determine ψ involve the momentum k only through E. Hence the K-matrix for the open channels holds also below the threshold and for a reasonable energy range below if the interactions are of suitably small range. However, it is usual to work with a K-matrix K(E) defined by (3.3) for the channels open at energy E, but to require the asymptotic form of the components of $(r\psi)$ for the closed channels to be damped, of the form $\exp(-|k|r)$. Let us denote the K-matrix for the n channels open below some new threshold ν by the notation $K^{(n)}$ and that for the (n+1) channels open above this threshold by $K^{(n+1)}$. If $K^{(n+1)}$ has the (n+1)x(n+1) matrix form

$$K^{(n+1)} = \begin{pmatrix} \gamma & \beta_\nu \\ \tilde{\beta}_\nu & \alpha_\nu \end{pmatrix} \tag{3.4}$$

above threshold, where γ is an nxn matrix, β_ν is 1xn and $\tilde{\beta}_\nu$ is the transpose of β_ν, it is a matter of simple algebra to deduce that, with the changed boundary condition, we have the nxn matrix form[7]

$$K^{(n)} = \gamma - |k_\nu|\tilde{\beta}_\nu\beta_\nu/(1 + |k_\nu|\alpha_\nu) \tag{3.5}$$

below threshold ν. We will note here that the scattering length $A_\nu = a_\nu + ib_\nu$ in the new channel, necessarily complex because of the transitions $\nu \to n$, is given in terms of (3.4) by the expression

$$A_\nu = \alpha_\nu + i\beta_\nu(1 - ik\gamma)^{-1}k\tilde{\beta}_\nu \tag{3.6}$$

where k denotes the nxn diagonal matrix of channel momenta. This is related with the complex phase shift δ_ν for elastic scattering in channel ν by the usual equation

$$k_\nu \cot\delta_\nu = 1/A_\nu . \tag{3.7}$$

The T-matrix for elastic scattering in channel ν, as we have defined it in Sec. 2 above, is then given by

$$T_{\nu\nu} = \sin\delta_\nu e^{i\delta_\nu} \tag{3.8a}$$

$$= k_\nu A_\nu/(1-ik_\nu A_\nu). \tag{3.8b}$$

Although A_ν is regular at the threshold and below it, $T_{\nu\nu}$ has a branch cut at the threshold due to the dependence of (3.8b) on the channel momentum k_ν.

We will now discuss briefly the application of these formulae to two situations of current physical interest.

<u>$\Lambda^*(1405)$.</u> At the K^-p threshold 27 MeV above this I=0 state, there are two channels open, $\bar{K}N$ and $\pi\Sigma$. Because K^- mesons are strongly produced with high energy proton accelerators and because they have a relatively long lifetime ($\sim 10^{-8}$ sec.), the

cross sections for K^-p scattering, charge-exchange and reaction processes at low energies have been measured in much detail. The S-wave interactions are very strong in this system and the scattering length (3.6) has a value about

$$A_0 \;=\; (-1.6 + i\,0.7)\text{fm}. \tag{3.9}$$

In fact, detailed analyses of the above-threshold data, with the effective range approximation that $K^{(2)}$ has energy dependence

$$(K^{(2)})^{-1} \;=\; (K_t^{(2)})^{-1} + \tfrac{1}{2}R_t k_\nu^2, \tag{3.10}$$

where the suffix t refers to threshold values, have given us best-fit values for the 2x2 matrices $K_t^{(2)}$ and R_t.[8]

The K-matrix thus determined can now be used below the $\bar{K}N$ threshold to predict $\pi\Sigma$ elastic scattering, using Eq. (3.5). In the present case, n=1; γ and β_ν are pure numbers, We note that, from Eq. (3.) the I=0 $\pi\Sigma$ phase shift is given by

$$\tan\delta_{\pi\Sigma} \;=\; k_{\pi\Sigma}(\gamma - \beta_\nu^2|k_\nu|/(1+\alpha_\nu|k_\nu|) \tag{3.11}$$

The threshold values for α_ν, β_ν and γ obtained from our best-fit $K_t^{(2)}$ matrix are -1.87 f., -0.95f. and -0.38f. Neglecting their energy dependence, we note that $\delta_{\pi\Sigma}$ passes through 90° at an energy corresponding to $|k_\nu| = -1/\alpha_\nu$ for the $\bar{K}N$ system, thus predicting an I=0 resonance at energy about $(m_K+M_N-1/(2m_{KN}\alpha_\nu^2)$ = 1415 MeV. If

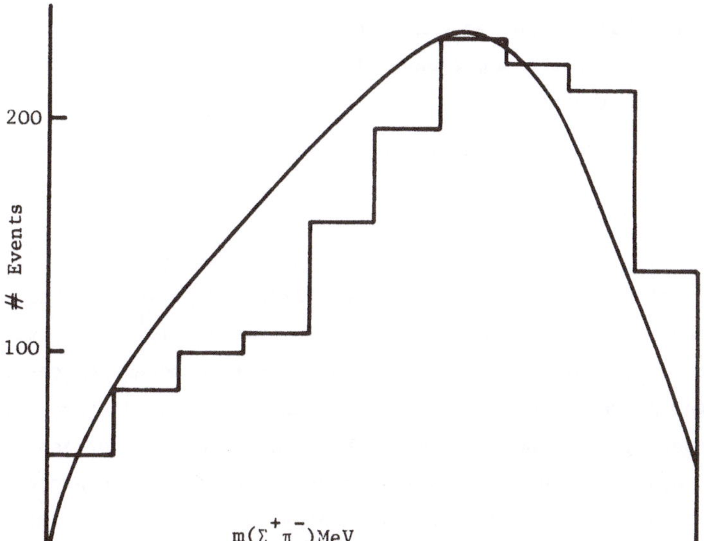

Fig. 10. Distribution of $(\Sigma^+\pi^-)$ c.m. energies in the reaction $K^-p \rightarrow \Sigma^-\pi^+\pi^-\pi^-$ at 4.2 GeV/c up to the K^-p threshold, is compared with prediction from Eq. (3.11) using the K-matrix parameters obtained from low energy K^-p interaction cross sections.

the effective range R_t is taken into account, this resonance energy moves to about 1395 MeV. This is precisely the mass region of the observed S-wave $\pi\Sigma$ $\Lambda^*(1405)$ resonance. Although we cannot carry out $\pi\Sigma$ scattering experiments directly, we can make use of Watson's approximation that, when a $\pi\Sigma$ S-wave final state is produced as a subsystem in a multiparticle final state, as in the reaction (2.41) for example, the dependence of the overall matrix element on the $\pi\Sigma$ c.m. energy $m(\pi\Sigma)$ is given by $\sim (\sin\delta_{\pi\Sigma}/k_{\pi\Sigma})$. Hence, if we plot the mass distribution for $\Sigma^+\pi^-$ pairs in the final state of reaction (2.41), this should be well-approximated by $\sin^2\delta_{\pi\Sigma}/k_{\pi\Sigma}$. Note that there is an additional factor $k_{\pi\Sigma}$ in the rate, which comes from the $\pi\Sigma$ phase space. On Fig. 10, we show the $m(\pi\Sigma)$ distribution predicted in this way, for the I=0 K-matrix derived from the analysis of K^-p scattering and reaction processes at and above the K^-p threshold, and compare it with preliminary data recently obtained by Hemingway[9] from a study of the reaction (2.41) at K^- momentum 4.2 GeV/c. This comparison has no free parameters. The prediction gives the general shape correctly, and it is clear that a good fit to both sets of data, those above and those below K^-p threshold, would obtain if the parameters were modified a little to bring the mean position of the $m(\pi\Sigma)$ distribution into accord with the data, a modification within the uncertainty with which these parameters have been determined.

We may note here that, although the $\Lambda^*(1405)$ resonance state does correspond to a simple pole in the T-matrix, a fact which we have not taken space to demonstrate above, the shape of the mass distribution is much distorted from a simple Breit-Wigner peak, owing to the finite width of the resonance and the consequent effects of the $\pi\Sigma$ and $\bar{K}N$ thresholds. The fact that the predicted distribution would be abnormal was clear already from the appearance of $|k_\nu|$ in Eq. (3.11) and this abnormality is clearly apparent in the data.

<u>A deuteron-like Hyperon-Nucleon state.</u> In the reaction

$$K^- + d \to \pi^- + p + \Lambda, \tag{3.12}$$

both in-flight and at-rest, there is observed a strong spike in the $m(\Lambda p)$ distribution approximately at the ΣN threshold, at about 2129 MeV. The data on this distribution for incident K^- momentum 700 MeV/c, due to Braun et al.[10], is plotted on Fig. 11. This spike is intimately related with the two-stage mechanism described by Fig. 12. The incident K^- meson interacts with one nucleon of the deuteron producing π^- meson and a Σ hyperon, the pion emerging with momentum roughly corresponding to this process. The Σ hyperon then reacts with the second nucleon, the process being $\Sigma N \to \Lambda N$ which releases about 80 MeV kinetic energy in the final Λp c.m. system.

The essential features of this process is given by the following simplified T-matrix

$$T(K^-d \to \Lambda p\pi^-) = T(\bar{K}N \to \pi\Sigma)\ F_d(\Delta q, k_\Sigma)\ T(\Sigma N \to \Lambda p) \tag{3.13}$$

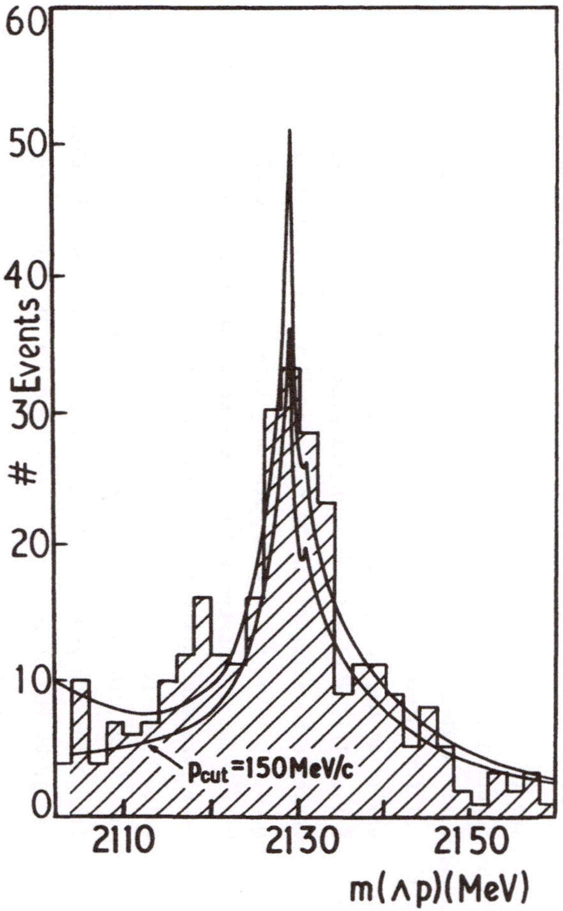

Fig. 11. Mass m(Λp) distribution for $K^-d\to\Lambda p\pi^-$ at 700 MeV/c, for forward pions ($\cos\theta_{K\pi lab}$ 0.9). Theoretical curves are for p_{cut} = 75 and 150 MeV/c for recoil proton.

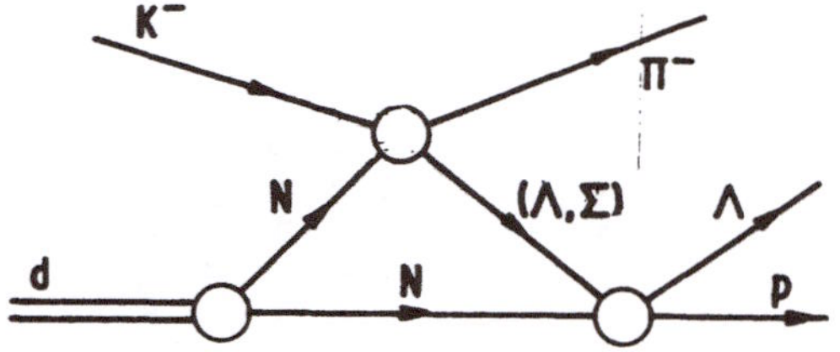

Fig. 12. Triangle graph shows two-stage mechanism for $K^-d\to\Lambda p\pi^-$ in threshold peak region.

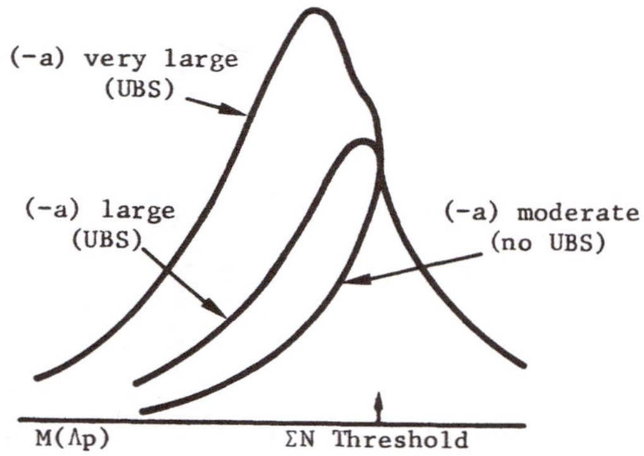

Fig. 13. Zero range model calculation for $K^-d\to\Lambda p\pi^-$ for interpretation of data of Fig. 11.

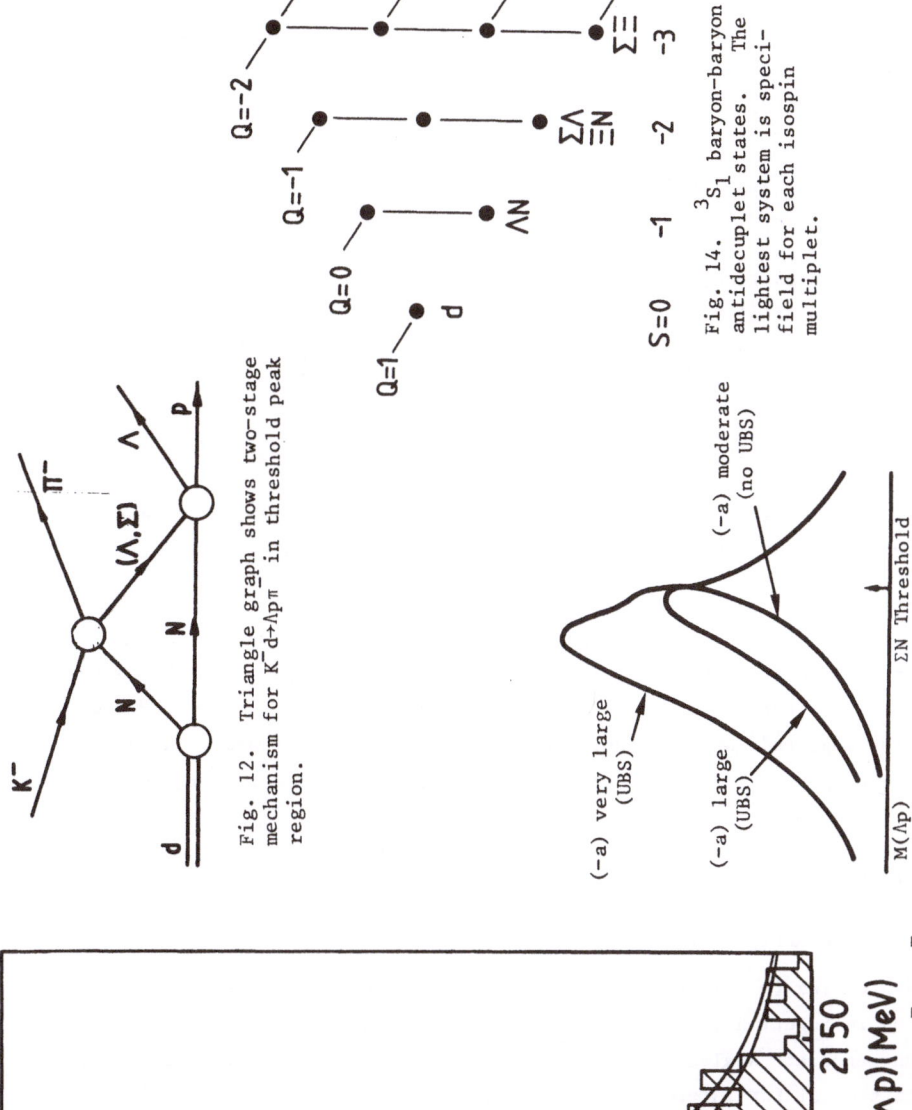

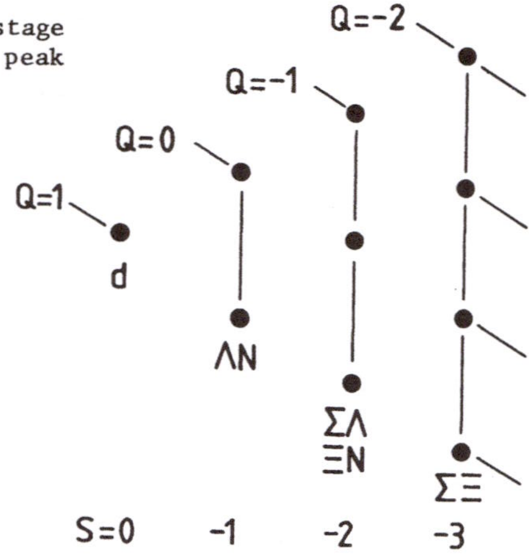

Fig. 14. $^3S_1$ baryon-baryon antidecuplet states. The lightest system is specified for each isospin multiplet.

The first factor is known empirically. The second factor depends on the $K^- \to \pi^-$ momentum transfer Δq, which is small in the data cited, and the c.m. momentum k_Σ in the intermediate ΣN system can be real or imaginary according as $m(\Lambda p)$ lies above or below the ΣN threshold. In the zero range approximation, this factor can be given explicitly as

$$F_d(0,k_\Sigma) = \int \frac{\exp(ik_\Sigma r)}{r} \psi_d(r)d^3\underline{r} \quad \propto \quad (\beta - ik_\Sigma)^{-1}(\gamma - ik_\Sigma)^{-1} \tag{3.14}$$

for a Hulthen wavefunction for the deuteron. Below the ΣN threshold, where $k_\Sigma = +i|k_\Sigma|$, the exponential factor depresses the integral increasingly as $m(\Lambda p)$ falls below $(m_\Sigma + m_N)$ and the rapid fall in the rate below the ΣN threshold is due mainly to this factor. The third factor has essentially the form

$$T(\Sigma N \to \Lambda p) = \beta^t_{\Sigma N}/(1 - ik_\Sigma A_\Sigma), \tag{3.15}$$

where A_Σ is the ΣN scattering length for $I = 1/2$, the isospin of the ΛN system. For $m(\Lambda p)$ below the ΣN threshold, this factor becomes

$$\beta^t_{\Sigma N}/\{(1 + |k_\Sigma|a_\Sigma) + i|k_\Sigma|b_\Sigma\}, \tag{3.16}$$

where $A_\Sigma = a_\Sigma + ib_\Sigma$. When $a_\Sigma < 0$, the real part of the denominator vanishes for $|k_\Sigma| = -1/a_\Sigma$, i.e. for $m(\Lambda p) = (m_\Sigma + m_N - 1/(2\mu_{\Sigma N}a_\Sigma^2))$ where $\mu_{\Sigma N}$ is the ΣN reduced mass, and the magnitude of this factor has its maximum value there. Above ΣN threshold, the rapid fall in rate arises mainly from the third factor, which contributes

$$(\beta^t_{\Sigma N})^2/\{(1 + k_\Sigma b_\Sigma)^2 + k_\Sigma^2 a_\Sigma^2\}, \tag{3.17}$$

since b_Σ is necessarily positive and is quite large, of order 1 fm.

The expression (3.13) provides a simple and illuminating model for the $m(\Lambda p)$ distribution in reaction (3.12), although it is not close enough to reality to be used for any accurate analysis of the data. We show on Fig. 13 the distributions given by this model for interesting parameter values. When a_Σ is negative and large in magnitude, there is an unstable bound state (UBS) not far below threshold. By this term we mean that if the transition K-matrix element $\beta_{\Sigma N}$ were placed zero, the K-matrix would predict a bound state in the new channel ν, here the ΣN channel. When $\beta_{\Sigma N}$ is re-introduced, this bound state becomes able to decay into the old channels n (here ΛN) with the release of energy, and we refer to it as an unstable bound state. It then appears as a resonance state in the old channels. We recall from above that this description held for the resonance $\Lambda^*(1405)$, as a UBS with respect to the $\bar{K}N$ channel. In the present case, the $m(\Lambda p)$ distribution will have a resonance peak not far from its mass value [note: the peak of the $m(\Lambda p)$ distribution would be shifted towards the ΣN threshold by the strong energy dependence of $|F_d(0,k_\Sigma)|^2$ in this region, already mentioned following Eq. (3.14)]. When a_Σ is smaller in magnitude, but still large enough to imply the existence of a UBS, the peak in $|T(\Sigma N \to \Lambda p)|^2$ due to the UBS may not be easily apparent, as is illustrated by the middle curve in Fig. 13; however the mean position of the peak observed in the m(Λp) distribution will then lie significantly below the ΣN threshold mass. When a_Σ has only a moderate magnitude, which does not imply the existence of a UBS, the peak will be an upward cusp and its mean position will lie rather close to the ΣN threshold. Although the situation is in practice complicated by the occurrence of two thresholds, Σ^+n and Σ^0p, it appears that the mean position of the peak observed is very close to the Σ^+n threshold mass and that the physical situation corresponds to the lower of the three possibilities illustrated on Fig. 13. The cusp itself is not seen directly, of course, owing to the finite resolution (full width at half maximum typically 5 MeV) of the experimental measurements. The present conclusion is there is no Λp resonance here, and no UBS. Yet this peak is really striking, while other cusps which have been seen in hadronic physics have been quite small and difficult to detect. What is the reason for this particular cusp to be so prominent?

The properties of the T-matrix over the ΣN threshold region have been calculated by a considerable number of authors using a variety of potential models for the hyperon-nucleon potentials, separable potentials, central potentials and spin-dependent potentials derived from various One-Boson-Exchange potentials. The data has also been analysed in a phenomenological way by fitting a K-matrix to all the data available on Σ^-p scattering, charge-exchange and reaction processes in the regime of low Σ^- lab. momentum and then examining the pole structure of the T-matrix obtained from this K-matrix. Some of these potentials, or fits to the data, imply the existence of an S-wave Λp resonance just below the Σ^+n threshold, but most do not. However all of them require or predict a pole in the T-matrix not far from the Σ^+n threshold.[11] The situation is illustrated on Fig. 14 where the pole locations are given from five investigations which have included a search for this pole, the sheet on which each is located being indicated by giving the path to reach it from the physical sheet. Most of these poles are not adjacent to the physical axis, not lying just inside the first unphysical sheet. The pole S is not even on the first unphysical sheet. The pole F lies reasonably close to the real axis but it is not close to the physical sheet, being separated from it by a branch cut. However, whichever sheet the pole lies on, it is relatively close to the Σ^+n branch point and that is what really matters if the rate of the reaction is to be large at the ΣN thresholds.

It seems very probable that this pole in the $^3S_1$ ΛN T-matrix has some relationship with the deuteron pole in the $^3S_1$ np T-matrix, through SU(3) symmetry. The simplest approach would be to assign the deuteron as the non-strange component of the baryon-baryon (BB) antidecuplet displayed on Fig. 15. However, the SU(3)-breaking effects in the BB system are very large (e.g. $\Delta m = m_\Sigma - m_\Lambda \approx 78$ MeV, and $m(\bar{K})/m(\pi) \approx 3.5$ for pseudoscalar meson exchange contributions to the BB potential) and such simple arguments are quite dubious, as we know from detailed calculations. The argument would be much stronger if it could be made at the quark level, but this requires the view that these BB systems are primarily "six quarks in a bag" objects.[12] However, in this situation, one is puzzled that the T-matrix pole corresponding to this state is not adjacent to the physical axis. Also the deuteron properties which we know so well do not seem to call for any appreciable "six quarks in a bag" component in its wavefunction.

An alternative approach is via potential models, using BB potentials calculated from One-Boson-Exchange (OBE). The deuteron properties are well accounted for by this approach, and explicit calculations by Brown, Downs and Iddings[13] (BDI) and by Nagels, Rijken & De Swart[14] have shown that they can give a good fit to all hyperon-nucleon interaction data. These fits were not predictive but were based on quite reasonable assumptions about SU(3)-breaking effects in the potentials, as well as in the Schrodinger equation used for the calculation of the T-matrix. It is not immediately obvious where the T-matrix pole will appear in such a calculation. It could appear anywhere between the ΛN and ΣN thresholds, or even below the ΛN threshold, corresponding to a stable bound state, or it might not appear at all, except far from this region and then only on a more remote sheet of the energy plane. However, the potential calculations mentioned just above are able to place the pole near the ΣN thresholds, for reasonable parameter choices, and this is what the data requires.

One test of these models will be the search for the corresponding $^3S_1$ state in the strangeness (-2) sector. With SU(3) symmetry, this will be the I=1 state of the antidecuplet shown on Fig. 15. A convenient production experiment would be to study the following proton-proton reactions

$$p + p \rightarrow \begin{cases} K^+ + K^+ + \Sigma^0 + \Lambda, & (3.18a) \\ K^+ + K^+ + \Sigma^+ + \Sigma^-, & (3.18b) \\ K^+ + K^+ + \Xi^- + p. & (3.18c) \end{cases}$$

If only the two K^+ mesons are measured, all three reactions (and others) will contribute to the "missing mass" distribution and this should lead to a peak at the s = -2

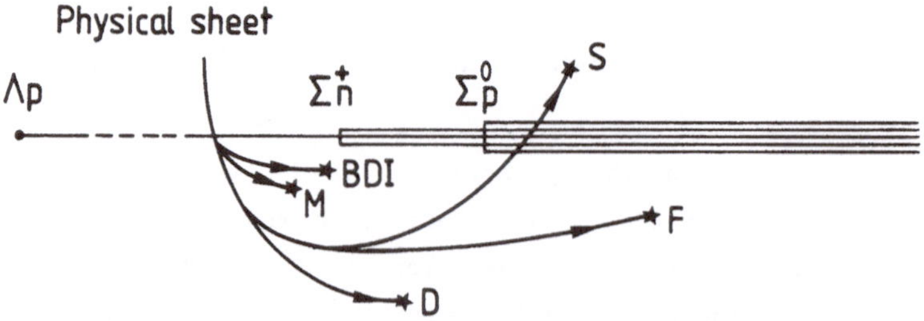

Fig. 15. Branch cuts for hyperon-nucleon T-matrix. Stars locate T-matrix poles found in five T-matrix investigations, and the lines identify the paths to each of these poles from the physical sheet.

resonant mass, if the anticipated resonance state exists. It would be advantageous to be able to distinguish between these final states, and ΞN is the I=1, s= -2 channel with the lowest threshold. The "six quarks in a bag" model would require that this state should exist, and would suggest, with the equal spacing rule appropriate for decuplets and antidecuplets, that the mass should be at about 2380 MeV, which is well above both ΞN and $\Sigma\Lambda$ thresholds. The OBE potential model has no firm prediction to make on this question, owing to the great sensitivity of any calculations with it to the hard-core radius to be adopted with this potential in a new sector.

4. CONCLUSION

We have seen that, conceptually (as for the Λ particle) or from experimental observation (as for $\Lambda^*(1520)$), isolated resonances do exist, narrow relative to the energy gap to the next resonance with the same quantum numbers, and even without appreciable background scattering. In such cases, and even with background scattering varying slowly with energy, the properties of the resonant state can be summarized concisely by the Breit-Wigner resonance amplitude and a small number of numerical parameters. This specification is accurate, concise and efficient.

More generally, we have found that the physical situation is more complicated than this. Resonances are broad and although the resonance may correspond to one resonance pole, more information is needed for an accurate description of it. The partial widths will vary with energy across the resonance, for example, as may also the background scattering. In particular, the resonance may straddle new thresholds and these often come in rather abruptly, causing strong distortion from the standard Breit-Wigner shape. Neighbouring resonances with the same quantum numbers overlap and there is interference between them.

Further, there appears to be some distinction between Breit-Wigner resonances and Unstable Bound State resonances, which is not yet clearly understood, the latter corresponding in large measure with stable bound states, like the deuteron. There are also effects observed which are as striking as any resonance in the bumps they produce in reaction cross sections for particular final states, but which are not to be represented as Breit-Wigner resonances at all, since the poles corresponding to them are not close to the physical axis, and no phase shift passes through 90°.

One has to conclude that our emphasis on resonance states is primarily a reflection of the fact that we are not able to make accurate calculations from our basic theories of particles and the forces between them. They represent an element of simplicity in the data on which we can focus our attention. However, whether or not some particular phenomena can be accounted for in terms of a resonance is not really the point of most fundamental importance, and arguments as to whether or not a pole important for the phenomena is or is not a resonance are not to the point.

If we had the power to calculate accurately the full consequences of our theory, we would not need to ask any such question; we would simply compare our predictive calculations with the data available for real physical energies and so decide whether or not our theory is capable of accounting for this data. Resonance is significant for us since strong, isolated resonances do govern many phenomena, and they provide a convenient standard pattern for comparisons with new data; but "resonance" would have much less significance for us if we had a more powerful ability to make quantitative predictions from our theories and to compare the data directly with these predictions.

REFERENCES

1. The identification and properties of all hadronic states mentioned in this lecture, stable or unstable, are given in the fullest detail in "Review of Particle Properties", by the Particle Data Group, which is scheduled for publication as vol. 138B of Physics Letters, to appear in April 1984.
2. G.C. Wick, Ann. Revs. Nucls. Sci 8 (1958) 1.
3. N. Hoshizaki, Prog. Theor. Phys. 60 (1978) 1796: ibid. 61 (1979) 129.
4. K. Kanai, A. Minaka, A. Nakamura and H. Sumiyoshi, "How to Search for Dibaryon Resonances Using Deuteron Targets", Proc. 2nd Meeting on Exotic Resonances (Eds. I. Endo, Y. Sumi, S. Wakaizumi and M. Yonezawa, Dept. of Physics, Hiroshima University, 1980) p.46.
5. B.J. Edwards and G.H. Thomas, Phys. Rev. D22 (1980) 2772.
6. B.J. Edwards, Phys. Rev. D23 (1981) 1978.
7. See, for example, R.H. Dalitz, and S.F. Tuan, Ann. Phys. (N.Y.) 3 (1960) 307.
8. A.D. Martin, Nucl. Phys. B179 (1981) 33.
9. R. Hemingway, private communication (1984).
10. O. Braun et al., Nucl. Phys. B124 (1977) 45. See also ref. 11.
11. R.H. Dalitz, Nucl. Phys. A354 (1981) 101c.
12. R.L. Jaffe, Phys. Rev. Letters 38 (1977) 175.
13. J.T. Brown, B.W. Downs and C.K. Iddings, Ann. Phys. (N.Y.) 60 (1970) 148.
14. M.M. Nagels, T.A. Rijken and J.J. De Swart, Phys. Rev. D20 (1979) 1633.

RESONANCES, RESONANCE FUNCTIONS AND SPECTRAL DEFORMATIONS

Erik Balsłev
Matematisk Institut
Aarhus Universitet
DK - 8000 Aarhus C

Introduction.

The present paper is aimed at an analysis of resonances and resonance states from a mathematical point of view. Resonances are characterized as singular points of the analytically continued Lippman-Schwinger equation, as complex eigenvalues of the Hamiltonian with a purely outgoing, exponentially growing eigenfunction, and as poles of the S-matrix. All these concepts refer to a comparison between an operator $H_1 = H_0 + U$ with a "background" potential U and the operator $H = H_1 + V$ obtained by perturbation of H_1 by a potential V. This approach emphasizes that the resonance is basically a scattering phenomenon involving asymptotic closeness of the time-evolution e^{-itH} to the "background" evolution e^{-itH_1}. The general scheme is given in section 1, and in sections 2-4 we analyse in detail the resonance functions of a pair of operators $(H_1, H_1 + V)$, where V is exponentially decaying. In section 2 and 3, $H_1 = H_0 (U = 0)$; in section 4 U can be taken as a dilation-analytic, short-range potential or, more generally, as any potential allowing an analytic continuation of the S-matrix of (H_0, H_1). Based on partial-wave analysis, the procedure is quite similar in both cases. The resolvent $R_1(k)$ is analytically continued as an operator $\tilde{R}_1(k)$ from a space of exponentially decaying to a space of exponentially growing functions (Lemmas 1.1, 2.1, 3.1, 4.4). A resonance z is characterized by the existence of an exponentially decaying solution Φ of the Lippman-Schwinger equation $\Phi + V\tilde{R}_1(k)\Phi = 0$. The resonance function $\psi = \tilde{R}_1(k)$ is an exponentially growing and purely outgoing solution of the Schrödinger equation $H\psi = z^2\psi$. Writing $\psi = fe^{i\varphi}$, we obtain in Theorem 2.2 $(U = 0, \ell = 0)$ and its analogues for $\ell \geqq 1$ (section 3) and $U \neq 0$ (section 4) a characterization of the pair (f, φ), which is precise in the sense, that given a point z and an amplitude function f, we can define uniquely the phase function φ and the potential V, such that $\psi = fe^{i\varphi}$ is a resonance function for $(H_1, H_1 + V)$ at the resonance z.

It is worth noticing, that the resonance function ψ does not have nodes. Anti - bound states, zero - energy resonances and bound states are discussed along the same lines in Theorem 2.6 $(U=0 , \ell=0)$ and its analogues for $\ell \geqq 1$ (section 3) and $U \neq 0$ (section 4). Here the phase function is zero, but there may be a finite number of nodes.

In section 5 we describe the various spectral deformation techniques, local as well as global (dilation - analytic). Their common feature is that the resonances are exposed as discrete eigenvalues of certain non-selfadjoint Hamiltonians with square - integrable eigenfunctions. This is what makes the method useful as a basis for computations. There seems to be an element of magic in these methods. The resonances appear and disappear, when the complex dilation parameter is turned on and off. In the case of an exponentially decaying potential, however, there is a simple connection between the resonance function ψ of section 1 and the square - integrable eigenfunctions of locally deformed Hamiltonians of section 5 A (Lemma 5.6). If the potential is <u>also</u> dilation - analytic, the global deformation (rotation) of the spectrum can be obtained as a limit of local deformations, and the mechanism is completely understood (section 5 C). For a sum of a dilation - analytic and an exponentially decaying potential the connection is not transparent, but local deformation <u>works</u>, in this case as the <u>only</u> method of defining resonances for the pair $(H_0 , H_0 + U + V)$ (section 5 B). This is very useful in connection with the results of section 4 (Lemma 5.7) to show that the resonances of $(H_0 + U , H_0 + U + V)$ found there are identical with the resonances of $(H_0 , H_0 + U + V)$, discarding resonances of $(H_0 , H_0 + U)$. For the n-body problem with dilation - analytic interactions (section 5 C), where the application of Theorem 5.8 has been very successful computationally, no connection has been established between the square-integrable eigenfunctions of the dilated Hamiltonians and a resonance function ψ of H itself. For the three - body problem below 0 with exponentially decaying interactions, the method described for the two - body problem in section 1 has been developed in [1] of sections 1 - 4, and ψ is well - defined. If the potentials are also dilation-analytic, it should be possible to establish the connection of ψ with the square-integrable eigenfunctions found by complex dilations.

The identification of resonances with poles of the S - matrix has been established in the two - body case (Lemmas 5.4, 5.4 - B, 5.10), a fact utilized in the analysis of section 4 (Lemma 4.1). For the n -body case partial results have been obtained (section 5. C).

The subject of resonances has been very fruitful from the point of view of interaction between mathematics and applications. Not only has the dilation - analytic method been used extensively as a basis for calculation of resonances, but the applications in quantum chemistry beyond its original domain of validity has inspired development of the mathematics of the Stark effect and of the Born - Oppenheimer approximation (section 5 C).

Acknowledgment.

The center for interdisciplinary research at Bielefeldt has provided a stimulating atmosphere for the present work. It is a pleasure to thank Ludwig Streit for the hospitality at Z.I.F. I would equally like to thank L. Streit and A. Grossmann for several fruitful discussions leading amongst other things to the formulation of a good question, which has acted as a motivation for this paper: What does a resonance function look like?

1. General theory of resonances and resonance functions.

Let $H = L^2(\mathbb{R}^3)$ with inner product $(\cdot,\cdot)$, linear in the second and antilinear in the first vector, and with norm $\|\cdot\|$. Let $H^2(\mathbb{R}^3)$ be the Sobolev space of order 2 with norm

$$\|f\|_{H^2} = \Big(\sum_{|\alpha| \leq 2} \|D^\alpha f\|^2 \Big)^{\frac{1}{2}} .$$

We denote by $B(H)$ and $C(H)$ the bounded and compact operators on H, respectively.

The free Hamiltonian H_0 is the self-adjoint operator with domain $\mathcal{D}(H_0) = H^2(\mathbb{R}^3)$ defined by

$$H_0 u = -\Delta u \quad , \quad u \in \mathcal{D}(H_0) .$$

The interaction V is assumed to be of the form

$$V = e^{-ar} W e^{-ar} \tag{1.1}$$

where W is an H_0 - compact, symmetric operator, i.e. $W(H_0 + 1)^{-1}$ is a compact, self - adjoint operator.

Examples. 1) W is multiplication by a real-valued function w in $L^2_{loc}(\mathbb{R}^3)$ such that $w(\bar{r}) \to 0$ for $|\bar{r}| \to \infty$.

2) $$W = \sum_{k=1}^{\infty} C_k |\phi_k\rangle\langle\phi_k| \,, C_k \in \mathbb{R} \,, \sum_{k=1}^{\infty} C_k^2 < \infty \,, (\phi_i, \phi_j) = \delta_i^j \,.$$

The Hamiltonian H is defined as a self-adjoint operator on $\mathcal{D}(H) = \mathcal{D}(H_0)$ by

$$Hu = -\Delta u + Vu \quad , \quad u \in \mathcal{D}(H_0) \,.$$

For the purpose of studying resonance functions we introduce the exponentially weighted spaces

$$\mathcal{H}_{\pm a} = \{ f \mid \| f \|_{\pm a} = \| e^{\pm ar} f \| < \infty \}$$

$$\mathcal{H}^2_{\pm a} = \{ f \mid \| f \|_{\pm a, 2} = \| e^{\pm ar} f \|_{H^2} < \infty \}$$

The operators $H_{0,-a}$ and H_{-a} acting in $\mathcal{H}_{-a}$ are defined on the domain

$$\mathcal{D}(H_{0,-a}) = \mathcal{D}(H_{-a}) = \{ u \in \mathcal{H}_{-a} \mid \Delta u \in \mathcal{H}_{-a} \}$$

by

$$H_{0,-a} u = -\Delta u \quad , \quad H_{-a} u = -\Delta u + Vu \,.$$

The free resolvent $R_0(k) \in \mathcal{B}(\mathcal{H})$ is defined for $k \in \mathbb{C}^+$ by

$$R_0(k) = (H_0 - k^2)^{-1} \,. \tag{1.2}$$

Under the condition (1.1) on V, the resolvent $R(k) \in \mathcal{B}(\mathcal{H})$, defined by

$$R(k) = (H - k^2)^{-1} \tag{1.3}$$

satisfies the 2nd resolvent equation

$$R(k) = R_0(k) (1 + V R_0(k))^{-1} \tag{1.4}$$

and is meromorphic in $\mathbb{C}^+$ with poles at discrete eigenvalues of H. Besides $R_0(k)$ and $R(k)$ we introduce the operators

$$R_0^a(k) = R_0(k) \mid \mathcal{H}_a \in \mathcal{B}(\mathcal{H}_a, \mathcal{H}^2_{-a}) \tag{1.5}$$

and

$$R^a(k) = R(k) \mid \mathcal{H}_a \in \mathcal{B}(\mathcal{H}_a, \mathcal{H}^2_{-a}) \,. \tag{1.6}$$

By condition (1.1), $V R_0^a(k)$ is an analytic, $\mathcal{C}(\mathcal{H}_a)$-valued function on $\mathbb{C}^+$, and we have

$$R^a(k) = R_0^a(k) (1 + V R_0^a(k))^{-1} \,. \tag{1.7}$$

We have denoted by $\mathbb{C}^+$ the upper half-plane $\{ k \mid \mathrm{Im} k > 0 \}$ and denote by $\mathbb{C}_a$ the half-plane

$$\mathbb{C}_a = \{k \mid \operatorname{Im} k > -a\} \quad .$$

The basis for the theory of resonances for the class of exponentially decaying potentials is the following well known result.

Lemma 1.1. $R_0^a(k)$ has an analytic continuation $\tilde{R}_0(k) \in B(H_a, H^2_{-a})$ to $\mathbb{C}_a$, given by

$$(\tilde{R}_0(k)f)(x) = \int_{\mathbb{R}^3} \frac{e^{ik|x-y|}}{4\pi|x-y|} f(y)dy \ . \tag{1.8}$$

From (1.1), (1.7), Lemma 1.1 and analytic Fredholm theory we obtain

Lemma 1.2. $V\tilde{R}_0(k)$ is analytic on $\mathbb{C}_a$ with values in $B(H_a)$ and $(1+V\tilde{R}_0(k))^{-1}$ is meromorphic in $\mathbb{C}_a$ with values in $B(H_a)$. Moreover, $R^a(k)$ has a meromorphic continuation $\tilde{R}(k) \in B(H_a, H^2_{-a})$ from $\mathbb{C}^+$ to $\mathbb{C}_a$ given by

$$\tilde{R}(k) = \tilde{R}_0(k)(1+V\tilde{R}_0(k))^{-1} \ . \tag{1.9}$$

The poles of $\tilde{R}(k)$ coincide with those of $(1+V\tilde{R}_0(k))^{-1}$.

We denote by $R^{a*}(k)$ the adjoint of $R^a(k)$ with respect to the duality between H_a and H_{-a} defined by

$$\langle u,v\rangle_{a,-a} = \int_{\mathbb{R}^3} \bar{u}\, v\, dx \ , \quad u \in H_a \ , \quad v \in H_{-a} \ .$$

For $k \in \mathbb{C}^+$ we have $R^*(k) = R(-\bar{k})$; this implies $R^{a*}(k) = R^a(-\bar{k})$ and hence by analytic continuation

$$\tilde{R}^*(k) = \tilde{R}(-\bar{k}) \ , \quad k \in \mathbb{C}_a \ . \tag{1.10}$$

Thus, the poles of $\tilde{R}(k)$ form a set Σ symmetric with respect to the imaginary axis.

For $z \in \Sigma$ the equation

$$\Phi + V\tilde{R}_0(z)\Phi = 0 \tag{1.11}$$

has a finite-dimensional space of solutions $\Phi \in H_a$.
For any solution $\Phi \in H_a$ of (1.11), let

$$\psi = \tilde{R}_0(z)\Phi \ . \tag{1.12}$$

Then $\psi \in H^2_{-a}$

$$\Phi = -V\psi \tag{1.13}$$

and

$$(H_{-a} - z^2)\psi = 0 \ . \qquad (1.14)$$

Equation (1.13) follows from the fact that, by analytic continuation, for $\Phi \in H_a$

$$(H_{0,-a} - z^2)\tilde{R}_0(z)\Phi = \Phi \ . \qquad (1.15)$$

A point $z \in \Sigma \cap \mathbb{C}^+$ lies on the imaginary axis and corresponds to a discrete eigenvalue of H with eigenfunction ψ.
A point $z \in \Sigma \cap (\mathbb{R} \setminus \{0\})$ corresponds to an eigenvalue of H embedded in the continuous spectrum; since it can be shown that if $\Phi + V\tilde{R}_0(z)\Phi = 0$ for such a z, then $\psi = \tilde{R}_0(z)\Phi \in H$.

Definition 1.3. A point $z \in \Sigma$ of the form $z = \alpha - i\beta$ with $\alpha, \beta > 0$ is called a resonance of the pair (H_0, H). If $\Phi \in H_a$ is a non-trivial solution of the equation (1.11), the function $\psi = \tilde{R}_0(z)\Phi$ is called a resonance function of the pair (H_0, H). The point $-\bar{z}$ is called a conjugate resonance.
A point $z \in \Sigma$ of the form $z = -i\beta$ with $\beta > 0$ is said to be a virtual pole of the pair (H_0, H). If $\Phi \in H_a$ is a non-trivial solution of (1.11), the function ψ is called a virtual state or anti-bound state for the pair (H_0, H).
If $0 \in \Sigma$, Φ is a solution of (1.11) for $z = 0$ and $\psi \notin H$, then we speak of a zero-energy resonance with resonance function ψ. If $\psi \in H$, then 0 is an eigenvalue of H with eigenfunction ψ. Thus, 0 may be both an eigenvalue of H and a resonance of (H_0, H).

The trace operator.

Using polar coordinates in momentum space, we identify $g \in L^2(\mathbb{R}^3_{\bar{k}})$ with $g(k,\cdot) \in L^2(\mathbb{R}^+_k, h; k^2 dk)$, where $h = L^2(S^2)$, S^2 the unit sphere in $\mathbb{R}^3$.
For $f \in H_a$, let $\hat{f} = \hat{f}(\bar{k})$ be the Fourier-transform of f.
The trace $\gamma(k)$ of f is defined for $k > 0$ by

$$\gamma(k)f = \hat{f}(k,\cdot) \in h \ . \qquad (1.16)$$

For $f \in H_a$, $\hat{f}(k,\cdot)$ has an analytic, h-valued extension to $T_a = \{k \in \mathbb{C} \mid |\mathrm{Im}\, k| < a\}$. Thus we have

Lemma 1.4. The trace operator $\gamma(k) \in \mathcal{B}(\mathcal{H}_a, h)$, defined by (1.16) for $k > 0$, has an analytic extension to T_a , given by the same expression (1.16) with the understanding that $\hat{f}(k,\cdot)$ is the analytic continuation of $\hat{f}$ from $\mathbb{R}^+$ to T_a .

The scattering matrix.

For scattering theory of the pair (H_0, H) , emphasizing the trace formalism, we refer to Kuroda [3]. The scattering matrix $S(k)$ is unitary on h for $k \in \overline{\mathbb{R}^+}$ and is given by

$$S(k) = 1 - \pi i\, k\, \gamma(k)\{V - V\tilde{R}(k)V\}\gamma^*(k) \tag{1.17}$$

where $\gamma^*(k) \in \mathcal{B}(h, \mathcal{H}_{-a})$ is the adjoint of $\gamma(k) \in \mathcal{B}(\mathcal{H}_a, h)$, with the inverse $S^{-1}(k)$ given by

$$S^{-1}(k) = 1 + \pi i\, \gamma(k)\{V - V\tilde{R}(-k)V\}\gamma^*(k) \tag{1.18}$$

We have the following result on analytic extension of the S-matrix.

Lemma 1.5. The S-matrix $S(k)$ has a $\mathcal{B}(h)$-valued analytic extension $\tilde{S}(k)$ to T_a with poles precisely at resonances and virtual poles. The map $\Phi \to \gamma(z)\Phi$ defines an isomorphism of $N(1 + V\tilde{R}_0(k))$ onto $N(\tilde{S}^{-1}(k))$ for $\mathrm{Re}\, k \geqq 0$, $\mathrm{Im}\, k < 0$.

Proof. By (1.17) and Lemmas 1.1 and 1.4 , $S(k)$ can be extended analytically to T_a with poles at most at Σ . The isomorphism, going back to [5], is established by a proof similar to the one used by A. Jensen (5 [26]) in the context of local deformation techniques, cf. Lemma 5.4. As a consequence of this isomorphism, every resonance and virtual pole is a pole of the S-matrix. As noted by A. Jensen (5 [27]), $\tilde{S}(k)$ can have no real poles.
The operator $\gamma^*(\bar{k}) \in \mathcal{B}(h, \mathcal{H}_{-a})$ is given explicitly by the formula

$$\gamma^*(\bar{k})\sigma = \int_{S^2} \sigma(\omega) f_\omega(k)\, d\omega \ , \quad \sigma \in h \tag{1.19}$$

where

$$[f_\omega(k)](x) = \exp(i\, k\, \omega\cdot x) \ .$$

In terms of $\gamma(k)$ the operator $\tilde{R}_0(k)$ is given by

Lemma 1.6. $\tilde{R}_0(k) - R_0(k) = \pi\, i\, k^{-1}\gamma^*(\bar{k})\gamma(k)$, $-a < \mathrm{Im}\, k < 0$. (1.20)

Using Lemma 1.6 we get the following representation of the resonance function $\psi = \tilde{R}_0(z)\Phi$ at the resonance z ;

Lemma 1.7. The resonance function ψ at the resonance z can be written in the form

$$\psi = \psi_1 + \psi_2 \ , \ \psi_1 = \pi i \, k^{-1} \gamma^*(\bar{z})\gamma(z)\Phi \ , \ \psi_2 = R_0(z)\Phi \tag{1.21}$$

where $\psi_1 \in N(H_{0,-a} - z^2)$ and $\psi_2 \in H_\beta$.

We have emphasized the following characterizations of a resonance of the pair (H_0,H) :

1) The analytically continued Lippman - Schwinger equation

$$\Phi + V\tilde{R}_0(z)\Phi = 0 \tag{1.22}$$

has a solution $\Phi \in H_a$, $\Phi \neq 0$.

2) The Schrödinger equation

$$-\Delta\psi + V\psi = z^2\psi \tag{1.23}$$

has a solution $\psi \in H_{-a}$ of the form

$$\psi = \tilde{R}_0(z)\Phi = \psi_1 + \psi_2 \tag{1.24}$$

where $\Phi \in H_a$ is a solution of (1.22), and

$$\psi_1 \in N(H_{0,-a}) \ , \ \psi_2 \in H_\beta \ .$$

3) The analytically continued scattering matrix $\tilde{S}(k)$ has a pole at z .

All these properties involve the pair (H_0,H) . While eigenvalues and eigenfunctions are characteristics of one self - adjoint operator, here H , the resonances and resonance functions are characteristics of a pair of operators, here (H_0,H) .

They are related to the evolution of scattering states of H and their asymptotic behaviour in time compared to the free evolution.

The S - matrix is an invariant characteristic of the pair of operators (H_0,H) , directly related to the scattering amplitude, scattering cross sections etc., and its poles close to the real axis have been associated with the experimentally observed local maxima of the scattering amplitude. It is therefore important to identify the resonances defined by other criteria with the poles of the S - matrix, as in Lemma 1.4.

Remark 1.8. Reflection on the method of this section shows that the only property of H_0 , which has been utilized, is the fact that $R_0^a(k)$ has an analytic continuation $\tilde{R}_0(k)$ from $\mathbb{C}^+$ to $\mathbb{C}_a$ as a $\mathcal{B}(H_a, H_{-a}^2)$-valued function. Thus, the theory immediately extends to a theory of resonances and resonance functions for a pair of self-adjoint operators (H_1, H) , which satisfies the conditions

(A) $R_1^a(k) = (H_1 - k^2)^{-1} \in \mathcal{B}(H_a, H_{-a}^2)$ has an analytic continuation $\tilde{R}_1(k)$ from $\mathbb{C}^+$ to a larger region $\mathcal{O}$, such that $\mathcal{O}_a = \{k \in \mathcal{O} \mid \operatorname{Re} k > 0, -a < \operatorname{Im} k < 0\} \neq \emptyset$

(B) $H = H_1 + V$, where V satisfies (1.1).

Typically,

$$H_1 = H_0 + U$$

where $R_1^a(k)$ can be proved under certain conditions on U to have an analytic continuation to such a region $\mathcal{O}_a$. In section 4 we shall treat this including the case, where U is a multiplicative, radial, dilation-analytic potential. The region $\mathcal{O}_a$ in this case is $\mathbb{C}_a \cap S_{\alpha'}$, where U is analytic in the angle $S_\alpha = \{k \mid |\operatorname{Arg} k| < \alpha\}$, and $\alpha' = \min\{\alpha, \frac{\pi}{2}\}$.

2. Analysis of s-wave resonance functions for (H_0, H) .

We assume from now on, that V is the operator of multiplication by a real-valued, measurable function $V(r)$ on $\mathbb{R}^+$ satisfying the following conditions:

(i) $V(r) = o(e^{-2ar})$ for $r \to \infty$

(ii) $V \in L^2_{loc}(\mathbb{R}^+)$, $\int_0^1 r^2 |V(r)|^2 dr < \infty$.

In this section we restrict the discussion to the case $\ell = 0$. The operators H_0 and H restricted to the subspace of spherically symmetric functions are unitarily equivalent via the map

$$f(r) \to g(r) = r f(r)$$

to the operators

$$H_0^0 = -\frac{d^2}{dr^2} , \quad H^0 = -\frac{d^2}{dr^2} + V(r) ,$$

acting in $L^2(0,\infty)$. Due to the conditions (ii) on V the operator $e^{2ar}V$ is H_0-compact, hence V is of the form considered in section 1, and the general theory applies.

We shall say that a function u is loc. a.c. on $\overline{\mathbb{R}^+}$, if u is absolutely continuous on $[0,R]$ for every $R > 0$.
Note that for $u \in \mathcal{D}(H^0)$

$u \in L^2(0,\infty)$, $u'' \in L^2(0,\infty)$, u and u' are loc. a. c. on $\overline{\mathbb{R}^+}$ and

$u(0) = 0$.
We introduce the spaces $h_{\pm a}$ and $h^2_{\pm a}$ defined by

$$h_{\pm a} = \{u \mid \|u\|_{\pm a} = \|e^{\pm ar}u\|_{L^2(\mathbb{R}^+)} < \infty\}$$

$$h^2_{\pm a} = \{u \mid \|u\|_{h^2_{\pm a}} = \|e^{\pm ar}u\|_{H^2(\mathbb{R}^+)} < \infty\}.$$

The operators $R^0_0(k) \in \mathcal{B}(L^2(\mathbb{R}^+))$ and $R^{0,a}_0(k) \in \mathcal{B}(h_a, h^2_{-a})$ are defined for $\operatorname{Im} k > 0$ by

$$R^0_0(k) = (H^0_0 - k^2)^{-1}, \quad R^{0,a}_0(k) = R^0_0(k)\,|h_a,$$

and $\tilde{R}^0_0(k)$ is the analytic continuation of $R^{0,a}_0(k)$ to $\mathbb{C}_a$.
By the standard construction of the Green's function we obtain the following representation of $\tilde{R}^0_0(k)$:

<u>Lemma 2.1</u>. $\tilde{R}^0_0(k)v$ is given for $v \in h_a$, $k \in \mathbb{C}_a$, by

$$(\tilde{R}^0_0(k)v)(r) = \frac{e^{ikr}}{k}\int_0^\infty \sin k\,t\,v(t)\,dt + \frac{e^{ikr}}{2ik}\int_r^\infty e^{-ikt}v(t)\,dt - \frac{e^{-ikr}}{2ik}\int_r^\infty e^{ikt}v(t)\,dt \qquad (2.1)$$

Let now $z = \alpha - i\beta$, $\alpha,\beta > 0$, be a resonance as given by Definition 1.3, and let $\Phi \in h_a$ be the solution of the equation

$$\Phi + V\tilde{R}^0_0(z)\Phi = 0, \qquad (2.2)$$

normalized by the condition

$$\frac{1}{z}\int_0^\infty \sin zt\,\Phi(t)\,dt = 1. \qquad (2.3)$$

The fact that the l. h. s. of (2.3) is non-zero, follows from Lemma 1.5, because $\frac{1}{z}\int_0^\infty \sin zt\,\Phi(t)\,dt$ is proportional to $\gamma(z)$, when Φ is radially symmetric.

The resonance function ψ is then given by

$$\psi(r) = e^{izr} + \frac{e^{izr}}{2iz}\int_r^\infty e^{-izr}\Phi(t)\,dt - \frac{e^{-izr}}{2iz}\int_r^\infty e^{izt}\Phi(t)\,dt \qquad (2.4)$$

and satisfies

$$-\psi'' + V\psi = z^2\psi \quad , \quad \psi(0) = 0 \qquad (2.5)$$

while

$$\Phi = -V\psi \qquad (2.6)$$

Simple estimates based on (2.1), using $\Phi \in h_a$, yield

$$\psi(r) = e^{izr} + o(e^{-ar}) \qquad (2.7)$$

By (i), (2.6) and (2.7)

$$\Phi(r) = o(e^{(\beta - 2a)r}) \qquad (2.8)$$

Iterating the estimates of the integrals in (2.1), using (2.8), gives

$$\psi(r) = e^{izr} + o(e^{(\beta - 2a)r}) \qquad (2.9)$$

Note that (2.9) is a decomposition of the resonance function ψ as a sum of the outgoing free solution of z and an asymptotically very small term, excluding an incoming free wave. This is different from the decomposition (1.21) of ψ as a sum of the regular free solution ψ_1 and a square-integrable function ψ_2 . For $\ell = 0$ (1.21) corresponds to setting

$$\psi(r) = 2i\sin zr + \psi_2(r) = e^{izr} + e^{-izr} + \psi_2(r) \qquad (2.10)$$

By (2.9) and (2.10)

$$\psi_2(r) = -e^{-izr} + o(e^{(\beta - 2a)r}) \qquad (2.11)$$

in agreement with the fact that $\psi_2 \in H_\beta$.

The decomposition (1.21), leading to (2.10) in the radial case for $\ell = 0$, is important because ψ_1 is a solution of the free Schrödinger equation in $\mathbb{R}^3$. It appears again in connection with spectral deformation techniques, see section 5.
The decomposition (2.9) is important, because it gives a characterization of the resonance function as the unique solution of (2.5) satisfying (2.9). This is the precise meaning of the statement, that a resonance is a point z at which the Schrödinger equation $H\psi = z^2\psi$ has a purely outgoing solution.
Starting from (2.5), (2.9) and (i), (ii) an analysis of resonance functions may be obtained as follows.

It can be shown, first of all, that ψ has no nodes, i.e.

$$\psi(r) \neq 0 \qquad \text{for} \quad r > 0 . \tag{2.12}$$

Using (2.12) we can write

$$\psi(r) = f(r)e^{i\varphi(r)} \qquad \text{for} \quad r \geqq 0 \tag{2.13}$$

where $f(r) > 0$ for $r > 0$ and the phase function φ can be chosen continuous on $\overline{\mathbb{R}^+}$.

Now insertion of (2.13) in (2.5) shows that the pair of functions (f , φ) satisfies the following pair of differential equations for $0 < r < \infty$

$$-f'' + Vf + \varphi'^2 f = E f \tag{2.14}$$

$$2f'\varphi' + f\varphi'' = \Gamma f \tag{2.15}$$

where

$$z^2 = E - i\Gamma , \quad E = \alpha^2 - \beta^2 , \quad \Gamma = 2\alpha\beta$$

Solving (2.15) for φ in terms of f and using $\psi(0) = 0$, (2.12), (2.14), (i), (ii) and the asymptotic estimate (2.9) we obtain the following result:

Theorem 2.2. A) Assume that V is a real-valued function on $\mathbb{R}^+$ satisfying (i) and (ii), and let $\psi = fe^{i\varphi}$ be the resonance function at the resonance z with $z^2 = E - i\Gamma$, normalized by setting

$$-\frac{1}{z}\int_0^\infty \sin zt\, V(t)\psi(t)\,dt = 1 . \tag{2.17}$$

Then the pair of functions (f , φ) satisfies the following conditions:

1) $f \in C^1(\overline{\mathbb{R}^+})$, f' is loc. a.c. on $\overline{\mathbb{R}^+}$

2) $f(0) = 0$, $f(r) > 0$ for $r > 0$, $f'(0) > 0$

3) $f(r) = e^{\beta r} + o(e^{(\beta - 2a)r})$

4) $f''(r) = \beta^2 e^{\beta r} + o(e^{(\beta - 2a)r})$

5) $f'' \in L^2(0,R)$ for every $R > 0$

6) $\int_0^\infty p(r)\,dr = \frac{1}{2\beta}$, where $p(r) = f^2(r) - e^{2\beta r}$

7) $\varphi \in C^2(\overline{\mathbb{R}^+})$, φ'' is loc. a.c. on $\overline{\mathbb{R}^+}$

8) $\varphi'(0) = 0$, $\varphi''(0) = \frac{\Gamma}{3}$

9) $$\varphi'(r) = \Gamma f^{-2}(r) \int_0^r f^2(t)\,dt$$

10) $$\varphi'(r) = \alpha + o(e^{-2ar})$$

11) $$\varphi(r) = \alpha r + \alpha \int_r^\infty \frac{p(t) + 2\beta \int_t^\infty p(s)\,ds}{e^{2\beta t} + p(t)}\,dt$$

12) $$\varphi(r) = \alpha r + o(e^{-2ar})$$

13) $$2f'\varphi' + f\varphi'' = \Gamma f$$

14) $$-f'' + Vf + \varphi'^2 f = Ef$$

B) Let f satisfy conditions 1) - 6), where $0 < \beta < a$, let $\alpha > 0$ and define φ by 11) and V by 14). Then V satisfies conditions (i) and (ii), and the function $\psi = fe^{i\varphi}$ is the resonance function for the pair (H_0, H) at the resonance $z = \alpha - i\beta$, normalized by (2.17).

The proof of Theorem 2.2 will be given elsewhere. Here we shall restrict ourselves to a discussion of the result. Conditions 1) - 5) on f and the corresponding conditions on φ are straightforward. In order to analyse condition 6) we consider the special case when V has compact support. In this case $\psi(r) = e^{ikr}$ outside the support of V, and the asymptotic conditions are trivially satisfied: $f(r) = e^{\beta r}$ and $\varphi(r) = \alpha r$ for $r > R$, when $V(r) = 0$ for $r > R$. Condition 6) in this case reduces to

$$\int_0^R p(r)\,dr = \frac{1}{2\beta}, \quad p(r) = f^2(r) - e^{2\beta r}. \tag{2.18}$$

To understand the significance of (2.18), notice that

$$p(0) = -1, \quad p'(0) = -2\beta, \quad p(R) = 0 \quad \text{(see Fig. 1)}$$

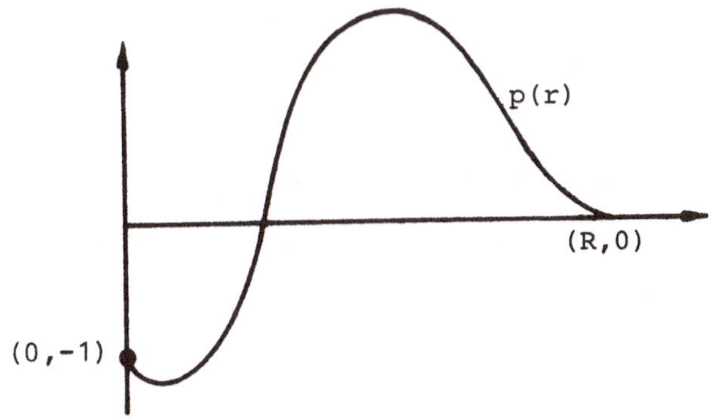

Fig. 1 Graph of $p(r)$

Thus, for β very small the amplitude function $f(r)$ must be very large on the support of V, it must exceed $e^{\beta r}$ so much on $[0,R]$ that (2.18) holds. This illustrates the increasing importance of a resonance and its resonance function, as the resonance gets closer to the real axis. (2.18) is related to the derivative of the phase shift at α and to time delay (cf. [4] (12.90)).
An important larger class of potentials is defined by the conditions (ii) and

(i') $\quad V(r) = o(e^{-ar})$ for every $a > 0$.

For this class of potentials we note the following property of resonance functions

<u>Corollary 2.3</u>. $V(r) = o(e^{-ar})$ for all $a>0$ if and only if

$$f(r) = e^{\beta r} + o(e^{-ar}) \qquad \text{for all } a>0$$

and

$$f''(r) = \beta^2 e^{\beta r} + o(e^{-ar}) \qquad \text{for all } a>0$$

in this case also

$$\varphi'(r) = \alpha + o(e^{-ar}) \qquad \text{for all } a>0 .$$

Regarding smoothness properties we have

<u>Corollary 2.4</u>. $V \in C^n(\mathbb{R}^+)$ if and only if $f \in C^{n+2}(\mathbb{R}^+)$;

if $V \in C^n(\mathbb{R}^+)$, then $\varphi \in C^{n+3}(\mathbb{R}^+)$;

$V \in C^\infty(\mathbb{R}^+)$ if and only if $f \in C^\infty(\mathbb{R}^+)$;

if $V \in C^\infty(\mathbb{R}^+)$, then $\varphi \in C^\infty(\mathbb{R}^+)$

Dilation-analytic potentials could also be discussed in connection with Theorem 2.2. It might be used to construct a dilation-analytic resonance function at a resonance z. In that connection we note the following

<u>Corollary 2.5</u>. If V is analytic in a set $\mathcal{O}$ with $\mathcal{O} \cap \mathbb{R}^+ \neq \emptyset$, then ψ is analytic in $\mathcal{O}$. If $\psi(\zeta) \neq 0$ for $\zeta \in \mathcal{O}$, then f and φ are analytic in $\mathcal{O}$. Conversely, if f is analytic in $\mathcal{O}$ and $f(\zeta) \neq 0$ for $\zeta \in \mathcal{O}$, then V is analytic in $\mathcal{O}$.

Virtual poles and zero - energy resonances.

We have so far restricted the discussion to resonance functions for (H_0,H) . Here the phase function φ plays an important role, although as we have seen φ is entirely determined by the amplitude f . Thus the phase function φ enters into the proof that ψ has no nodes and in establishing the important condition 6).
The phase function is asymptotic to the free phase function αr . In the case of virtual states, when $\alpha = 0$, the "free outgoing wave" is $e^{\beta r}$ with phase function 0 . This simplifies the analysis considerably, since we can choose in this case $\varphi \equiv 0$. On the other hand, the possibility of nodes of f can not be ruled out. It is convenient to treat the other cases when $\alpha = 0$ at the same time, i.e. zero - energy resonances $(\beta = 0)$ and bound states corresponding to discrete eigenvalues $(\beta < 0)$. We obtain the following result.

Theorem 2.6. A) Assume that V is a real - valued, measurable function on $\mathbb{R}^+$ satisfying the conditions (i) and (ii). Let $-i\beta \in \Sigma$, and let Φ be a solution of

$$\Phi + V\tilde{R}_0^0(-i\beta)\Phi = 0$$

with

$$\tilde{R}_0^0(-i\beta)\Phi$$

and

$$\frac{1}{-i\beta}\int_0^\infty \sin(-i\beta t)\Phi(t)\,dt = 1 \ .$$

Then ψ is a real - valued function on $\overline{\mathbb{R}^+}$ satisfying the following conditions

1) $\psi \in C^1(\overline{\mathbb{R}^+})$, ψ' is loc. a.c. on $\overline{\mathbb{R}^+}$

2) $\psi(0) = 0$, $\psi'(0) \neq 0$

3) ψ has at most a finite number of positive zeros $r_1 \ldots r_k$ $r_1, \ldots, r_k$:

$$\psi(r_i) = 0 \ , \ \psi'(r_i) \neq 0 \ , \quad i = 1, \ldots, k \ .$$

4) $\psi'' \in L^2_{loc}(\mathbb{R}^+)$, and for some $\delta > 0$

$$\int_0^\delta |\psi''(r)|^2 dr < \infty \ , \quad \int_{r_i-\delta}^{r_i+\delta} |\psi''(r)|^2 |r-r_i|^{-2} dr < \infty$$

5) $$\psi(r) = e^{\beta r} + o(e^{(\beta - 2a)r})$$
and
$$\psi''(r) = \beta^2 e^{\beta r} + o(e^{(\beta - 2a)r}) \quad \text{for} \quad \beta \neq 0$$
$$\psi(r) = 1 + o(r\,e^{-2ar})$$
and
$$\psi''(r) = o(e^{-2ar}) \quad \text{for} \quad \beta = 0 .$$

B) Let $-a < \beta < a$ and let ψ be a real-valued function on $\overline{\mathbb{R}^+}$ satisfying conditions 1) - 5). Define V by

$$V(r) = \frac{\psi''(r)}{\psi(r)} - \beta^2 .$$

Then V satisfies (i) and (ii), and ψ is a virtual state $(\beta > 0)$ of (H_0, H), zero-energy resonance function $(\beta = 0)$ of (H_0, H) or bound state $(\beta < 0)$ of H at the point $-i\beta \in \Sigma$.

3. Analysis of resonance functions of (H_0, H) for $\ell \geq 1$.

In this section we extend the results of section 2 to the case of angu-angular momentum $\ell \geq 1$.
The restrictions of H_0 and H to the subspace of $L^2(\mathbb{R}^3)$ characterized by angular momentum quantum number ℓ and magnetic quantum number m (m fixed between $-\ell$ and ℓ) are unitarily equivalent via the map

$$f(r) \to g(r) = r\, f(r)$$

to the operators

$$H_0^\ell = -\frac{d^2}{dr^2} + \frac{\ell(\ell+1)}{r^2} , \quad H^\ell = H_0^\ell + V ,$$

acting in $L^2(0,\infty)$ and defined under the assumptions (i) and (ii) of section 2 on V as for $\ell = 0$ without the condition $u(0) = 0$.
The operators $R_0^\ell(k) \in \mathcal{B}(L^2(\mathbb{R}^+))$ and $R_0^{\ell,a}(k) \in \mathcal{B}(h_a, h_{-a}^2)$ are defined for $\operatorname{Im} k > 0$ by

$$R_0^\ell(k) = (H_0^\ell - k^2)^{-1} , \quad R_0^{\ell,a}(k) = R_0^\ell(k)\,|\,h_a .$$

Moreover, $\tilde{R}_0^\ell(k)$ is the analytic continuation of $R_0^{\ell,a}(k)$ to $\mathbb{C}_a$.
Replacing $e^{\pm ikr}$ in Lemma 2.1 by the Riccati-Hankel functions $u_{\ell\pm}(kr)$, we get

Lemma 3.1. $\tilde{R}_0^{\ell}(k)v$ is given for $v \in h_a$, $k \in \mathbb{C}_a$ by

$$(\tilde{R}_0^{\ell}(k)v)(r) = \frac{1}{k} e^{-i\frac{\pi\ell}{2}} u_{\ell+}(kr) \int_0^{\infty} u_{\ell}(kt)v(t)dt$$
$$+ \frac{u_{\ell+}(kr)}{2\,ik} \int_r^{\infty} u_{\ell-}(kt)v(t) - \frac{u_{\ell-}(kr)}{2\,ik} \int_r^{\infty} u_{\ell+}(kt)v(t)dt \qquad (3.1)$$

Here $u_{\ell}(kt)$ is the regular solution of the equation

$$-u'' + \frac{\ell(\ell+1)}{r^2} u = k^2 u \qquad (3.2)$$

and the outgoing and incoming solutions $u_{\ell\pm}(kr)$ are given by

$$u_{\ell\pm}(kr) = P_{\ell}(\pm(kr)^{-1})e^{\pm ikr} ,$$

where P_{ℓ} is a certain polynomial of degree ℓ, normalized by $P_{\ell}(0) = 1$ (cf. [4]).

Let now z be a resonance and $\Phi \in h_a$ a solution of the equation

$$\Phi + V\tilde{R}_0^{\ell}(z)\Phi = 0 \qquad (3.3)$$

normalized by

$$\frac{1}{z} e^{-i\frac{\pi\ell}{2}} \int_0^{\infty} u_{\ell}(zt)\Phi(t)dt = 1 \qquad (3.4)$$

and let

$$\psi = \tilde{R}_0^{\ell}(z)\Phi \qquad (3.5)$$

Then we obtain from (3.1) as for $\ell = 0$

$$\psi(r) = u_{\ell+}(zr) + o(e^{(\beta - 2a)r}) \qquad (3.6)$$

From (i), (ii), (3.6), the equation

$$-\psi'' + \frac{\ell(\ell+1)}{r^2}\psi + V\psi = z^2 \qquad (3.7)$$

and the regularity condition

$$\psi(r) \sim C\,r^{\ell+1} \quad \text{for } r \to 0 \qquad (3.8)$$

we obtain the result analogous to Theorem 2.2.

It is to be noted that the resonance function ψ has <u>no nodes</u>. The free asymptotic amplitude and phase functions $e^{\beta r}$ and αr are replaced by $f_{\ell+}(r)$ and $\varphi_{\ell+}(r)$, where we set

$$u_{\ell+}(zr) = f_{\ell+}(r)e^{i\varphi_{\ell+}(r)} \qquad (3.9)$$

For example, 3) and 10) are replaced by

3') $f(r) = f_{\ell+}(r) + o(e^{(\beta - 2a)r})$

10') $\varphi'(r) = \varphi'_{\ell+}(r) + o(e^{-2ar})$

and the important condition 6) becomes for $\ell \geqq 1$

6') $\int_0^r f^2(t) + \int_r^\infty \{f^2(t) - f^2_{\ell+}(t)\}dt = \Gamma^{-1}\varphi'_{\ell+}(r) f^2_{\ell+}(r)$ for any $r > 0$.

Given f satisfying 1') - 6'), φ' is obtained by solving 13) (which is the same for all ℓ), and V is defined by

$$V = \frac{f''}{f} - \varphi'^2 + E - \frac{\ell(\ell+1)}{r^2} . \tag{3.10}$$

Remark 3.2. Condition 6') is more restrictive than 6) for $\ell = 0$. Thus, if $u_{\ell+}(zr)$ has a node r_0 and $V(r) = 0$ for $r > r_0$, then 6') can not be satisfied with $r = r_0$, so z can not be a resonance for any such V. Otherwise, the result for $\ell \geqq 1$ is quite similar to the one derived for $\ell = 0$, taking into account the different asymptotic behaviour of $u_{\ell\pm}(zr)$ for $r \to 0$. The resonance function ψ at z for a given ℓ is characterized by the precise asymptotic behaviour of its amplitude f and phase φ as the amplitude and phase function of the free outgoing wave $u_{\ell\pm}$. Given an amplitude function f and a point z, the phase function φ is derived and an exponentially decaying potential V is constructed producing the resonance function $\psi = fe^{i\varphi}$ at the resonance z. The discussion of section 2 carries over to $\ell \geqq 1$ with some modification due to the replacement of 6) by the more complicated condition 6'). Corollaries 2.3. - 2.5 hold identically for all ℓ.

Virtual states and zero-energy bound states.

The analysis of virtual states corresponding to $z = -i\beta$, $\beta > 0$ and bound states corresponding to $z = -i\beta$, $\beta < 0$, is also analogous to the case $\ell = 0$. However, for $\ell \geqq 1$ there is no possibility for zero-energy resonances, whereas zero-energy bound states may occur. Thus, Theorem 2.3 is valid for $0 < |\beta| < a$ with $e^{\beta r}$ replaced by $u_{\ell+}(-i\beta r)$, V replaced by $V + \frac{\ell(\ell+1)}{r^2}$ and the conditions on ψ at 0 replaced by

$$\psi(r) \sim cr^{\ell+1}, \quad \psi'(r) \sim c(\ell+1)r^\ell \quad \text{for } r \to 0. \tag{3.11}$$

For $\beta = 0$ condition 5) of Theorem 2.3 is replaced by

$$\psi(r) = r^{-\ell} + o(r^{1-\ell} e^{-2ar})$$
$$\psi''(r) = \ell(\ell+1)r^{-\ell-2} + o(r^{-\ell} e^{-2ar}).$$

Thus $\psi \in L^2(0,\infty)$ for $\ell \geqq 1$ in contrast to the case $\ell = 0$, and we have a zero-energy bound state if $0 \in \Sigma$.

4. Resonances of $(H_0 + U , H_0 + U + V)$ for $\ell = 0$.

In section 2 we have given a detailed analysis of s-wave resonance wave functions for the pair $(H_0 , H_0 + V)$, where $V = o(e^{-2ar})$. In this section we extend the theory to the more general case, where H_0 is replaced by an operator $H_1 = H_0 + U$ with the property that $R_1^{0,a}(k)$ has an analytic continuation $\tilde{R}_1^0(k)$ to a domain $\mathcal{O}_a$, as suggested in Remark 1.8. The existence of this analytic continuation is closely related to the possibility of continuing the S-matrix of the pair (H_0 , H_1) analytically, which has been proved under rather general conditions on U , using spectral deformation techniques, see Lemmas 5.4, 5.4 B and 5.10. In this section we shall show first, how the existence of the analytic continuation $\tilde{R}_1^0(k)$ follows from the continuation of the S-matrix of $(H_0 , H_0 + U)$, and then by the method of section 2 discuss resonances and resonance functions of the pair $(H_0 + U , H_0 + U + V)$.

We make use of the well-known partial-wave analysis, referring to [4] for general background. We summarize the main results needed for our purpose. The basic equation is

$$-v'' + Uv = k^2 v \tag{4.1}$$

The potential U is assumed to be a real-valued function satisfying

(i) $\quad U \in L^2_{loc}(\mathbb{R}^+) , \quad \int_0^1 r^2 |U(r)|^2 dr < \infty$

(ii) $\quad \int_1^\infty |U(r)| dr < \infty$

We also consider the following stronger condition

(ii') $\quad \int_1^\infty r |U(r)| dr < \infty$

The self-adjoint operator $H_1 = -\frac{d^2}{dr^2} + U(r)$ is defined as in sections 2 and 3.

We set

$$R_1(k) = (H_1 - k^2)^{-1} \in \mathcal{B}(L^2(\mathbb{R}^+)) \qquad \text{for } \operatorname{Im} k > 0$$

$$R_1^a(k) = R_1(k) \mid h_a \in \mathcal{B}(h_a , h_{-a}^2) \qquad \text{for } \operatorname{Im} k > 0 .$$

Under conditions (i) and (ii) the equation (4.1) has the following solutions:

The <u>regular</u> solution $u_0(k,r)$ is defined for $k \neq 0$, $r \geq 0$ by the conditions

$$u_0(k,0) = 0 \ , \quad u_0'(k,0) = 1 \tag{4.2}$$

The function $u_0(k,r)$ is for every $r \geq 0$ analytic in k for $k \neq 0$. The outgoing solution $u_+(k,r)$ is defined for $k \neq 0$, $\operatorname{Im} k \geq 0$, by the condition

$$u_+(k,r)e^{-ikr} \longrightarrow 1 \quad \text{for} \quad r \to \infty \tag{4.3}$$

The function $u_+(k,r)$ is for every $r \geq 0$ analytic in k for $\operatorname{Im} k > 0$ and continuous in k for $k \neq 0$, $\operatorname{Im} k \geq 0$.
The incoming solution $u_-(k,r)$ is defined for $k \neq 0$, $\operatorname{Im} k \leq 0$, by the condition

$$u_-(k,r)e^{ikr} \longrightarrow 1 \quad \text{for} \quad r \to \infty \tag{4.4}$$

The function $u_-(k,r)$ is for every $r \geq 0$ analytic in k for $\operatorname{Im} k < 0$ and continuous in k for $k \neq 0$, $\operatorname{Im} k \leq 0$.

In particular, the Jost function $F(k) = u_+(k,0)$ is analytic for $\operatorname{Im} k > 0$ and continuous for $k \neq 0$, $\operatorname{Im} k \geq 0$. The connection between u_0, u_+ and u_- is given for k real, $k \neq 0$, by

$$u_0(k,r) = \frac{1}{2ik}\left[\, F(-k)u_+(k,r) - F(k)u_-(k,r)\right] \tag{4.5}$$

The S-matrix $S(k)$ is given for $k \in \mathbb{R}$, $k \neq 0$, by

$$S(k) = \frac{F(-k)}{F(k)} \tag{4.6}$$

From (4.5) and (4.6) we get for $k \in \mathbb{R}$, $k \neq 0$

$$u_+(k,r) = \frac{2ik}{F(-k)}\, u_0(k,r) + S^{-1}(k)u_-(k,r) \tag{4.7}$$

If U satisfies (ii'), then $u_0(k,r)$ is entire analytic in k, $u_+(k,r)$ is continuous for $\operatorname{Im} k \geq 0$, $u_-(k,r)$ is continuous for $\operatorname{Im} k \leq 0$ and $S(k)$ is given by (4.5) for $k \in \mathbb{R}$.

Suppose that U satisfies (i) and (ii).
Let $\mathcal{O}$ be a domain contained in $\mathbb{C}_a \setminus \{0\}$, intersecting $\mathbb{R}^+$ in an interval I.

Lemma 4.1. The following statements are equivalent:

1) The Jost function $F(k)$ has an analytic continuation from $\mathbb{C}^+$ to $\mathcal{O}$.
2) The S-matrix $S(k)$ has an analytic extension from I to $\mathcal{O} \cap \mathbb{C}^-$ with poles at the zeros of $F(k)$.
3) For $r \geq 0$, $u_+(k,r)$ has an analytic continuation from $\mathbb{C}^+$ to $\mathcal{O}$.

If U satisfies (ii'), $k=0$ should be included in 1) - 3), except that $S(k)$ is regular at 0 even if $F(0)=0$.

It is important to know the asymptotic behaviour for $r \to \infty$ of the analytic continuation of $u_+(k,r)$. To obtain this we first recall the following estimate (cf.[4]).

$$|u_0(k,r) - k^{-1}\sin kr| \leqq C e^{|\mathrm{Im}\,k|r} \frac{r}{1+|k|r} \int_0^r dr' |U(r')| \frac{r'}{1+|k|r'}$$

$$\leqq C' e^{|\mathrm{Im}\,k|r} \qquad (4.8)$$

where C' is independent of k.
By (4.8),

$$|u_0(k,r)e^{-ikr}| \leqq C'' \quad \text{for} \quad \mathrm{Im}\,k \leqq 0 \qquad (4.9)$$

From (4.3), (4.4), (4.7) and (4.9) we obtain

<u>Lemma 4.2</u>. Suppose that $u_+(k,r)$ for $r \geqq 0$ has an analytic continuation from $\mathbb{C}^+$ to $\mathcal{O}$. Let K be a compact subset of $\mathcal{O}$. Then

$$|u_+(k,r)e^{-ikr}| \leqq C(K) \quad \text{for} \quad k \in K,\ r \geqq 0 \qquad (4.10)$$

Now we obtain from (4.3) and (4.10) by means of Vitali's convergence theorem (cf. [6]).

<u>Lemma 4.3</u>. Under the conditions of Lemma 4.2 we have

$$u_+(k,r)e^{-ikr} \xrightarrow[r\to\infty]{} 1, \text{ uniformly for } k \in K$$

From Lemmas 4.2 and 4.3 we obtain the following result on analytic contionation of $R_1^a(k)$, analogous to Lemma 2.1.

<u>Lemma 4.4</u>. Under the conditions of Lemma 4.2, the operator-valued function $R_1^a(k) \in \mathcal{B}(h_a, h^2_{-a})$ has an analytic continuation $\tilde{R}_1(k)$ from $\mathbb{C}^+$ to $\mathcal{O} \cap \mathbb{C}_a$, given by

$$(\tilde{R}_1(k)v)(r) = u_+(k,r)\frac{1}{F(k)}\int_0^\infty u_0(k,t)v(t)dt$$
$$+ \frac{1}{2ik}u_+(k,r)\int_r^\infty u_-(k,t)v(t)dt - \frac{1}{2ik}u_-(k,r)\int_r^\infty u_+(k,t)v(t)dt \qquad (4.11)$$

In $\mathcal{O}_a = \{k \in \mathcal{O} \mid \mathrm{Re}\,k > 0,\ -a < \mathrm{Im}\,k < 0\}$, $\tilde{R}_1(k)$ has the same poles as the S-matrix, at the zeros of the Jost function, which are precisely the resonances of the pair (H_0, H_1).

By Lemma 4.1, the conclusion of Lemma 4.4 holds if $S(k)$ has an analytic extension from I to $\mathcal{O}$. By Lemmas 5.10 and 5.4 B this holds with $I = \mathbb{R}^+$ and $\mathcal{O} = S_\alpha \cap \mathbb{C}_a$, if

a) U is S_α-dilation analytic and $U(r) = o(r^{-1})$ for $r \to \infty$.

b) $U = U_1 + U_2$, where U_1 is S_α-dilation-analytic and $U_1(r) = o(r^{-2})$ and $U_2 = o(e^{-2ar})$ for $r \to \infty$.

Based on Lemma 4.4 we can now apply the method of section 2 to analyze the resonances and resonance functions of the pair (H_1, H_1+V), where $V = o(e^{-2ar})$, in the domain $\mathcal{O}_a \setminus \mathcal{R}(H_0,H_1)$, where $\mathcal{R}(H_0,H_1)$ is the set of resonances of (H_0,H_1).
We briefly indicate the results. Assume that U satisfies (i) and (ii) of this section, that the conditions of Lemma 4.2 are satisfied, and that V satisfies (i) and (ii) of section 2.
Let the self-adjoint operators

$$H_0 = -\frac{d^2}{dr^2}, \quad H_1 = H_0 + U, \quad H = H_1 + V$$

with the boundary condition $u(0) = 0$ be defined as in section 2.

Let $z = \alpha - i\beta \in \mathcal{O}_a \setminus \mathcal{R}(H_0,H_1)$ be a resonance of the pair (H_1,H) with resonance function ψ, i.e. the equation

$$\Phi + V\tilde{R}_1(z)\Phi = 0 \tag{4.12}$$

has a solution $\Phi \in h_a$, normalized by setting

$$\frac{1}{F(z)} \int_0^\infty u_0(z,t)\Phi(t)\,dt = 1, \tag{4.13}$$

$$\psi = \tilde{R}_1(z)\Phi, \quad \Phi = -V\psi. \tag{4.14}$$

Then

$$\psi \in h^2_{-a}, \quad \psi(0) = 0$$

and

$$\left(-\frac{d^2}{dr^2} + U + V\right)\psi = z^2\psi \tag{4.15}$$

It is first of all proved that ψ has no nodes, $\psi(r) \neq 0$ for $r > 0$.
We can then write ψ in the form

$$\psi(r) = f(r)e^{i\varphi(r)}$$

with a continuous phase function φ.
By (4.11) and (4.13),

$$\psi(r) = u_+(z,r) + \frac{u_+(z,r)}{2ik}\int_r^\infty u_-(z,t)v(t)\,dt - \frac{u_-(z,r)}{2ik}\int_r^\infty u_+(z,t)v(t)\,dt \tag{4.16}$$

From (4.16) we get as in section 2, using Lemma 4.2

$$\psi(r) = u_+(z,r) + o(e^{(\beta - 2a)r}) \tag{4.17}$$

Then we can prove a result analogous to Theorem 2.2, replacing $e^{\beta r}$ by $f_+(r)$ and αr by $\varphi_+(r)$, where we set

$$u_+(z,r) = f_+(r)e^{i\varphi_+(r)} \tag{4.18}$$

For example, 3) and 10) are replaced by

3") $f(r) = f_+(r) + o(e^{(\beta - 2a)r})$

10") $\varphi(r) = \varphi_+(r) + o(e^{-2ar})$.

Otherwise, only 6) and 11) require modification. Thus, 6) is replaced by

6") $\int_0^\infty [f^2(r) - f_+^2(r)]dr = \frac{\varphi_+'(0)f_+^2(0)}{\Gamma}$

Again, given an amplitude function f satisfying 1") - 6"), the phase function φ and the potential V are determined such that $\psi = fe^{i\varphi}$ is a resonance function for the pair $(H_1, H_1 + V)$ at the resonance $z = \alpha - i\beta$.

In this way we obtain a complete characterization of resonance functions for $(H_1, H_1 + V)$, where $V = o(e^{-2ar})$, associated with resonances in the domain $\mathcal{O}_a \setminus R(H_0, H_1)$ of analytic continuation of $\tilde{R}_1(k)$. It has turned out to be a necessary condition for z to be a resonance of $(H_1, H_1 + V)$ for any V with support in $(0, r_0)$, that $u_+(z,r)$ has no nodes in $(0, r_0)$. In fact, it can be proved using the differential equations 13) and 14) of Theorem 2.2, which also hold for (f_+, φ_+) on any interval free from nodes of u_+, that u_+ has <u>at most one</u> node. Moreover, $u_+(k,r)$ has a node at most for k on certain smooth curves in $\mathcal{O}$. This approach emphasizes the point of view, that resonances and resonance functions are characteristics of <u>pairs of operators</u>. However, the free Hamiltonian H_0 plays a special role, and the question arises, whether the resonances of (H_1, H) are also resonances of (H_0, H) and vice versa. This is proved in section 5, using local deformation technique, see Lemma 5.7 and Corollary 5.8.
Theorem 2.6 on anti-bound states, zero-energy resonances and bound states also generalize to the case, where H_0 is replaced by H_1 , provided the domain $\mathcal{O}_a$ includes the relevant points on the imaginary axis. Note that anti-bound states may occur, if U is S_α - dilation - analytic with $\alpha > \frac{\pi}{2}$, whereas the dilation - analytic technique does not allow zero to be included in $\mathcal{O}_a$ and hence does not make it possible to study zero-energy resonances for $(H_1, H_1 + V)$.

The results of this section can all be extended to $\ell \geq 1$, just as it was done for $U = 0$ in section 3.

5. Spectral deformation techniques.

A variety of spectral deformation techniques have been applied to the study of resonances, both from a theoretical and a practical computational point of view. A common feature of these methods is the deformation of the continuous spectrum, displaying resonances as discrete eigenvalues of certain non-selfadjoint operators with square-integrable eigenfunctions, representing the resonance function. Local deformation techniques have been applied to the two-body problem and are based on a local complexification of the momentum variable. Global deformation techniques have been successful also for many-body problems and are based on a rotation of the continuous spectrum around fixed thresholds. We do not intend here to give a complete survey of this subject, but refer to the fairly comprehensive bibliography of the mathematical literature on resonances and spectral deformation techniques. For the applications of the complex dilation method to quantum chemistry and physics we refer to [25] and the recent review articles [24], [29], [34], which contain extensive bibliographies.
The aim of this section is to establish the connection between the various spectral deformation techniques and the analysis of resonances and resonance functions given in the previous sections and to explain how the resonances and resonance functions characterized by the properties 1) - 3) of section 1 appear as discrete eigenvalues with square-integrable eigenfunctions through spectral deformation.

A. Local spectral deformation technique for exponentially decaying potentials.

Complex deformations of the three-dimensional momentum space has been studied by Nuttall [30] and L. E. Thomas [42]. In [6] the method of local deformations was formulated in polar coordinates and used to study the S-matrix. We briefly indicate the method and the main results from [6].
Let Q be the operator of convolution by the function $(k^2 + a^2)^{-2} = c\,\mathcal{F}(e^{-ar})$, and let Y be a compact operator from $L^{2,1}$ to $L^{2,-1}$ where

$$L^{2,\pm 1} = L^2(\mathbb{R}^+, h; (1+k^2)dk), \quad h = L^2(S^2)$$

are the weighted L^2-spaces written in polar coordinates.
The interaction V is supposed to be of the form $V = QYQ$. Note that the potential $V(r)$ of section 1 in momentum space is of this form.
Let Γ be a C^1-curve contained in $\{k \in \mathbb{C} \mid \mathrm{Re}\, k \geqq 0, -a < \mathrm{Im}\, k \leqq 0\}$ and starting from the origin, as indicated in Fig. 2.

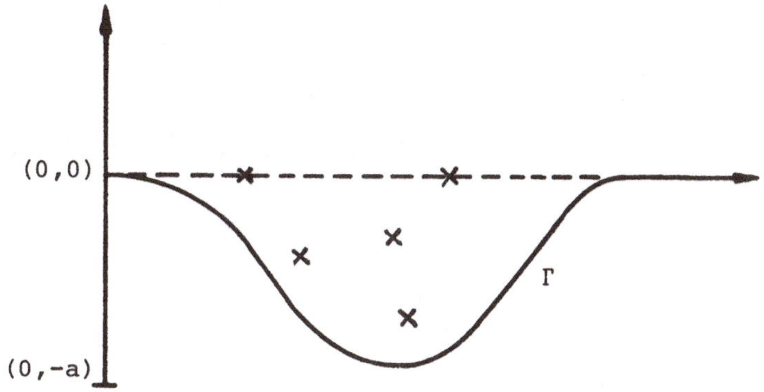

Fig. 2. Local deformation Γ

Let $L^{2,1}_\Gamma = L^2(\Gamma, h; (1+|k|^2)|dk|)$ and define the operators ${}_\Gamma Q \in \mathcal{B}(L^{2,-1}, L^{2,-1}_\Gamma)$ and $Q_\Gamma \in \mathcal{B}(L^{2,+1}_\Gamma, L^{2,+1})$ by

$$({}_\Gamma Qf)(k,\omega) = \int_{\mathbb{R}^+} dk' \int_{S^2} d\omega' [k^2+k'^2-2kk'\omega\cdot\omega']^{-2} k'^2 f(k',\omega') \tag{5.1}$$

$$(Q_\Gamma f)(k,\omega) = \int_\Gamma dk' \int_{S^2} d\omega' [k^2+k'^2-2kk'\omega\cdot\omega']^{-2} k'^2 f(k',\omega') \tag{5.2}$$

Let $H_{0\Gamma} \in \mathcal{B}(L^{2,+1}_\Gamma, L^{2,-1}_\Gamma)$ be defined by $H_{0\Gamma} f = k^2 f$,

$$V_\Gamma = {}_\Gamma Q\, Y\, Q_\Gamma \in \mathcal{B}(L^{2,+1}_\Gamma, L^{2,-1}_\Gamma)$$

$$H_\Gamma = H_{0\Gamma} + V_\Gamma \in \mathcal{B}(L^{2,+1}_\Gamma, L^{2,-1}_\Gamma) .$$

For the identification of these operators and their spectra for $\Gamma = \mathbb{R}^+$ with the operators and their spectra defined in the usual way through quadratic forms we refer to [6]. The operators V_Γ are linked through their action on the dense set of functions analytic in $\{k \mid \mathrm{Re}\, k > 0, |\mathrm{Im}\, k| < a\}$.

If χ is such a function and χ_Γ its restriction to Γ, we have that $V_{\Gamma'}\chi_{\Gamma'}$ is the analytic continuation of $V_\Gamma \chi_\Gamma$ to Γ'.

Lemma 5.1. a) The essential spectrum $\sigma_e(H_\Gamma)$ is the set $\overline{\Gamma}^2$.

b) The non-real, discrete spectrum $\sigma_d(H_\Gamma)$ lies between $\mathbb{R}^+$ and Γ^2, and $\sigma_d(H_{\Gamma_2}) \subseteq \sigma_d(H_{\Gamma_1})$ for Γ_1 below Γ_2 .

Definition 5.2. The point z is a resonance of (H_0, H) if there is a Γ with z between $\mathbb{R}^+$ and Γ such that $z^2 \in \sigma_d(H_\Gamma)$.

Let $\Phi_\Gamma \in L_\Gamma^{2,-1}$ be a solution of the Γ - distorted Lippman - Schwinger equation for z between $\mathbb{R}^+$ and Γ ,

$$\Phi_\Gamma + V_\Gamma R_{0\Gamma}(z)\Phi_\Gamma = 0 \tag{5.3}$$

Then $\psi_\Gamma = R_{0\Gamma}(z)\Phi_\Gamma$ is the Γ - resonance function at the resonance z.

Lemma 5.3. Let z be a resonance of (H_0, H) , and let Φ_Γ be a solution of (5.3). Then Φ_Γ has an analytic extension to $\{k \mid |\mathrm{Im}\, k| < a\}$ with values in h , and for z between $\mathbb{R}^+$ and Γ'

$$\Phi_{\Gamma'} + V_{\Gamma'} R_{0\Gamma'}(z)\Phi_{\Gamma'} = 0 \tag{5.4}$$

Lemma 5.4. The S - matrix has an analytic extension $\tilde{S}(k)$ to $\{k \mid |\mathrm{Im}\, k| < a\}$. The poles in $\{k \mid \mathrm{Re}\, k > 0 , -a < \mathrm{Im}\, k < 0\}$ coincide with the resonances.

The fact that every resonance is a pole of the S - matrix was proved by A. Jensen [26] by showing that $N(1 + V_\Gamma R_{0\Gamma}(z))$ is isomorphic to $N(\tilde{S}^{-1}(z))$ via the map $\Phi_\Gamma \to \Phi(z)$.

From Lemmas 1.5 and 5.4 follows the identity of the resonances of Definitions 1.3 and 5.2. The connection between the methods of sections 1 and 5 is established as follows. Using the residue theorem, we obtain from Lemma 1.3 by deformation of Γ into $\mathbb{R}^+$, crossing the resonance z ,

Lemma 5.5. Let z be a resonance and ϕ_Γ the solution of (5.3) and $\psi_\Gamma = R_{0\Gamma}(z)\phi_\Gamma$ the Γ - resonance function at z . Then

$$\phi + VR_0(z)\phi + V\pi i z^{-1}\gamma^*(\overline{z})\gamma(z)\phi = 0 \tag{5.5}$$

where ϕ is the restriction to $\mathbb{R}^+$ of the analytic continuation of ϕ_Γ to $\{k \mid |\mathrm{Im}\, k| < a\}$.

Comparing (5.5) with (1.21), we conclude

Lemma 5.6. Let z be a resonance, and let Φ, ψ and ψ_1 be the functions defined in (1.22), (1.24) and Φ_Γ, ψ_Γ the functions of Definition 5.3. Then Φ_Γ and ψ_Γ are the analytic continuations to Γ of the Fourier transforms of Φ and ψ_1 respectively.

Thus, the connection between the methods of sections 1 and 5 is very simple. The resonance function ψ of Definition 1.3 is a sum of two functions, $\psi_1 \in N(H_{0,-a})$ and $\psi_2 \in H_\beta$. By deformation of the continuous spectrum below the resonance, the function ψ_1 disappears as a residue term, and only ψ_2 is retained; the analytic continuation of its Fourier transform to Γ - with a simple pole at z - appears as a square-integrable eigenfunction of H_Γ with eigenvalue z^2. Note that the resonance function can be recovered from any of the square-integrable functions ψ_Γ as follows. Let ψ_0 be the analytic continuation of ψ_Γ to $\mathbb{R}^+$, set $\Phi_0 = (k^2 - z^2)\psi_0$ and $\Phi = F^{-1}\Phi_0$. Then $\psi = \tilde{R}_0(z)\Phi$ is the resonance function at z.

B. Local spectral deformation technique for more general potentials.

A dilation-analytic potential like $r^{-\alpha}$, $0 < \alpha < 2$, does not in an obvious way permit local spectral deformations, since its Fourier transform $c\,r^{\alpha-3}$ has a singularity at 0. However, by cutting off the forward direction of scattering we obtain operators which allow local spectral deformation, and by letting the cut-off go to 0 we obtain analogues of some of the results of A.

For the most general formulation of this theory we refer to [14], where the local spectral deformation technique is developed for a class of integral operators in momentum space comprising operators of convolution by the Fourier transform of a dilation-analytic, radial potential as well as the class of operators considered in A. An alternative approach, using analytic continuation in the parameter of the group generated by a function of the dilation generator, was suggested by I. Sigal [36]. In [14], Operators V_Γ and H_Γ are defined, and they are linked by analytic continuation of their action on analytic functions. Lemma 5.1 and Definition 5.2 are maintained under these general conditions. We shall refer to these generalizations as Lemma 5.1 - B and Definition 5.2 - B. Lemma 5.3 is not valid in general, since ϕ_Γ does not extend analytically across z. However, we have the following result.

Lemma 5.3 - B. Suppose that U is an S_α -dilation-analytic, radial potential and that $V = QYQ$ satisfies the condition of section 5 A. Let $H = H_0 + U + V$, let z be a resonance of (H_0, H), and let ϕ_Γ be a solution of (5.3) with z between $\mathbb{R}^+$ and Γ,

$$\Gamma \subset \mathcal{O} = \{k \mid \operatorname{Re} k > 0, -a < \operatorname{Im} k \leqq 0\} \cap S_\alpha .$$

Then ϕ_Γ has an analytic extension to

$$\mathcal{O} \setminus \{\zeta \in \mathcal{O} \mid \zeta = ze^{i\varphi}, \varphi \geqq 0\},$$

and for z between $\mathbb{R}^+$ and Γ' (5.4) holds.
Lemma 5.4 has been generalized under the conditions of Lemma 5.3 - B with the further assumption that for some $\varepsilon > 0$, $R > 0$

$$|U(r)| < cr^{-2-\varepsilon} \quad \text{for } r > R ;$$

we refer to this result as Lemma 5.4 - B. This shows that the resonances defined by local spectral deformation are genuine resonances.
Lemma 5.5 and 5.6 do not generalize, since they are based on Lemma 5.3.

The above results provide the basis for a further discussion of the results of section 4. Extending an argument of A. Jensen [28] to the present situation, we obtain

Lemma 5.7. Let U and V be as in Lemma 5.3 - B.
Suppose that the pair $(H_0, H_0 + U)$ does not have a resonance at z. Then $(H_0, H_0 + U + V)$ has a resonance at z if and only if $(H_0 + U, H_0 + U + V)$ has a resonance at z. Moreover, if z is such a resonance and ϕ_Γ is a solution of

$$\phi_\Gamma + (U_\Gamma + V_\Gamma) R_{0\Gamma}(z) \phi_\Gamma = 0 , \tag{5.6}$$

then

$$\chi_\Gamma = (1 + U_\Gamma R_{0\Gamma}(z)) \phi_\Gamma \tag{5.7}$$

is a solution of

$$\chi_\Gamma + V_\Gamma R_{1\Gamma}(z) \chi_\Gamma = 0 \tag{5.8}$$

where

$$R_{1\Gamma}(z) = (H_{0\Gamma} + U_\Gamma - z^2)^{-1} .$$

Proof. This follows from Definitions 5.2 and 5.2-B and the identity

$$1 + (U_\Gamma + V_\Gamma) R_{0\Gamma}(z) = (1 + V_\Gamma R_{1\Gamma}(z))(1 + U_\Gamma R_{0\Gamma}(z)) \tag{5.9}$$

It is an important consequence of Lemma 5.7, that the resonances discussed in section 4 are in general independent of the decomposition of the potential as a sum $U + V$. We formulate this as

Corollary 5.8. Let U_i and V_i satisfy the conditions of Lemma 5.3-B with $U_1 + V_1 = U_2 + V_2$, let $H_i = H_0 + U_i$, and suppose that z is not a resonance of (H_0, H_i) , $i = 1,2$. Then z is a resonance of $(H_1, H_1 + V_1)$ if and only if z is a resonance of $(H_2, H_2 + V_2)$.

C. Dilation - analytic potentials.

The method of B applies in particular to radial, dilation - analytic potentials. In this case the analyticity in an angle and the decay properties of the Fourier transform of the potential allows the curves Γ to be replaced by the half-lines $e^{i\varphi}\mathbb{R}^+$. This follows from Cauchy's integral theorem applied to the integral kernel, taking the limit over a sequence of curves Γ_n , such that Γ_n contains the line segment $e^{i\varphi}[\frac{1}{n}, n]$, as indicated in Fig. 3.

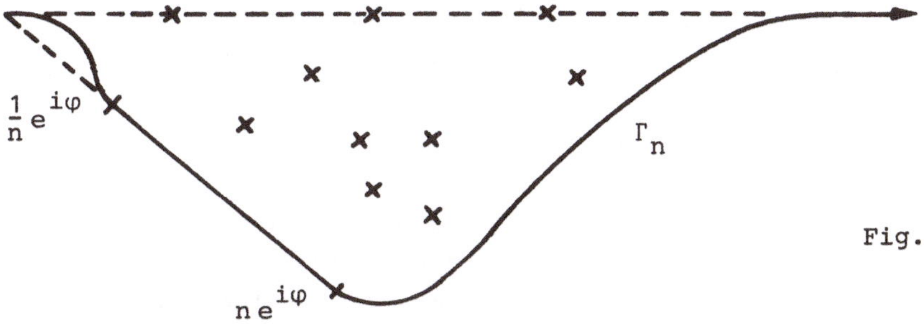

Fig. 3.

Fig. 3. Rotation as a limit of local deformations Γ_n

Thus, Lemma 5.1 and Definition 5.2 are valid with the curves Γ replaced by $e^{i\varphi}\mathbb{R}^+$, $-\alpha < \varphi < 0$.
This establishes the connection between the local and global deformation techniques in the case of dilation-analytic, radial potentials. The operators $H_{e^{i\varphi}\mathbb{R}^+}$ are identical with the operators $H(e^{i\varphi})$ defined by the method of complex dilations in momentum space. As an abstract method this dilation-analytic technique can be applied either in position- or momentum representation. Originally it was introduced in position space ([1],[15]) with the obvious application to Coulomb potentials and Yukawa potentials. The importance of the dilation-analytic method lies in its possibility of extension to the n - body problem, where first of all no other method so far applies, and secondly

the dilated operators are sufficiently simple to be useful in numerical calculations. We formulate the basic result of [15] as follows

Let $\{U(\rho) \mid \rho > 0\}$ be the one-parameter unitary group of dilations in $L^2(\mathbb{R}^3)$ defined by

$$(U(\rho)f)(\bar{r}) = \rho^{3/2} f(\rho\bar{r}) .$$

A symmetric operator V in $L^2(\mathbb{R}^3)$ is said to be S_α-dilation-analytic, if V is Δ-compact and the operator-valued function

$$V(\rho) = U(\rho)\, V\, U(\rho^{-1}) : \mathbb{R}^+ \to C(H^2(\mathbb{R}^3), L^2(\mathbb{R}^3))$$

has an analytic extension to the angular region

$$S_\alpha = \{\zeta = \rho e^{i\varphi} \mid \rho > 0 , |\varphi| < \alpha\}$$

For example, the operator of multiplication by a function $V(r)$, which is the restriction of a function $V(r\, e^{i\varphi})$ to $\mathbb{R}^+$, is dilation analytic, if $V(r\, e^{i\varphi}) = 0(r^{-\frac{3}{2}+\varepsilon})$ for $r \to 0$ and $V(r\, e^{i\varphi}) \to 0$ for $r \to \infty$, $|\varphi| < \alpha$ (cf. [5] for a more general result).

Let

$$H = H_0 + \sum_{i<j=1}^{n} V_{ij}$$

by the n-body Hamiltonian with the center-of-mass separated out and with Δ_{ij}-compact, S_α-dilation-analytic two-body interactions.

A self-adjoint, analytic family of type A in the sense of Kato is defined for $\zeta \in S_\alpha$ by

$$H(\zeta) = \zeta^{-2} H_0 + \sum_{i<j=1}^{n} V_{ij}(\zeta)$$

The operators $H(e^{i\varphi})$ are unitarily equivalent for φ fixed, and discrete eigenvalues are analytic in ζ and therefore constant unless absorbed by the cintinuous spectrum. These simple facts and the Weinberg-van Winter equation are some of the ingredients in the induction proof [15] of the following result, cf. Fig. 4.

<u>Theorem 5.9</u>. a) The essential spectrum of $H(\zeta)$ is the set of half-lines $\{\lambda + e^{-2i\varphi}\mathbb{R}^+\}$, where λ runs over all thresholds of the system (including non-real thresholds due to resonances of subsystems).

b) The real, discrete eigenvalues of $H(\zeta)$ are the discrete eigenvalues of H and the embedded eigenvalues of H, which are not thresholds.

c) The non-real, discrete eigenvalues of $H(\zeta)$ are φ-independent unless absorbed by the continuous spectrum and are contained in the sector bounded by $\lambda_e + \mathbb{R}^+$ and $\lambda_e + e^{-2i\varphi}\,\mathbb{R}^+$, where λ_e is the smallest threshold of the system.

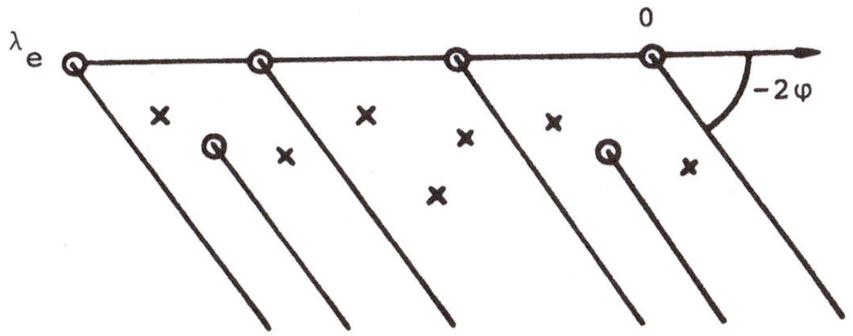

Fig. 4. Spectrum of $H(e^{i\varphi})$

Definition 5.10. A point μ is a resonance, if μ is a discrete eigenvalue of $H(e^{i\varphi})$ for some $\varphi \in (0,\alpha)$ (and hence for all φ in some interval). The eigenfunction ψ of the operator $H(e^{i\varphi})$ corresponding to the eigenvalue μ is called the φ - resonance function associated with the resonance μ.

Whereas examples of resonances of pairs $(H_0, H_0 + U)$ of two - body operators, where U is dilation-analytic, are scarce, and it is a non-trivial problem to prove existence of resonances for concrete examples, the situation is quite different for the many - body problem. This has to do with the fact that many - body resonances can be produced by perturbation of eigenvalues embedded in the continuum.

A good illustration of this difference is provided by the Coulomb potential. Let

$$H = -\Delta + \alpha r^{-1}, \quad \alpha \neq 0 .$$

The operator-valued function

$$H(\zeta) = -\zeta^{-2}\Delta + \alpha\zeta^{-1}r^{-1}$$

is analytic for $\zeta \in \mathbb{C} \setminus \{0\}$, and

$$H(e^{i\pi}) = -\Delta - \alpha r^{-1}$$

is self-adjoint. If H had a resonance μ in the sense of Definition 5.10, by the standard dilation-analytic argument μ would be an eigenvalue of $H(e^{i\pi})$, a contradiction. Thus, the two - body operator with Coulomb potential has no resonances.

In contrast to this, consider the operator of the Helium atom with infinitely heavy nuclens,

$$H(\gamma) = -\Delta_1 - \Delta_2 - \frac{1}{r_1} - \frac{1}{r_2} + \gamma\frac{1}{r_{12}}$$

with a coupling constant γ. With a suitable choice of units, $\gamma = 1$ corresponds to the actual Helium atom, and $\gamma = 0$ corresponds to a system of two Hydrogen atoms. The operator $H(0)$ has for each threshold λ except the lowest one a sequence $\lambda_n \to \lambda$ of eigenvalues embedded in the continuum. Through complex dilation each λ_n becomes a discrete eigenvalue of $H(e^{i\varphi}, 0)$ for $0 < \varphi < \frac{\pi}{2}$. For small positive γ, each λ_n moves into the lower half-plane as a discrete eigenvalue of $H(e^{i\varphi}, \gamma)$, i.e. a resonance in the sense of Definition 5.10. For a detailed discussion, we refer to Simon [38]. This example clearly shows, that resonances in two - body and many - body systems are distinctly different phenomena. Resonances neccessarily appear in a multi-channel situation, if analytic continuation is at all possible. In the two - body case, even if analytic continuation is assured, additional features of the potential are required to produce resonances.

Theorem 5.9 has been very useful from a computational point of view. From a theoretical point of view the understanding of resonances and resonance functions is still very incomplete.
First of all, the resonances of Definition 5.10 should be identified with poles of the S - matrix. Some efforts in this direction have been made. For the two - body problem the following result was proved in [11]:

Lemma 5.11. Let V be a dilation-analytic, short-range interaction, i.e. $V(1 + r^2)^{\frac{1}{2} + \varepsilon}$ is dilation-analytic for some $\varepsilon > 0$. Then the S - matrix has an analytic extension from $\mathbb{R}^+$ to S_α with poles precisely at resonances.
This result has been to some extent generalized to the three - body problem [12] and to the n - body problem below the smallest three - cluster threshold [13].
Diagonal elements S_λ of the S - matrix are continued analytically on the part of the Riemann surface attached to (λ, λ'), where S_λ describes elastic scattering within the channel defined by the threshold λ, and λ' is the next threshold. The poles of the analytic continuation $\tilde{S}_\lambda$ are resonances between the cuts $\lambda + e^{-2i\varphi}\mathbb{R}^+$ and $\lambda' + e^{-2i\varphi}\mathbb{R}^+$ for φ in some interval. Conversely, such a resonance μ is a pole of $\tilde{S}_\lambda$ unless μ becomes an embedded eigenvalue of $H(e^{i\varphi_0})$ for that value φ_0 of φ such that μ lies on the half-line $\lambda + e^{-2i\varphi_0}\mathbb{R}^+$. In this latter case μ re-appears as a resonance on the next Riemann-sheet, and $\tilde{S}_\lambda(\zeta)$ is regular at μ.

If the potentials are exponentially decaying, it was proved in 1-4 [1] that for the three - body problem with $\lambda < 0$, the total S - matrix can be analytically continued with poles precisely at resonances. This indicates that the dilation-analytic result of [12] is incomplete due to the method. From the point of view of scattering theory the complex dilation method selects the particular diagonal elements $\tilde{S}_\lambda$. It remains an open problem, whether such embedded eigenvalues of the dilated operators do occur. If such special resonances exist, which are duplicated on two different sheets of the Riemann surface, they are of mathematical interest, whatever might be their physical significance. In the two - body case and on the lowest cut they are excluded [11] [13]). The problem of identification of resonances with poles of the S - matrix for the many - body problem has also been discussed in [21], [31], [35].

The advantage of spectral deformation techniques is that the resonances become discrete eigenvalues with square-integrable eigenfunctions of a non-selfadjoint Hamiltonian. In the case of exponentially decaying potentials we have explained in sections 2 and 5 A, how these eigenfunctions are related to the resonance functions. A similar understanding of the resonance functions has not been achieved in the dilation-analytic case, even for the two - body problem.

The resonance eigenfunctions at the resonance μ are known to be analytic in $\zeta = \rho e^{i\varphi}$ for $\rho > 0$ and φ in the interval for which μ is a discrete eigenvalue of $H(e^{i\varphi})$. In the case of an embedded eigenvalue of $H(e^{i\varphi})$ for $\varphi > 0$, the eigenfunction continues analytically as an eigenfunction at the same resonance on the next Riemann sheet. Also, by the method of Combes and Thomas [16], including the boost group in the dilation-analytic theory, it follows that the eigenfunctions of $H(e^{i\varphi})$ lie in an exponentially weighted L^2-space.

Let us mention finally two important examples of a fruitful interaction between pure mathematics and applications, the Stark effect and the Born - Oppenheimer approximation. The dilation-analytic technique was applied for numerical calculation of resonances for atoms in a constant electric field (Stark effect), although the theory did not cover such potentials. The solution of this problem was given by I. Herbst [22] and extended by Herbst and Simon [23] to the n - body problem. The key result is the surprising fact, that the spectrum of the dilated $(-\Delta + \varepsilon x)$ - operator has an empty spectrum and hence by perturbation theory the dilated $(-\Delta + \varepsilon x - \frac{1}{r})$ - operator has a discrete spectrum, identified with the resonances of the Stark effect.

It is further proved that these resonances as $\varepsilon \to 0$ converge to the eigenvalues of the Hydrogen atom. Thus, the resonances are of a different character from the two - body resonances as well as the many - body resonances discussed above. Whereas the latter are produced by perturbation of continuum - eigenvalues by a small interaction between the electrons, the Stark effect resonances are produced by perturbation of discrete eigenvalues by a small electric field. Alternative approaches to the Stark effect are due to S. Graffi and V. Grecchi [18], [19] and K. Yajima [46] and the Stark effect has been treated by K. Yajima [47] and by S. Graffi, see the talk of S. Graffi at this conference.

Another case of sucessful application of dilation-analytic techniques outside its original domain of validity is the calculation of molecular resonances using the Born - Oppenheimer approximation. A solution of this problem was proposed by B. Simon [40], who suggested the use of exterior dilation, i.e. to keep all coordinates fixed up to a certain distance and dilate only outside this distance (outside the fixed nuclei). For an application of exterior scaling to a two - body resonance problem, see the talk of R. Seiler at this conference.

Acknowledgement

The Center for Interdisciplinary Research at Bielefeld has provided a stimulating atmosphere for the present work. I would like to express my thanks to Prof. L. Streit for the hospitality at ZiF, to the organizers of this conference, Profs. S. Albeverio, L. Ferreira and L. Streit, and to Profs. L. Streit and A. Grossmann for fruitful discussions.

References.

Sections 1 - 4:

1. E. Balslev, Aarhus University preprint ser. 1981/82 no. 3, to appear in Adv. in Appl. Math.

2. C. L. Dolph, J. B. McLeod and D. Thoe, J. Math. Anal. Appl. 16 (1966), 311 - 332.

3. S. T. Kuroda, An Introduction to Scattering Theory, Aarhus University lecture notes ser. no. 51, 1978.

4. R. Newton, Scattering Theory for Waves and Particles, Springer-Verlag, 2nd edition 1982.

5. N. Schenk and D. Thoe, Rocky Mountain J. Math. 1 (1971), 89 - 125.

6. E. Titchmarsh, The Theory of Functions, Oxford University Press 1939.

Section 5:

1. F. Aguilar and J. M. Combes, Comm. Math. Phys. 22 (1971), 269-279.

2. S. Albeverio and R. Høegh-Krohn, Univ. Bochum preprint, to appear in J. Math. Anal. Appl.

3. J. Avron and I. W. Herbst, Comm. Math. Phys. 52 (1977), 239 - 254.

4. D. Babbitt and E. Balslev, Comm. Math. Phys. 35 (1974), 173 - 179.

5. D. Babbitt and E. Balslev, J. Funct. Anal. 18, 1 (1975), 1 - 14.

6. D. Babbitt and E. Balslev, J. Math. Anal. Appl. 54, 2 (1976) 316 - 347.

7. E. Balslev, Ann. Phys. 73, 1 (1972), 49 - 107.

8. E. Balslev, Rep. Math. Phys. 5 (1974) 219-279, 393 - 413.

9. E. Balslev, Arch. Rat. Mech. Anal. 59 , 4 (1975), 343 - 357.

10. E. Balslev, Comm. Math. Phys. 52 (1977), 127 - 146.

11. E. Balslev, J. Funct. Anal. 29 , 3 (1978), 375 - 396.

12. E. Balslev, Ann. Inst. H. Poincaré 32, 2 (1980), 125 - 160.
Aarhus Univ. preprint no. 26 (1978).

13. E. Balslev, Comm. Math. Phys. 77 (1980), 173 - 210.

14. E. Balslev, Mittag - Leffler Inst. Publ. Series 1982,
to appear in J. Funct. Anal.

15. E. Balslev and J. M. Combes, Comm. Math. Phys. 22 (1971), 280 - 294.

16. J. M. Combes and L. E. Thomas,
Comm. Math. Phys. 24 (1973), 251 - 270.

17. M. Courbage, Lett. Math. Phys. 2 (1978), 451 - 457.

18. S. Graffi and V. Grecchi, Comm. Math. Phys. 62 (1978), 83 - 96.

19. S. Graffi and V. Grecchi, Comm. Math. Phys. 79 (1981), 91 - 109.

20. A. Grossman and A. Tip, J. Phys. A: Math. Gen. 13 (1980),
3381 - 3397.

21. G. A. Hagedorn, Comm. Math. Phys. 65, 2 (1979), 181 - 201.

22. I. W. Herbst, Comm. Math. Phys. 64 (1979), 279 - 298.

23. I. W. Herbst and B. Simon, Comm. Math. Phys. 80 (1981), 181 - 216.

24. Y. K. Ho, Phys. Rev. 99, 1 (1983).

25. Int. J. Quantum Chemistry 14 (1978).

26. A. Jensen, J. Math. Anal. Appl. 59 (1977), 503 - 513.

27. A. Jensen, Manuscripta Math. 25 (1978), 61 - 77.

28. A. Jensen, Ann. Inst. Henri Poincaré XXXIII, 2 (1980), 209 - 223.

29. B. R. Junker, Adv. Atomic and Molecular Phys. 18 (1982), 207-263.

30. J. Nuttall, J. Math. Phys. 8 (1967), 873 - 877.

31. J. Nuttall and S. E. Singh, Comm. Math. Phys. 72, 1 (1980), 1-13.

32. J. Rauch, J. Funct. Anal. 35 (1980), 304 - 315.

33. M. Reed and B. Simon, Methods of Modern Mathematical Physics IV, Academic Press, London, New York, ch.XII 6, XIII 10 (1978).

34. W. P. Reinhardt, Ann. Rev. Phys. Chem. 33 (1982), 223.

35. I. M. Sigal, Trans. Amer. Math. Soc. 270 , 2 (1982), 409 - 437.

36. I. M. Sigal, Weizman Inst. preprint 1983.

37. B. Simon, Comm. Math. Phys. 27 (1972), 1 - 10.

38. B. Simon, Ann. Math. 97 (1973), 247 - 274.

39. B. Simon, Math. Ann. 207 (1974), 133 - 138.

40. B. Simon, Phys. Lett. 71 A (1979), 2 - 2, 211 - 214.

41. E. C. Sudarshan, G. Parravicini and Y. Gorini, J. Math. Phys. 21 , 8 (1980), 2208 - 2226.

42. L. E. Thomas , Helv. Phys. Acta 45 (1973), 1057 - 1065.

43. R. A. Weder, Ann. Inst. Henri Poincaré, 20 (1974), 211 - 220.

44. R. A. Weder, J. Math. Phys. 15 (1974), 20 - 24.

45. C. van Winter, J. Math. Anal. Appl. 47 (1974), 633 - 670 ; 48 (1974), 368 - 399 ; 49 (1975), 88 - 123.

46. K. Yajima, J. Fac. Sci. University Tokyo, I; Sect. I A, 26 (1979), 377 - 390 ; II ; Sect. I A, 28 (1981), 1 - 15.

47. K. Yajima, Comm. Math. Phys. 87 (1982), 331 - 352.

On the Shape Resonance

J.M. Combes and P. Duclos

Université de Toulon et du Var
and Centre de Physique Théorique, CNRS
Marseille Cedex 9, France

R. Seiler

Fachbereich Mathematik,
Technische Universität Berlin
D-1000 Berlin 12

Introduction

Before we present our main results about shape resonances we should like to make two remarks. The first one concerns the historical origin of shape resonances, the second one the connection between lifetime of a state and poles of the S-matrix.

1. In 1928/29 Gamov published the celebrated articles "Zur Quantentheorie der Atomkerne" and "Zur Quantentheorie der Atomzertrümmerung" [1] [1]. The two articles contain the answer to the following questions

a) Why do nuclei radiate?
b) If they radiate, why do they not radiate much more?
c) Why is the range of lifetime for nuclei so large (largest:smallest $\sim 10^{16}$)?

Gamov explained these phenomena by applying the Schrödinger equation to the nuclear system. (Previously it had been used to describe electrons of atoms and molecules). He considers a charged particle in a potential V(x), $0 \leq x < \infty$, which is attractive for x small and has a Coulomb repulsion for x large (Fig. 1).

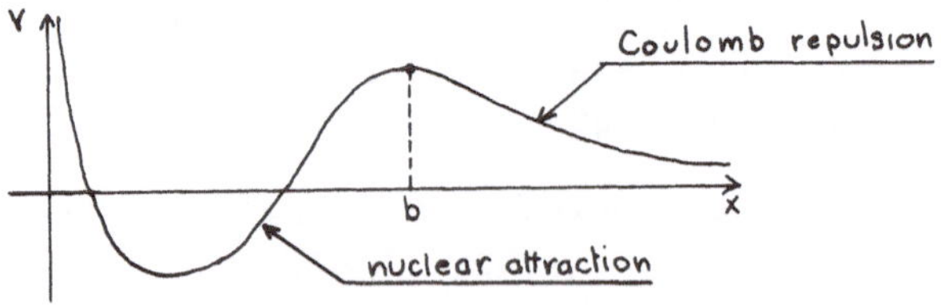

Fig 1

Since the precise form of the repulsive part of the potential was not known he computed just the case of a simple square well, $V(x) = 0$ $(0 \leq x < a,\ b < x < \infty)$, $V(x) = V$ $(a \leq x \leq b)$. The solutions of the Schrödinger equation vanishing at zero behave for large x and $E > 0$ in the well known manner

$$\psi(k,x) = \frac{i}{2k}\{f(k)e^{-ikx} - f(-k)e^{+ikx}\}\ ,\ k = \sqrt{E}\ .$$

In analogy to the boundary condition describing bound states $f(k) = 0$, $\operatorname{Im} k > 0$, Gamov imposed the boundary condition $f(k) = 0$, $\operatorname{Im} k < 0$, $\operatorname{Re} k > 0$, for the decaying (Gamov) state. After telling the reader not to worry about the exponential increase in x of such states he computes their life time. Generalizing the expression to potentials of the type depicted in Fig. 1 he gets for the inverse of life time (the width)

$$\text{const.}\ \exp - \frac{2m}{\hbar^2}\int \sqrt{V - E}\ dx\ ,$$

where the integral ranges over all x with $V(x) \geq E$. It has all the qualitative features necessary to answer the questions raised at the beginning. In particular it is exponentially small in the potential hence difficult to compute.

2. The connection between life time and poles of the S-matrix: Since there is no satisfactory version of this connection we shall describe it on a heuristic level for the simplest physically interesting model; 2 particles without spin interacting via a radially symmetric potential. It would be nice to see progress in understanding this question better. First, let us relate the poles of the resolvent on the second Riemann sheet with the exponentially slow decay in time. Consider the one particle Schrödinger operator $H = -\Delta + V$. We know that for a regular potential of compact support the function $\langle\phi,(H-z)^{-1}\phi\rangle$ is meromorphic on the Riemann surface of $\sqrt{z}$, for all ϕ regular and decreasing faster than $\exp - \alpha|x|$ for all $\alpha > 0$, [2]. Consider the expectation value of the time evolution operator in a state ϕ orthogonal to all bound states of H (there are finitely many). It is related to the resolvent by Laplace transformation

$$\langle\phi,e^{-iHt}\phi\rangle = \frac{1}{2\pi i}\int_\Gamma dz\, e^{-izt}\langle\phi,(H-z)^{-1}\phi\rangle\ .$$

The contour of integration is depicted in Fig. 2a)

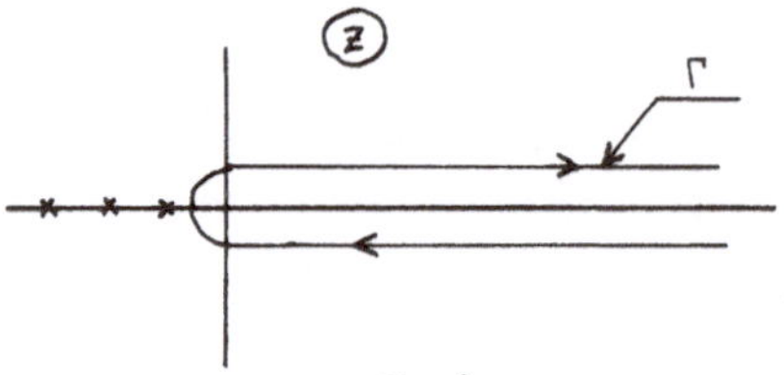

Fig 2a

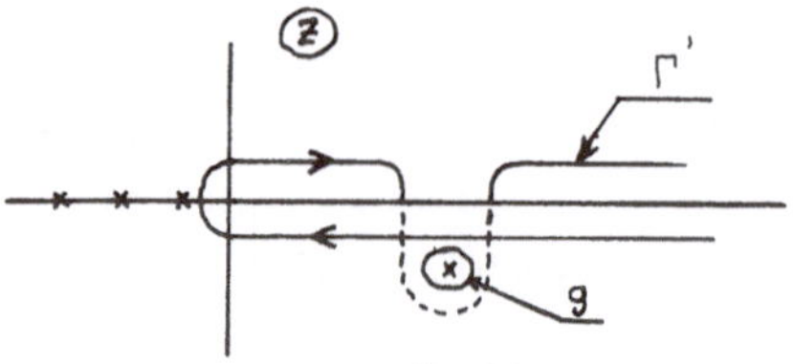

Fig 2b

Due to analyticity of the integrand we are allowed to deforme the contour Γ into Γ' (Fig. 2b). Assume there is a Pole g on the second sheet. Then we get the result

$$<\phi, e^{-iHt}\phi> = <\phi, P_g \phi> e^{-igt} + R(t)$$

where P_g is the "Projection on a Gamov state" [2)] and R(t) is the integral over Γ'. The above formula shows that life time is given by the inverse of the imaginary part of g, provided R(t) is small. It is an important problem to prove under what conditions R(t) is small in an appropriate time interval (t_0, t_1). On general grounds we know that t_0 must be larger than zero and t_1 finite (see e.g. [3] for a discussion of this problem in a special case). Second, we shall relate the poles of $<\phi, (H-z)^{-1}\phi>$ with poles of the S-matrix: The integral kernal of the resolvent in a sector of fixed angular momentum can be expressed by the fundamental solutions of the Schrödinger equation u(k,x), f(k,x), $u(k,0) = 0$, $u'(k,x) = 1$, $f(k,x)e^{-ikx} \to 1$ $(x \to \infty)$,

$$(H-z)^{-1}(x,y) = \frac{1}{f(k)} \begin{cases} (k,x)\ f(k,y), & x < y \\ f(k,x)\ u(k,y), & x \geq y \end{cases} \qquad (k = \sqrt{z},\ \mathrm{Im}\, k \geq 0)$$

where $f(k) = f(k, x=0)$ still denotes the Jost function [4].
This formula proves again that the expectation value $<\phi, (H-z)^{-1}\phi>$ is meromorphic in $k \in \mathbb{C}$ and has singularities at zeros of f(k) for the ϕ's described above. The Jost function f(k) is related to the S-matrix by [4] $S(k) = \frac{f(-k)}{f(k)}$.
Hence the poles of S are identical to the poles of $<\phi, (H-z)^{-1}\phi>$ (ϕ regular and compact support); for Im k < 0, the poles coincide with the values of k for which the Gamov boundary condition in the solution of the Schrödinger equation is satisfied.
This identification between resonances and poles of the analytic continuation of expectation values $<\phi, (H-z)^{-1}\phi>$ for certain ϕ's can

also be made for other classes of potentials than those of compact support, in particular for "dilation analytic potentials" [5]. In [6] it is shown that the poles which will be obtained by the methods described in section III coincide again with zeros of the Jost function hence with resonances.

II. The Shape Resonance

Now we shall describe a simplified version of the model we have analysed in [6]. Consider a particle on the positive half axis whose dynamics is given by the Schrödinger operator $H(k) = -k^4\Delta + V \quad (k > 0)$. The potential is supposed to satisfy the conditions A.

A1. V is a positive twice differentiable function up to a finite number of points.

2. $\lim_{x\to\infty} V(x) = 0$

3. There exists a positive number b such that
   - i) the absolute minimum of V on [0,b] is taken at only one point x_m in the interior of the interval and $V(x_m) = 0$, $V''(x_m) > 0$. [3)]
   - ii) V is strictly positive on $[b,\infty]$. [4)]

In the following we shall analyse properties of the operator H(k) for $k \to 0$. However we shall omit k in all the formulae where there is no danger of confusion.

Later some further technical conditions will be imposed on the potential. V looks typically like the potential function used by Gamov (Fig. 1). Our approach to the shape resonance problem is the following: We want to compare the dynamics of H with the one given by H^D, where H^D has the same differential symbol but a Dirichlet boundary condition at b (Fig. 1) i.e. the domain of definition contains only functions vanishing at x = b. H^D splits into the direct sum $H^D = H_{int} \oplus H_{ext}$.

H_{int} and H_{ext} act on functions defined on the interval $(0,b)$ and (b,∞) respectively. The spectrum of H^D is the union of the spectra of H_{int} and H_{ext}. It has therefore lots of point spectrum in the continuum (Fig. 3a) since H_{int} has pure point spectrum; here it is important to notice that for k small the lowest part of this point spectrum depends mostly on the shape of the potential near its minimum, whereas an infinity of eigenvalues depends on the boundary conditions at 0 and b.

Fig 3a

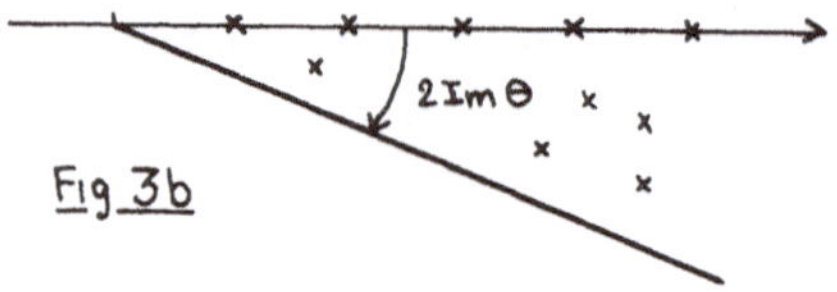

Fig 3b

Starting from H^D we consider the transition to H as a perturbation. It is expected that the lowest part of the point spectrum changes into resonances which we call accordingly "Shape Resonances". The precise statement of this expectation, the proof of it and development of a computational procedure for computing the resonant energies (poles of resolvent in the second sheet) is the purpose of our analysis [6]. Let us stress that we shall only be able to make statements for k sufficiently small. This means the mass of the particles considered have to be sufficiently large. Since k^2 is proportional to $\hbar$ this is also the semiclassical regime. There is a large literature on the problem of shape resonances. To our knowledge only Ashbaugh and Harrell II [7] treated a general class of potentials with mathematical rigour. However their restrictions on the potentials are very strong (they are necessarily of compact support). Furthermore their method does not seem to be of simple computational nature. There is an interesting nonrigorous article by Jona-Lasinio, Martinelli and Scopola on the subject where they apply ideas of probability theory [19] (large deviation technics). Those ideas were successful in the analysis of tunneling for the double well potential [8]. In our analysis the exterior scaling technic introduced by Simon was very helpful [9]. We shall describe the method in short terms: Consider the mapping

$$U_\theta : R_+ \to R_+ \qquad (\theta \in \mathbb{R})$$

$$x \to \begin{cases} x, & (x \leq b) \\ b + e^{\theta}(x-b), & x > b \end{cases}$$

U_θ induces on $L^2(R_+)$ a unitary mapping U_θ. The operator valued function $H_\theta := U_\theta H U_\theta^{-1}$ is analytic in the strip $S_\alpha = \{\theta \,|\, |\mathrm{Im}\,\theta| < \alpha\}$ provided V, restricted to the exterior interval (b,∞), is dilation analytic in S_α. This follows from Krein's formula [10] relating the resolvent of H_θ and H_θ^D, since H_θ^D is analytic in S_α by construction. 5)
The spectrum of H_θ^D looks typically like the one depicted in Fig. 3b.

For our general result the potential has to satisfy the following technical conditions B [6]:

B1. $V'(x) < 0$ for $b \leq x < \infty$

2. There is an $\alpha > 0$ such that V is dilation analytic in $S_\alpha = \{\theta | \theta \in \mathbb{C}, |\text{Im}\,\theta| < \alpha\}$ as an operator on $L^2(b,\infty)$. [7]

3. The L^1-norm of $(x-b)V(x)$ on $[b,\infty)$ is finite

4. There exists $\delta \in (0,1]$ such that

$$|V''(x)| \leq \text{const}\; x^{3\delta-4} |V'(x)|^\delta \;, \quad (x > b) \;.$$

Our main result is summarized in the following theorem:

<u>Theorem</u>: Let V satisfy conditions A and B and let $E^D(k) \in \sigma(H_{int}(k))$ be the n-th spectrum valued function, $k \in (0,k_0)$ for a $k_0 > 0$, counted from below. Then there exists a $k_1 \in (0,k_1)$, a complex number θ with $\text{Im}\,\theta < \alpha$ and a spectrum valued function $E(k) \in \sigma(H_\theta(k))$ $(k \in (0,k_1))$ such that

$$E(k) = E^D(k) + \sum_{n=1}^{\infty} \frac{t^n}{n!} a_n \qquad (0 < k < k_1).$$

The tunneling parameter t and the coefficients a_n are k dependant. The a_n's $(n > 1)$ are polynomially bounded in k^{-1} and $a_1(k) = O(k^2)$;

$$t(k) = o(\exp - 2\beta k^{-2} \int_b^\infty \sqrt{V(s)ds}\,) \qquad (\beta < 1) \quad .$$

$E(k)$ is the unique spectral point of $H_\theta(k)$ in an appropriate neighbourhood of $E^D(k)$.

We conclude this section by a series of remarks

1. There exists a dense set of vectors ϕ such that $\langle\phi, (H - z)^{-1} \phi\rangle$ has a meromorphic extension into the second sheet of $\sqrt{z}$ and has a pole at $E(k)$. Accordingly $E(k)$ is a shape resonance energy.

2. A potential which satisfies the conditions of the theorem is e.g.

$$V(x) = \begin{cases} (x-1)^2 & , \; x \leqq 2 \\ (x-1)^{-4} & , \; x \geqq 2 \end{cases}$$

Another example is the potential $V(x) = (x-1)^2 e^{-x^2}$ similar to the one considered by Brändas and Korsch in this conference.

3. Definitions of the tunneling parameter t and the coefficients $\{a_n\}_{n=1}^{\infty}$ will be given later (section III, Box 11 and remark 13 respectively).

III Structure of Proof

In this last section we present the structure of proof in diagramatic form (Fig. 4). The boxes contain declarations of used objects and/or statements separated by a broken line. They are linked by arrows indicating the flow of the argument. The main line of thought is marked by a double line. The operator valued analytic function H_θ is defined by the steps 1 to 4, 14 and 15. The theorem in step 5 is equivalent to the main theorem mentioned previously. Remarks about the individual steps are added subsequently. 7 to 9 are technical results necessary to get 10 which is technically the most important result.

Remarks

1) H and H^D have been introduced earlier. According to our approach we consider H^D as the point of reference. In particular we want to analyse the influence of the perturbation by the boundary condition $H^D \to H$ on the discrete eigenvalue E^D of H^D with eigenvektor Φ^D. Notice that Φ^D is supported in (0,b) due to the assumptions on the potential.

2) For $k \to 0$ $E^D(k)$ converges to zero like k^2; because the particle is getting heavier and heavier the harmonic approximation gets predominant [11]. It is technically useful to introduce scaled objects

   $H(k) = k^2 U(k) h(k) U(k)^{-1}$, $(k > 0)$

   U(k) unitary dilation by $1/k$ around the minimum of V.

   $\Phi^D(k) = U(k)\phi^D(k)$.

   $E^D(k) = k^2 e^D(k)$.

3) r and r^D are the resolvents of h and h^D respectively considered at a point $z = -a, a > 0$. The analysis of the perturbation by a Dirichletboundary condition is conveniently discussed in terms of resolvents because of Krein's identity

   $r = r^D + |\tau\rangle\langle\tau|$

   where the vector τ is given by

$$\tau(x) := (r(k^{-1}b, k^{-1}b))^{-1/2}\, r(k^{-1}b,x)$$
$$= (r(k^{-1}b, k^{-1}b))^{+1/2}\, D_1\{r_{int}(k^{-1}b,x) - r_{ext}(k^{-1}b,x)\}.$$

D_1 denotes the derivative with regard to the first argument.

4) Exterior scaling to the right of $k^{-1}b$ is now introduced. [8)] $u_\theta r^D u_\theta^{-1}$ is analytic in S_α because $r^D = r_{int} \oplus r_{ext}$ and $u_\theta = \mathbb{1} \oplus u_{\theta,ext}$ [9)]. The analogous statement about r is more delicate. It turns out that τ is an U_θ analytic vector. So we can define r_θ by extending Krein's formula to complex values of $\theta \in S_\alpha$. This defines h_θ (Box 14) and H_θ (Box 15). h_θ^D and H_θ^D can either be defined along the same lines or directly. Notice that the quadratic form generated by h_θ has a θ-dependant domain. It can be characterised explicitly [16].

5) Our main result stated in section II can be reformulated as a statement about resolvents and eigenvalues of microscopic operators. The resulting theorem is equivalent to the theorem stated previously. The rule of substitution is: $E^D(k) \in \sigma(H_{int}(k))$ is replaced by $f^D(k) \in \sigma(r_{int}(k))$, $E(k) \in \sigma(H_\theta(k)$ by $f(k) \in \sigma(r_\theta(k))$, $\{a_n\}_{n=1}^\infty$ by $\{\sigma_n\}_{n=1}^\infty$. The coefficients a_n can be expressed in terms of the σ_n's and vice versa. Their behaviour in the limit $k \to 0$ is the same.

6) The resolvents of r and r^D are related by the Weinstein-Aronszajn formula

$$(r-z)^{-1} = (r^D - z)^{-1} + \frac{(r^D - z)^{-1}\,\tau\rangle\langle\tau\,(r^D - z)^{-1}}{w(z)}$$

where w(z) denotes the W-A determinant. It can be shown that the zero set of the meromorphic function w(z) coincides with the spectrum of r_θ in an appropriatly chosen neighbourhood $N \subset \mathbb{C}$ of f^D (N will have to depend on k). This step reduces the proof of the theorem to the analysis of the zero set of w.

7) The numerical range of $r_{ext,\theta}$ can of course be analysed in terms of $h_{ext,\theta}$. This operator is analytic in θ and can be written as a second order polynomial and remainder in θ. The numerical range of the second order part can be estimated more or less explicitly (assumption B1 and B3 are relevant). The remainder can be treated perterbatively.

8) This technical result is proved à la Mourre [20], (Assumption B4 is relevant).

9) Here we use a comparison technic similar to the one used previously for the estimate of τ itself (Prop. 6 and 7 of Part I in ref. [11]).

10) $\widehat{(r^{D}-z)}^{-1}$ is the resolvent reduced with respect to the projection onto ϕ^{D}. It is enough to prove $\langle\tau|\widehat{(r^{D}-z)}^{-1}\tau\rangle\to 0$ $(z\in N,\ k\to 0)$. The scalar product contains two terms, one from the interior one from the exterior. The first one geos to zero because the spacing of the eigenvalues of r^{D}_{int} is of the order one and $\|\tau\|\to 0$, $(k\to 0)$. The second one is difficult to control. By simple algebra the problem can be reformulated in terms of $f(z) := \langle\tau,(h_{ext}-z)^{-1}\tau\rangle$. It can be analytically continued through the real axis, $f(z) := \langle\tau_{\theta},(h_{ext,\theta}-z)^{-1}\tau_{\theta}\rangle, \theta\in S_{\alpha}$. The difficulty in proving statement 10 is to show that f(z) has no poles in a neighbourhood of e^{D}. This is done as follows: f(z) can be splitt into two parts using the resolvent equation,

$$f(z) = f_1(z) + f_2(z)$$

$$f_1(z) := \langle\tau,(h_{ext}-(e^{D}+i0))^{-1}\tau\rangle$$

$$f_2(z) := (z-e^{D})\langle\tau_{\theta},(h_{ext,\theta}-z)^{-1}(h_{ext,\theta}-e^{D})^{-1}\tau_{\theta}\rangle$$

The first term can be controled by statements 8 and 9. The second one by 7 and an appropriate choice of N.

11) The exponential bound on t(k) is proven in the same way as in the case of a multiple well potential (by a complex boost method, [12]). $t(k)>0$ because τ can not be orthogonal on an eigenstate of r^{D}.

12) This statement follows from the identity

$$w(z) = (f^{D}-z)^{-1}\sigma(z)^{-1}\tilde{w}(z),$$

statement 10 and $t>0$ (statement 11).

13) The zeros of $\tilde{w}(z)$ in the neighbourhood of f^{D} can be computed by the formula of Lagrange

$$f = f^{D} + \sum_{n=1}^{\infty}\frac{t^{n}}{n!}\sigma_n$$

$$\sigma_n := \left(\frac{d}{dz}\right)^{n-1}(\sigma(z))^{n}\Big|_{z=f^{D}+i0}$$

This final part is analogous to the double well case [12]. The coefficients $\{a_n\}_{n=1}^{\infty}$ are related to $\{\sigma_n\}_{n=1}^{\infty}$ by the following equations:

$$E - E^{D} = k^{2}\left(\frac{1}{f}-\frac{1}{f^{D}}\right) = \sum_{n=1}^{\infty}\frac{t^{n}}{n!}a_n .$$

To get the a_n's explicitly one has to invert a formal power series (see e.g. [17]).

Summary

We computed poles of the S-matrix as the poles of some expectation values of the resolvent through a perturbation by a Dirichlet boundary condition. They are exponentially close to the real axis. Our results are valid if k is small (large masses, quasiclassical regime).

The perturbation theory is based upon the equation

$r_\theta = r_\theta^D + \pi_\theta$.

The resonant energies are given in terms of a convergent power series expansion in the tunneling parameter t. This parameter is exponentially small in k^{-2}.

Acknowledgement

Pierre Duclos thanks the Deutsche Forschungsgemeinschaft for its financial support by which this work was partially supported and Eric Mourre of the Centre de Physique Théorique, Marseille, for profitable discussion about results of L.A.P. in the classical limit, box 8.

Footnotes

1) After this talk was delivered we checked again the historical origin of shape resonances. It turned out that there are two articles by Gurney and Condon [18] treating the decay of nuclei extensively along the same lines as Gamov reaching the same conclusions. The first one was published in the Septembre issue of Nature in 1928, the second one in Phys. Rev. 33 in 1929. The work of Gurney and Condon was done independently and at the same time. (Gamov's first article reached the editors of Zeitschrift für Physik August 2, 1928, Gamov was in Göttingen, Gurney and Condon in Princeton).

2) P_g is not a bounded operator on Hilbert space; it can most easily be understood in terms of nested Hilbert spaces or by introducing a semidefinite sesquilinear form as explained by Hoegh-Krohn in his talk three days ago. In the later sense projections onto different Gamov states are orthogonal.

3) The low lying spectrum of H in the case $V(x_{min}) < 0$ is discrete, This situation has already been investigated recently in [11],[12] (see also [13]-[15]).

4) We expect that a similar result to the one given in our main theorem still hold if one consider either the situation where $V(x_{min}) > 0$ or/and the case where V is only bounded below on (b,∞) with still $V(\infty) = 0$. However we have not yet worked out the technical complications arising in these cases.

5) The analyticity of H_θ has been proved earlier by Graffi and Yajima [16].

6) Condition B.3 is in fact stronger than necessary. One may consider a V which obeys $\exists\, \varepsilon_0 \in [0,2]$ such that $\forall\, x > b \quad \varepsilon_0 V(x) + x V' \leq 0$. A condition of that type insures that $H_{ext}(k = 1)$ has no resonances in a sector around $\mathbb{R}^+$ with opening angle $\alpha(\theta)$ such that $tg\, \alpha(\theta) = \varepsilon_0\, \theta + 0(\theta^2)$. This condition is stronger than those resulting from the theorem for non existence of positive eigenvalues.

7) See for instance the talk by Balslev in this conference.

8) For simplicity of notations we will not indicate explicitly the dependance of this exterior scaling on k.

9) Notice that u_θ is the image of the exterior scaling operator U_θ introduced in Section II under the dilation U(k) (Remark 2).

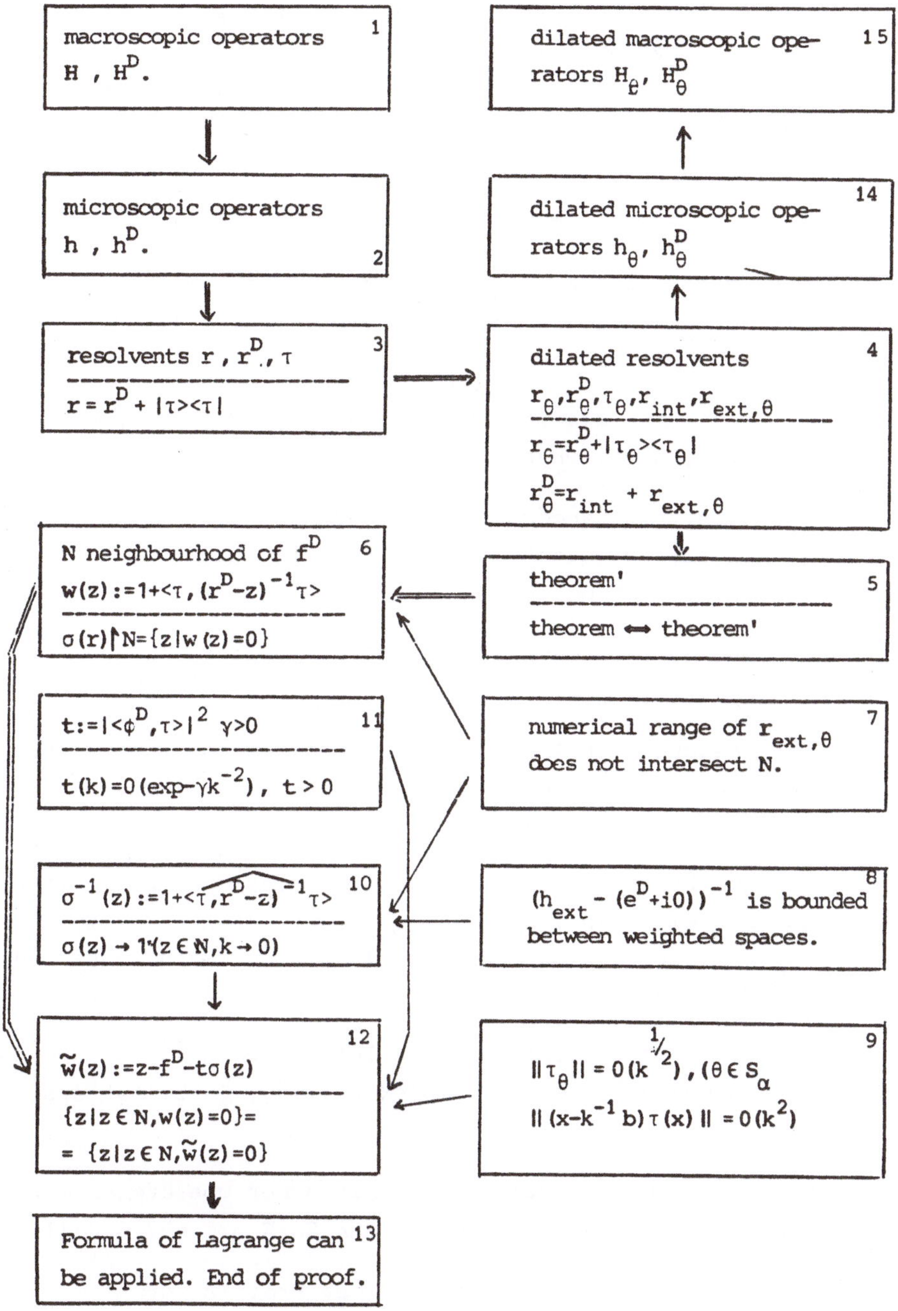
macroscopic operators H , H^D. 1
microscopic operators h , h^D. 2
resolvents r , r^D, τ 3
$r = r^D + |\tau><\tau|$
dilated resolvents 4
$r_\theta, r_\theta^D, \tau_\theta, r_{int}, r_{ext,\theta}$
$r_\theta = r_\theta^D + |\tau_\theta><\tau_\theta|$
$r_\theta^D = r_{int} + r_{ext,\theta}$
theorem' 5
theorem ⟺ theorem'
N neighbourhood of f^D 6
$w(z) := 1 + <\tau, (r^D - z)^{-1}\tau>$
$\sigma(r)\restriction N = \{z | w(z) = 0\}$
numerical range of $r_{ext,\theta}$ 7
does not intersect N.
$(h_{ext} - (e^D + i0))^{-1}$ is bounded 8
between weighted spaces.
$\|\tau_\theta\| = O(k^{1/2})$, ($\theta \in S_\alpha$ 9
$\|(x - k^{-1}b)\tau(x)\| = O(k^2)$
$\sigma^{-1}(z) := 1 + <\tau, \widehat{r^D - z)^{-1}}\tau>$ 10
$\sigma(z) \to 1$ ($z \in N, k \to 0$)
$t := |<\phi^D, \tau>|^2$ $\gamma > 0$ 11
$t(k) = O(\exp - \gamma k^{-2})$, $t > 0$
$\tilde{w}(z) := z - f^D - t\sigma(z)$ 12
$\{z | z \in N, w(z) = 0\} =$
$= \{z | z \in N, \tilde{w}(z) = 0\}$
Formula of Lagrange can 13
be applied. End of proof.
dilated microscopic ope- 14
rators h_θ, h_θ^D
dilated macroscopic ope- 15
rators H_θ, H_θ^D

Fig. 4

[1] G.Gamov: Zur Quantentheorie der Atomkerne, Zeitschrift f. Phys. 51, 204-212 (1928) and
Zur Quantentheorie der Atomzertrümmerung, Zeitschrift f. Phys. 52, 510-515 (1929).

[2] A. Grossman, T.T.Wu: Schrödinger Scattering Amplitude III, Journal of Math. Phys. 3, 684-689 (1962) and ref. therein.

[3] M.H. Mittelmann and A. Tip: Decay of an adiabatically prepaired state, J. Phys. B At. Mol. Phys. 17, 571-576, (1984).

[4] R. Jost: Über die falschen Nullstellen der Eigenwerte der S-Matrix, Helv. Phys. Acta 20, 256 (1947).

[5] B. Simon: Resonances in N-body quantum systems with dilation analytic potentials and the foundations of time dependant perturbation theory, Ann. Math. 97, 247-274 (1973).

[6] J.M. Combes, P. Duclos, R. Seiler: Shape Resonances, to appear.

[7] M.S. Ashbough, E.R. Harrel II: Perturbation Theory for Shape Resonances and large Barrier Potentials, Commun. Math. Phys. 83, 151-170 (1982).

[8] G. Jona-Lasinio, F. Martinelli and E. Scoppola: New Approach to the Semi-classical Limit of Quantum Mechanics I. Multiple Tunnelings in one Dimension. Comm. Math. Phys. 80, 223-254 (1981).

[9] B. Simon: The definition of Molecular Resonances Curves by the Method of Exterior Complex Scaling, Phys. Letters 71 A, 211-214 (1979).

[10] M. Krein: Über Resolventen hermitescher Operatoren mit Defektindex (m,m), Dokl. Akad. Nauk SSSR 52, 657-660 (1946).

[11] J.M. Combes, P. Duclos, R. Seiler: Krein's formula and one dimensional multiple well. J. Funct. Anal. 52, 257-301 (1983).

[12] J.M. Combes, P. Duclos, R. Seiler: Convergent Expansions for Tunneling, Commun. Math. Phys. 92, 229-245, (1983).

[13] B. Simon: Semiclassical analysis of low lying eigenvalues I. Nondegenerate minima: asymptotic expansions, Ann. Inst. Henri Poincaré 38, 295-307 (1983).

[14] B. Helffer et D. Robert: Asymptotique des niveaux d'énergie pour des hamiltoniens à un degré de Liberté. Duke Math. Journ. 49, 853-868 (1982).

[15] B. Helffer, J. Sjöstrand: Multiple Well in the semiclassical limit I. Preprint 83 T 25 Orsay.

[16] S. Graffi, K. Yajima: Exterior scaling and the AC-Stark Effect in a Coulomb Field. Comm. Math. Phys. 89, 277-301, (1983).

[17] H. Cartan: Théorie Elémentaire des fonctions analytique d'une ou plusieurs variables complexes, Edition Scientifiques Hermann, Paris (1961).

[18] R.W. Gurney, E.U. Condon: Quantum Mechanics and Radioactive Disintegration, Phys. Rev. 33, 127-132 (1929) and Nature 122,439 (1928).

[19] G. Jona-Lasinio, F. Martinelli, E. Scoppola: Decaying Quantum-Mechanical States: An informal Discussion within Stochastic Mechanics. Lettere al Nuovo cimento 34, 13-17, (1982).

[20] E. Mourre: Absence of Singular Continuous Spectrum for certain Self-Adjoint Operators. Commun. Math. Phys. 78, 391-408 (1981).

PERTURBATION THEORY FOR RESONANCES IN TERMS OF FREDHOLM DETERMINANTS

F. Gesztesy*,**
Zentrum für interdisziplinäre Forschung
Universität Bielefeld
D-4800 Bielefeld 1, FR Germany

I. Introduction

This lecture essentially consists of two parts. In the first part (Sections II, III, IV and VI) we give a unified treatment of bound states and resonances associated with general Hamiltonians in terms of Fredholm determinants. By taking a suitable limiting procedure in which the range of the interaction tends to zero these Hamiltonians give rise to certain models (so-called point interactions) whose bound states and resonances can be computed explicitly. These models are described in the second part (Sections V and VII).

Section II contains the general theory and relates bound states and resonances of Hamiltonians H defined as quadratic forms to poles in the complex momentum plane. Based on results of [46] the multiplicity of bound states and resonances is then discussed in terms of the multiplicity of the zero of certain (modified) Fredholm determinants. Finally, by introducing an additional complex parameter λ, we study perturbation theory for bound states and resonances of H_λ. In particular, if for some fixed λ_0 H_{λ_0} has bound states and/or resonances we show how to use (modified) Fredholm determinants in order to get Puiseux expansions in $(\lambda-\lambda_0)$ for the bound states and/or resonances of H_λ near those of H_{λ_0}. This is achieved by extending previous results of [7]. Concrete applications to Schrödinger operators with local, short-range potentials in dimensions 1, 2 and 3 are studied in Section III. Extensions of this approach to nonlocal short-range interactions are sketched in Section IV. Coulomb-type interactions with local or nonlocal short-range potentials are discussed in Section VI. Based on [9] we describe spectral and scattering properties of point interactions in 1, 2 and 3 dimensions in Section V and treat Coulomb plus

---

*Alexander von Humboldt Research Fellow.

**On leave of absence from Institute for Theoretical Physics, University of Graz, Austria.

point interactions in Section VII.

For other methods not covered by this lecture see e.g. [25], [27], [82], [85] and the references therein.

II. Bound States and Resonances - An Abstract Approach

We describe a unified treatment of bound states and resonances of Hamiltonians defined as quadratic forms. These results form the basis for the concrete applications in the following sections.

Let $\mathcal{B}(\mathcal{H})$ and $\mathcal{B}_\infty(\mathcal{H})$ denote the spaces of bounded and compact operators on a separable Hilbert space $\mathcal{H}$. Similarly $\mathcal{B}_p(\mathcal{H})$, $p \geq 1$ denotes the set of compact operators whose singular values are in ℓ_p. Next let H_o be a semi-bounded self-adjoint operator in $\mathcal{H}$ and assume

(I) E and F are closed operators in $\mathcal{H}$ which are infinitesimally bounded with respect to $|H_o|^{1/2}$.

We then introduce in $\mathcal{H}$

$$H = H_o \dot{+} E^* F \tag{2.1}$$

by the method of forms ([54], [81]) (note that H need not be self-adjoint) and define

$$K(k) = F(H_o - k^2)^{-1} E^* \ , \quad \text{Im}k > 0, \quad k^2 \notin \sigma(H_o). \tag{2.2}$$

(If no confusions arise we always identify operators of the type $(H_o - k^2)^{-1} E^*$ and $[E(H_o - \bar{k}^2)^{-1}]^*$ etc.)

We start with

Lemma 2.1. Suppose assumption (I) holds. Then

a) $$(H - k^2)^{-1} = (H_o - k^2)^{-1} - (H_o - k^2)^{-1} E^* F (H_o - k^2)^{-1} \ , \quad \text{Im}k > 0, \quad k^2 \notin \sigma(H) \cup \sigma(H_o). \tag{2.3}$$

If in addition $K(k) \in \mathcal{B}_\infty(\mathcal{H})$ for all $\text{Im}k > 0$, $k^2 \notin \sigma(H_o)$ then

$$(H - k^2)^{-1} = (H_o - k^2)^{-1} - (H_o - k^2)^{-1} E^* [1 + K(k)]^{-1} F (H_o - k^2)^{-1} \ , \quad \text{Im}k > 0 \ , \quad k^2 \notin \sigma(H) \cup \sigma(H_o). \tag{2.4}$$

b) Let $K(k) \in \mathcal{B}_\infty(\mathcal{H})$ for all $\mathrm{Im}k > 0$, $k^2 \notin \sigma(H_o)$ and let $E_o = k_o^2 \notin \sigma(H_o)$, $\mathrm{Im}k_o > 0$. Then H has the eigenvalue E_o with geometric multiplicity n_o if and only if $K(k_o)$ has the eigenvalue -1 with the same geometric multiplicity n_o. In particular, if $K(k_o)\phi_o = -\phi_o$, $\phi_o \in \mathcal{H}$ then $\psi_o = (H_o - k_o^2)^{-1}E^*\phi_o$ fulfills $H\psi_o = E_o\psi_o$, $\psi_o \in \mathcal{D}(H)$. Conversely, if $H\tilde{\psi}_o = E_o\tilde{\psi}_o$, $\tilde{\psi}_o \in \mathcal{D}(H)$ then $\tilde{\phi}_o = -F\tilde{\psi}_o \in \mathcal{H}$ fulfills $K(k_o)\tilde{\phi}_o = -\tilde{\phi}_o$ and $\tilde{\psi}_o = (H_o - k_o^2)^{-1}E^*\tilde{\phi}_o$.

c) If $F(|H_o| + c)^{-1/2}$ or $E(|H_o| + c)^{-1/2}$ is compact for some $c > 0$ then

$$\sigma_{ess}(H) = \sigma_{ess}(H_o).$$

A proof of Lemma 2.1 appeared e.g. in [60] (cf. also [55], [81]). For a generalized version see [8].

Next we give an abstract discussion of resonances of H based on [7]. We restrict ourselves to the case where $K(k)$ is compact and analytic for all $\mathrm{Im}k > 0$, $k^2 \notin \sigma(H_o)$. In addition we assume that K can be analytically continued into some subset of the lower k-plane and remains compact there. More precisely we assume

(II) Let $\Omega \subset \mathbb{C}$ be open and connected, $\Omega \supset \{k \in \mathbb{C} \,/\, \mathrm{Im}k > 0,\ k^2 \notin \sigma(H_o)\}$, $k_o \in \Omega$ for some k_o with $\mathrm{Im}k_o < 0$. Assume that $K: \Omega \to \mathcal{B}_\infty(\mathcal{H})$ is analytic.

By Lemma 2.1 b) there exists a one-to-one correspondence between eigenvalues $E_1 = k_1^2$, $\mathrm{Im}k_1 > 0$, $k_1 \in \Omega$ of H and the eigenvalue -1 of $K(k_1)$. For resonances we introduce

<u>Definition 2.2</u>. Assume hypotheses (I) and (II). Then $k_2 \in \Omega$, $\mathrm{Im}k_2 < 0$ is called a resonance of H if and only if $K(k_2)$ has an eigenvalue -1.

We note

<u>Lemma 2.3</u>. In addition to hypotheses (I) and (II) assume that $K: \Omega \to \mathcal{B}_p(\mathcal{H})$ for some $p \in \mathbb{N}$. Then if for some $k_o \in \Omega$ $K(k_o)$ has an eigenvalue -1 $[1 + K(k)]^{-1}$ has a norm convergent Laurent expansion around $k = k_o$

$$[1 + K(k)]^{-1} = \sum_{m=-M}^{\infty} K_m(k - k_o)^m \quad \text{for some } M \in \mathbb{N}. \tag{2.5}$$

Here for each m $K_m \in \mathcal{B}(\mathcal{H})$ and for $-M \le m \le -1$ K_m is of finite rank. Moreover, $-1 \in \sigma(K(k_o))$ if and only if $\det_p [1 + K(k_o)] = 0$ and the geometric multiplicity of the eigenvalue -1 of $K(k_o)$ coincides with the multiplicity of the zero of the (modified) Fredholm determinant $\det_p[1+K(k)]$ at $k = k_o$ if and only if $M = 1$. In particular, if H is self-adjoint and $k_o \in \Omega$, $\mathrm{Im}k_o > 0$ then $M = 1$.

Remark 2.4. That M need not be one (e.g. if resonances collide) has been discussed in [70] and [72]. In fact, if $M > 1$ then $\nu(k_o)$, the order of the zero of $\det_p[1+K(k)]$ at $k=k_o$, strictly dominates the geometric multiplicity of the eigenvalue -1 of $K(k_o)$. (In general $\nu(k_o) \geq M$)([46]). On the other hand, $[1+K(k)]^{-1}$ has a simple pole at $k=k_o$ if and only if $[1+K(k_o)+(k-k_o)K'(k_o)]^{-1}$ has a simple pole at $k=k_o$. In addition, if the geometric and algebraic multiplicity of the eigenvalue -1 of $K(k_o)$ coincide, then $[1+K(k)]^{-1}$ has a simple pole at $k=k_o$ if and only if $P(k_o)K'(k_o)$ (or equivalently if $K'(k_o)$) is injective on $\mathrm{Ker}[1+K(k_o)]$ ([46]).
(Here $P(k_o) = -(2\pi i)^{-1} \oint_{|z+1|=\varepsilon} dz[K(k_o)-z]^{-1}$, $0 < \varepsilon$ small enough, denotes the projection onto the algebraic eigenspace of $K(k_o)$ to the eigenvalue -1 and $\mathrm{Ker}[T]$ denotes the kernel of some $T \in \mathcal{B}(\mathcal{H})$.)

A generalization of Lemma 2.3 based on results of [46] appeared in [8]. For earlier results on multiplicities of bound states of Schrödinger operators in $\mathbb{R}^3$ cf. [67]. Modified Fredholm determinants are e.g. treated in [29], [84] and [86].

In concrete applications it frequently happens (cf. Sections V, VI, VII) that bound states and resonances of H_o decouple from those of H in the sense that poles of $(H_o-k^2)^{-1}$ (resp. $K(k)$) are not poles of $(H-k^2)^{-1}$ (resp. $F(H-k^2)^{-1}E^*$). As a first result in this direction we state

Lemma 2.5. a) In addition to hypothesis (I) assume that $K(k) \in \mathcal{B}_\infty(\mathcal{H})$ for all $\mathrm{Im}k > 0$, $k^2 \notin \sigma(H_o)$.

b) Let $k_o^2 < 0$ ($\mathrm{Im}k_o > 0$) be an eigenvalue of H_o such that for $0 < |k-k_o|$ small enough

$$(H_o - k^2)^{-1} \underset{k\to k_o}{=} (k^2-k_o^2)^{-1} \sum_{j=1}^{N} (\tilde{\Psi}_{oj},\cdot)\Psi_{oj} + M(k), \quad N \in \mathbb{N} \tag{2.6}$$

where $\tilde{\Psi}_{oj}, \Psi_{oj} \in \mathcal{H}$, $1 \leq j \leq N$ and M is analytic near $k=k_o$. Then $FM(k)E^*$ is analytic and compact near $k=k_o$.

c) Suppose that $-1 \notin \sigma(FM(k_o)E^*)$.

d) Assume that

$$F\Psi_{oj} \neq 0, \quad E\tilde{\Psi}_{oj} \neq 0, \quad 1 \leq j \leq N', \quad N' \leq N$$

and in case $N' < N$ assume that

$$F\Psi_{0\ell} = E\tilde{\Psi}_{0\ell} = 0 , \quad N' + 1 \le \ell \le N.$$

e) Suppose the existence of the inverse of the matrix $(E\tilde{\Psi}_{0j},[1+FM(k_0)E^*]^{-1}F\Psi_{0\ell})$, $1 \le j,\ell \le N$ denoted by

$$(E\tilde{\Psi}_0,[1+FM(k_0)E^*]^{-1}F\Psi_0)^{-1}_{j\ell} .$$

Then, for $0 < |k-k_0|$ small enough

$$(H-k^2)^{-1} \underset{k\to k_0}{=} \begin{cases} O(1) & \text{if } N' = N, \\ (k^2-k_0^2)^{-1} \sum_{j=N'+1}^{N} (\tilde{\Psi}_{0j},\cdot)\Psi_{0j} + O(1) & \text{if } N' < N. \end{cases} \qquad (2.7)$$

In particular if and only if $N' = N$ $(H-k^2)^{-1}$ is analytic near $k = k_0$ and hence the first order pole of $(H_0-k^2)^{-1}$ at $k = k_0$ cancels in Eq. (2.4).

<u>Proof</u>. Inserting Eq. (2.6) and

$$[1+K(k)]^{-1} = \{(k^2-k_0^2)^{-1} \sum_{j=1}^{N'} (E\tilde{\Psi}_{0j},\ )F\Psi_{0j} + 1 + FM(k)E^*\}^{-1}$$

$$\underset{k\to k_0}{=} [1+FM(k)E^*]^{-1}$$

$$- \sum_{j,\ell=1}^{N} (E\tilde{\Psi}_0,[1+FM(k)E^*]^{-1}F\Psi_0)^{-1}_{j\ell}([1+EM(k)^*F^*]^{-1}E\tilde{\Psi}_{0\ell},\cdot)\cdot$$

$$\cdot\ [1+FM(k)E^*]^{-1}F\Psi_{0j}$$

$$+ (k^2-k_0^2) \sum_{j,\ell=1}^{N'} (E\tilde{\Psi}_0,[1+FM(k)E^*]^{-1}F\Psi_0)^{-2}_{j\ell} ([1+EM(k)^*F^*]^{-1}E\tilde{\Psi}_{0\ell},\cdot) \cdot$$

$$\cdot\ [1+FM(k)E^*]^{-1}F\Psi_{0j} + O((k^2-k_0^2)^2) \qquad (2.8)$$

into Eq. (2.4) directly proves the result (2.7).

Lemma 2.5 can be extended to resonances as follows. By hypotheses (I) and (II)

$$F(H-k^2)^{-1}E^* = 1 - [1+K(k)]^{-1}, \quad \mathrm{Im}k > 0, \quad k^2 \notin \sigma(H) \cup \sigma(H_0). \qquad (2.9)$$

(cf. Eq. (2.4)). Thus the left hand side of Eq. (2.9) has a compact meromorphic continuation into Ω, denoted by $L(k)$. As a consequence we obtain

Lemma 2.6. a) Assume hypotheses (I) and (II).

b) Let $k_o \in \mathbb{C}$ and suppose that for $0 < |k - k_o|$ small enough

$$K(k) \underset{k\to k_o}{=} (k-k_o)^{-1} \sum_{j=1}^{N} (\tilde{\Phi}_{oj},.)\Phi_{oj} + R(k), \quad N \in \mathbb{N}$$

where $\tilde{\Phi}_{oj}, \Phi_{oj} \in H$, $1 \le j \le N$ and R is analytic near $k = k_o$.

c) Assume that $-1 \notin \sigma(R(k_o))$.

d) Suppose the existence of the inverse of the matrix $(\tilde{\Phi}_{oj},[1+R(k_o)]^{-1}\Phi_{o\ell})$, $1 \le j,\ell \le N$ denoted by

$$(\tilde{\Phi}_o,[1+R(k_o)]^{-1}\Phi_o)^{-1}_{j\ell} \quad .$$

Then $L(k)$ (the meromorphic continuation of $F(H-k^2)^{-1}E^*$ into Ω) is analytic near $k=k_o$

$$\begin{aligned} L(k) &= 1 - [1+K(k)]^{-1} \\ &\underset{k\to k_o}{=} 1 - [1+R(k_o)]^{-1} \\ &\quad + \sum_{j,\ell=1}^{N} (\tilde{\Phi}_o,[1+R(k_o)]^{-1}\Phi_o)^{-1}_{j\ell}([1+R(k_o)^*]^{-1}\tilde{\Phi}_{o\ell},.)[1+R(k_o)]^{-1}\Phi_{oj} \\ &\quad + O(k-k_o). \end{aligned}$$

Proof. Analogously to that of Lemma 2.5.

These results can be generalized in various directions. As an example we finally give

Lemma 2.7. a) Assume hypotheses (I) and (II).

b) Let $k_o \in \mathbb{C}$ and suppose that for $0 < |k-k_o|$ small enough

$$K(k) \underset{k\to k_o}{=} (k-k_o)^{-1} (\tilde{\Phi}_o,.)\Phi_o + R(k)$$

where $\tilde{\Phi}_o, \Phi_o \in H$ and R is analytic near $k=k_o$.

c) Assume that $-1 \in \sigma(R(k_o))$ and that for $0 < |\varepsilon|$ small enough

$$[1+\varepsilon+R(k_o)]^{-1} = \varepsilon^{-1}(\tilde{\varphi}_o,.)\varphi_o + J(\varepsilon)$$

where $\tilde{\varphi}_o, \varphi_o \in H$ and J is analytic near $\varepsilon = 0$.

d) Let $(\tilde{\varphi}_0, \Phi_0) \neq 0$, $(\tilde{\Phi}_0, \varphi_0) \neq 0$ and also $(\tilde{\varphi}_0, R'(k_0)\varphi_0) \neq 0$.

Then

$$L(k) = 1 - [1 + K(k)]^{-1} \underset{k \to k_0}{=} O(1)$$

is analytic near $k = k_0$.

Proof. Insertion of

$$[1 + R(k)]^{-1} \underset{k \to k_0}{=} (k-k_0)^{-1} (\tilde{\varphi}_0, R'(k_0)\varphi_0)^{-1} (\tilde{\varphi}_0, .)\varphi_0 + O(1)$$

into

$$\{(k-k_0)^{-1}(\tilde{\Phi}_0, .)\Phi_0 + 1 + R(k)\}^{-1} = [1 + R(k)]^{-1}$$

$$- \frac{([1+R(k)^*]^{-1}\tilde{\Phi}_0, .)[1+R(k)]^{-1}\Phi_0}{k-k_0 + (\tilde{\Phi}_0, [1+R(k)]^{-1}\Phi_0)}, \quad 0 < |k-k_0| \text{ small enough} \quad (2.10)$$

proves that all pole terms on the right hand side of Eq. (2.10) indeed cancel.

Lemmas 2.5 - 2.7 together with Eqs. (2.4) and (2.9) show that under certain additonal assumptions bound states and resonances of H are in a one-to-one correspondence with eigenvalues -1 of $K(k)$ even if $K(k)$ is meromorphic in some open, connected domain $\Omega' \subset \mathbb{C}$. We also note that in general $N' = N$ when local interactions such as those in Sections III and VI are considered. For nonlocal interactions (see Sections IV, VI) it is simple to construct examples with $N' < N$ such that k_0^2 is an eigenvalue for both H_0 and H. For questions concerning the multiplicity of eigenvalues k_0^2 of H which might also be eigenvalues of H_0 we refer to [44].

Concerning multiplicities of resonances we introduce

Definition 2.8. In addition to hypotheses (I) and (II) assume that for all $k \in \Omega$ $K(k) \in \mathcal{B}_p(\mathcal{H})$ for some $p \in \mathbb{N}$. If $k_0 \in \Omega$, $\text{Im} k_0 < 0$ denotes a resonance of H its multiplicity is defined to be that of the zero of the (modified) Fredholm determinant $\det_p[1+K(k)]$ at $k = k_0$.

By Lemma 2.1b) and the last statement of Lemma 2.3 the multiplicity of bound states of H automatically coincides with that of the zero of the corresponding (modified) Fredholm determinant if H is self-adjoint.

Finally we turn to perturbations of eigenvalues and resonances and introduce the following assumption

(III) Let $\Lambda \subset \mathbb{C}$ be open and connected and suppose E_λ, F_λ fulfill hypothesis (I) for all $\lambda \in \Lambda$. Moreover, assume $K_\lambda(k) = F_\lambda(H_0-k^2)^{-1}E_\lambda^*$ to be (norm) analytic in $\Omega \times \Lambda$ and for some $p \in \mathbb{N}$ $K_\lambda(k) \in \mathcal{B}_p(\mathcal{H})$ for all $(k,\lambda) \in \Omega \times \Lambda$.

In analogy to Eq. (2.1) we then define $H_\lambda = H_0 \dotplus E_\lambda^* F_\lambda$. It turns out that if $k_0 \in \Omega$ corresponds to a bound state ($\mathrm{Im}k_0 > 0$) or to a resonance ($\mathrm{Im}k_0 < 0$) of H_{λ_0} for some fixed $\lambda_0 \in \Lambda$ then, for λ in a neighbourhood of λ_0, H has bound states resp. resonances $k(\lambda) \in \Omega$ with $k(\lambda) = k_o + o(\lambda-\lambda_o)$. The functions $k(\lambda)$ are given by the solutions of $\det_p[1+K_\lambda(k)] = 0$ near (k_0,λ_0). More precisely we have

<u>Theorem 2.9</u>. Assume hypotheses (I)-(III). If for some $(k_0,\lambda_0) \in \Omega \times \Lambda$ $K_{\lambda_0}(k_0)$ has an eigenvalue -1 such that the multiplicity of the zero of $\det_p[1+K_{\lambda_0}(k)]$ at $k=k_0$ equals m_0 then, for $|\lambda-\lambda_0|$ small enough, there exist m_0 (not necessarily distinct) functions $k_\ell(\lambda) \in \Omega$, $1 \le \ell \le s$ which are all the solutions of $\det_p[1+K_\lambda(k)]=0$ for (k,λ) near (k_0,λ_0). They are given by Puiseux expansions of the type

$$k_\ell(\lambda) = k_0 + \sum_{j=1}^{\infty} k_{j,\ell}\,(\lambda-\lambda_0)^{j/m_\ell},\ m_\ell \ge 1,\ \ 1 \le \ell \le s,\ \sum_{\ell=1}^{s} m_\ell = m_0 . \tag{2.11}$$

If e.g. k_0 corresponds to a simple bound state or resonance of H_{λ_0} (i.e. if $m_0 = 1$) then $k(\lambda)$ is analytic near $\lambda = \lambda_0$

$$k(\lambda) \underset{\lambda\to\lambda_0}{=} k_0 + k_1(\lambda-\lambda_o) + O(\lambda-\lambda_o)^2), \tag{2.12}$$

$$k_1 = -\frac{(\tilde{\phi}_0)[\frac{\partial}{\partial\lambda} K_\lambda(k_0)]\big|_{\lambda=\lambda_0}\phi_0)}{(\tilde{\phi}_0,[\frac{\partial}{\partial k} K_{\lambda_0}(k)]\big|_{k=k_0}\phi_0)} \tag{2.13}$$

where

$$K_{\lambda_0}(k_0)\phi_0 = -\phi_0,\ K_{\lambda_0}(k_0)^*\tilde{\phi}_0 = -\tilde{\phi}_0,\quad \phi_0,\tilde{\phi}_0 \in \mathcal{H}\ . \tag{2.14}$$

If $\mathrm{Im}k_o > 0$, $\lambda_o \in \Lambda \cap \mathbb{R}$, and H_λ is self-adjoint for λ in a real neighbourhood of λ_0, the additional constraint $E_\ell(\lambda) = k_\ell(\lambda)^2 < 0$, $1 \le \ell \le s$ for $\lambda \in \mathbb{R}$, $|\lambda-\lambda_0|$ small enough, in fact leads to $m_\ell = 1$

i.e., $k_\ell(\lambda)$, $1 \le \ell \le s$ are analytic near $\lambda = \lambda_o$.

Proof. Eq. (2.11) can be derived as follows. Let

$$\mathrm{Ker}[1+K_{\lambda_o}(k_o)] = \{\phi_j,\ 1 \le j \le n_o\}\ ,\quad \mathrm{Ker}[1+K_{\lambda_o}(k_o)^*] = \{\tilde{\phi}_j,\ 1 \le j \le n_o\}$$

and define corresponding biorthogonal sets ([72])

$$\{\chi_j,\ 1 \le j \le n_o\}\ ,\quad \{\tilde{\chi}_j\ ,\ 1 \le j \le n_o\}$$

such that

$$(\phi_j,\tilde{\chi}_\ell) = (\chi_j,\tilde{\phi}_\ell) = \delta_{j\ell}\quad ,\quad 1 \le j,\ell \le n_o.$$

If we denote

$$Q = \sum_{j=1}^{n_o} (\tilde{\chi}_j,\cdot)\chi_j$$

then $[1+K_{\lambda_o}(k_o)+Q]^{-1} \in \mathcal{B}(\mathcal{H})$ and hence for (k,λ) in a sufficiently small neighbourhoof of (k_o,λ_o)

$$1 + K_\lambda(k) = [1+K_\lambda(k)+Q]\{1 - [1+K_\lambda(k)+Q]^{-1}Q\}.$$

Taking e.g. $p = 1$ this leads to

$$\det[1+K_\lambda(k)] = \det[1+K_\lambda(k)+Q]\det\{1-[1+K_\lambda(k)+Q]^{-1}Q\}$$

and hence we get that

$$\det[1+K_\lambda(k)] = 0 \quad \text{for} \quad (k,\lambda) \quad \text{near} \quad (k_o,\lambda_o) \tag{2.15}$$

if and only if

$$\det\{1-[1+K_\lambda(k)+Q]^{-1}Q\} = 0 \quad \text{for } (k,\lambda) \quad \text{near} \quad (k_o,\lambda_o). \tag{2.16}$$

If e.g. $p = 2$ it follows from

$$\det_2\{[1+A][1+B]\} = \det_2[1+A]\det_2[1+B]e^{-\mathrm{Tr}AB}$$

$$= \det_2[1+A]\ \det[1+B]e^{-\mathrm{Tr}[(1+A)B]} \quad \text{for} \quad A \in \mathcal{B}_2(\mathcal{H}),\quad B \in \mathcal{B}_1(\mathcal{H})$$

that again $\det_2[1+K_\lambda(k)] = 0$ for (k,λ) near (k_o,λ_o) if and only if Eq. (2.16) holds. Similarly this can be proved for general $p \in \mathbb{N}$. Since Q is of rank n_o and hence $\det\{1-[1+K_\lambda(k)+Q]^{-1}Q\}$ is analytic near (k_o,λ_o) ([45]) one can pass to a matrix representation and then apply the Weierstraß preparation theorem to arrive at Eq. (2.11).

If $m_o = 1$ (and hence $n_o = 1$) we may choose

$$Q = (\tilde{\phi}_o, \cdot)\phi_o$$

and thus Eq. (2.15) holds if and only if

$$\det\{1-[1+K_\lambda(k)+Q]^{-1}Q\} = 1-(\tilde{\phi}_o)[1+K_\lambda(k)+Q]^{-1}\phi_o) = 0. \qquad (2.17)$$

Expanding in norm

$$[1+K_\lambda(k)+Q]^{-1} \underset{\substack{k\to k_0\\ \lambda\to\lambda_0}}{=} [1+K_{\lambda_0}(k_o)+Q]^{-1}$$

$$- [1+K_{\lambda_0}(k_o)+Q]^{-1}[(\lambda-\lambda_o)\dot{K}_{\lambda_0}(k_o)+(k-k_o)K'_{\lambda_0}(k_o)][1+K_{\lambda_0}(k_o)+Q]^{-1}$$

$$+ O((\lambda-\lambda_o)^2) + O((k-k_o)^2) + O((\lambda-\lambda_o)(k-k_o)) \qquad (2.18)$$

(where prime denotes $\frac{\partial}{\partial k}$ and dot abbreviates $\frac{\partial}{\partial\lambda}$) and noting

$$[1+K_{\lambda_0}(k_o)+Q]^{-1}\phi_o = (\tilde{\phi}_o,\phi_o)^{-1}\phi_o$$

Eq. (2.13) follows after insertion of Eq. (2.18) into Eq. (2.17). (Note that $(\tilde{\phi}_o,\phi_o) \neq 0$ and $(\tilde{\phi}_o,K'_{\lambda_0}(k_o)\phi_o) \neq 0$ by the end of Remark 2.4 since $m_o = 1$.) The last part of Theorem 2.9 is known as Rellich's theorem ([54, 75]).

Remark 2.10. That $(\tilde{\phi}_o,K'_{\lambda_0}(k_o)\phi_o)$ in Eq. (2.13) cannot vanish can be seen explicitly in the special case where $K_\lambda(k) = \lambda K(k)$ and $H_\lambda = H_o \dotplus \lambda E^*F$ is self-adjoint for $\lambda \in \mathbb{R}$: E.g. if $\lambda_o \in \mathbb{R}\backslash\{0\}$ and $\mathrm{Im}k_o > 0$ ($k_o = i\sqrt{-E_o}$, $E_o = k_o^2 < 0$) then

$$(\tilde{\phi}_o,K'_{\lambda_0}(k_o)\phi_o) = 2k_o\lambda_o(\tilde{\phi}_o,F(H_o-k_o^2)^{-1}E^*\phi_o) = 2k_o\lambda_o^{-1}\|\psi_o\|^2 \neq 0$$

where $\psi_o = \lambda_o(H_o-k_o^2)^{-1}E^*\phi_o \in \mathcal{D}(H)$ fulfills $H\psi_o = E_o\psi_o$.

In particular Eq. (2.13) simplifies to

$$k_1 = \lambda_o(2k_o)^{-1}\|\psi_o\|^{-2}(\tilde{\phi}_o,K_{\lambda_0}(k_o)\phi_o) = (2k_o)^{-1}\|\psi_o\|^{-2}(\tilde{\phi}_o,\phi_o)$$

$$= (2k_o)^{-1}\|\psi_o\|^{-2}(u\psi_o, v\psi_o)$$

a well known result in first order perturbation theory ([54], [75]).

Theorem 2.9 is an extended verison of theorem 3.1 of [7] which deals Schrödinger operators in $L^2(\mathbb{R}^3)$.

For the definition of resonances in the context of an abstract analytic scattering theory we refer to [50].

III. Local Short-Range Potentials in Dimensions 1,2 and 3

In this section we apply the abstract results of Section I to Schrödinger operators with local short-range potentials. We concentrate on dimension 1,2 and 3 and emphasize the common structure as well as particular differences depending on the number of dimensions.

Let

$$H_o = -\Delta \ , \quad \mathcal{D}(H_o) = H^{2,2}(\mathbb{R}^3), \quad 1 \le n \le 3 \tag{3.1}$$

denote the unperturbed Hamiltonian in $L^2(\mathbb{R}^n)$, $1 \le n \le 3$ and introduce its resolvent G_k by

$$G_k = (-\Delta - k^2)^{-1} \ , \quad \mathrm{Im}k > 0 \ . \tag{3.2}$$

Depending on n the kernel $G_k(x,x')$ of G_k is given by

$$G_k(x,x') = \begin{cases} (i/2k)\ e^{ik|x-x'|} \ , & n = 1, \\ -(i/4)\ H_o^{(1)}\ (k|x-x'|), \quad x \neq x', & n = 2, \\ (4\pi|x-x'|)^{-1}\ e^{ik|x-x'|}, \quad x \neq x', & n = 3; \ \mathrm{Im}k > 0 \end{cases} \tag{3.3}$$

where $H_o^{(1)}(z)$ denotes the Hankel function of order zero and first kind [1]. Next we define the interaction V. Let

$$e^{2a|x|}\ V \in L^1(\mathbb{R}) \quad \text{for some} \quad a > 0, \quad n = 1, \tag{3.4}$$

$$V \in L^{1+\delta}\ (\mathbb{R}^2), \quad e^{2a|x|}\ V \in L^1(\mathbb{R}^2) \quad \text{for some} \quad \delta, a > 0, \ n = 2, \tag{3.5}$$

$$e^{2a|x|}\ V \in R \quad \text{for some} \quad a > 0, \quad n = 3 \tag{3.6}$$

where R denotes the class of Rollnik potentials (i.e. $\int_{\mathbb{R}^6} d^3x\ d^3x'\ |V(x)||V(x')|e^{2a[|x|+|x'|]}|x-x'|^{-2} < \infty$). Then the Schrödinger operator H in $L^2(\mathbb{R}^n)$, $1 \le n \le 3$ is defined by the method of forms ([54], [81])

$$H = -\Delta \dotplus V. \tag{3.7}$$

In particular by identifying

$$E^* = |V|^{1/2} \equiv v \ , \quad F = |V|^{-1/2} V \equiv u, \quad K(k) = uG_k v$$

and observing $\sigma(H_o) = \sigma_{ess}(H_o) = [0,\infty)$, Lemma 2.1 applies. For a proof that assumption (I) holds and that $K(k) \in \mathcal{B}_\infty(L^2(\mathbb{R}^n))$, $1 \le n \le 3$ for all $\mathrm{Im}k > 0$ see e.g. [8], [16], [57], [58], [75], [81] and [83].

Next we turn to resonances. Let Ω_n, $1 \le n \le 3$ be defined as follows

$$\Omega_1 = \{k \in \mathbb{C}\backslash\{0\} / \ \mathrm{Im}k > -a\} \ , \tag{3.8}$$

$$\Omega_2 = \{k \in \mathbb{C}\backslash\{0\} \quad \mathrm{Im}k > -a, \quad -\pi < \arg k < \pi\}, \tag{3.9}$$

$$\Omega_3 = \{k \in \mathbb{C} \ / \ \mathrm{Im}k > -a\} \ . \tag{3.10}$$

For $n = 2$, Ω_2 could be replaced by Ω_2', the entire logarithmic Riemann surface restricted to $\mathrm{Im}k > -a$

$$\Omega_2' = \{k \in \mathbb{C} \setminus \{0\} \ / \ \mathrm{Im}k > -a, -\infty < \arg k < \infty\}. \tag{3.9'}$$

Then the decomposition ([16], [81], [83])

$$uG_k v = \begin{cases} A_1(k)k^{-1} + B_1(k) \ , & n = 1, \\ A_2(k)\ln k + B_2(k) \ , & n = 2, \\ A_3(k), & n = 3 \ ; \quad \mathrm{Im}k > 0 \end{cases} \tag{3.11}$$

where A_n, B_n, $1 \le n \le 3$ are analytic with respect to k in $\mathrm{Im}k > -a$ shows that $uG_k v$ actually has a $\mathcal{B}_p(L^2(\mathbb{R}^n))$-valued analytic continuation into Ω_n, where $p = 1$ for $n = 1$ and $p = 2$ for $n = 2,3$. As a consequence all conditions of Lemma 2.3 are fulfilled.

Moreover, if V is real-valued, the following property of the free Green's function, viz.

$$\overline{G_k(x,x')} = G_{e^{i\pi}\bar{k}}(x,x') \ , \quad n = 1,2,3 \tag{3.12}$$

shows that resonances of H lie symmetrically with respect to the negative imaginary axis for $n = 1$ and 3. For $n = 2$ different sheets of the logarithmic Riemann surface are connected. Of course bound states of H exclusively lie on the nonnegative imaginary axis for $1 \le n \le 3$ in case V is real.

In the special case where $n = 1$ and V is real-valued more detailed information than contained in Lemma 2.3 is available: First of all $\det[1+uG_k v]$ coincides with the Jost function ([86]). As a consequence all negative bound states of H have multiplicity one since by ODE-techniques the zeros of the Jost function on the positive imaginary axis are necessarily simple ([21]).

If $k_o \in \Omega_n$, $-a < \mathrm{Im}k_o < 0$ corresponds to a resonance of H then the resonance function $\psi(k_o,x)$ defined as

$$\psi(k_o,x) = \int_{\mathbb{R}^n} d^nx' \, G_{k_o}(x,x') \, v(x')\phi_{k_o}(x') \, , \quad 1 \le n \le 3 \tag{3.13}$$

where

$$uG_{k_o} v\phi_{k_o} = -\phi_{k_o} \, , \quad \phi_{k_o} \in L^2(\mathbb{R}^n)$$

obeys $\psi(k_o)$, $\nabla\psi(k_o) \in L^2_{loc}(\mathbb{R}^n)$ as well as

$$(-\Delta + V) \, \psi(k_o) = k_o^2 \, \psi(k_o)$$

in the sense of distributions.

Next we describe the connection between poles of the off-shell scattering operator and bound states and resonances of H. We start with $n = 1$. Under the assumption (3.4) the on-shell scattering matrix in one dimension reads ([21])

$$S(k) = \begin{pmatrix} S_{++}(k) & S_{+-}(k) \\ & \\ S_{-+}(k) & S_{--}(k) \end{pmatrix} \, , \quad k > 0 \tag{3.14}$$

where S_{++} (resp. S_{--}) represents the transmission coefficient from the right (resp. from the left) and S_{+-} (resp. S_{-+}) represents the reflection coefficient from the right (resp. from the left).

Explicitly we have ([16])

$$S_{\varepsilon_1\varepsilon_2}(k) = \delta_{\varepsilon_1\varepsilon_2} + (2ik)^{-1}(v \, e^{i\varepsilon_1 kx}, (1+uG_kv)^{-1} u \, e^{i\varepsilon_2 kx'}) \tag{3.15}$$

$$\varepsilon_j = \pm 1, \quad j = 1,2.$$

Moreover, using Jost function techniques, one can show ([21], [65]) that $S(k)$ has a meromorphic continuation into $\tilde{\Omega}_1 = \{k \in \Omega_1 / |\mathrm{Im}k| < a\}$ and that nonreal poles $k_o \in \tilde{\Omega}_1$ of $S(k)$ coincide with bound states (if $\mathrm{Im}k_o > 0$) or resonances (if $\mathrm{Im}k_o < 0$) of H.

Poles on the real line need a separate discussion. In fact, one can show, assuming V to be real-valued, that a pole in $(1 + uG_kv)^{-1}$ occurs at most at $k = 0$ (H cannot have positive embedded eigenvalues) where one obtains the Laurent expansion ([16])

$$(1 + uG_k v)^{-1} = \sum_{n=-M}^{\infty} t_n k^n , \quad M = 0 \text{ or } 1, \qquad 0 < |k| \text{ small enough.} \tag{3.16}$$

If $M = 1$, H has a zero-energy resonance ([16]) (it never has a zero-energy bound state [59]). In any case $S(k)$ never has poles on the real axis. In particular,

$$\text{if } M = 0: \quad S_{\substack{++\\--}}(k) \underset{k\to 0}{=} O(k), \qquad S_{\substack{+-\\-+}}(k) \underset{k\to 0}{=} O(1), \tag{3.17}$$

$$\text{if } M = 1: \quad S(k) \underset{k\to 0}{=} O(1). \tag{3.18}$$

The discussion for $n = 2$ and 3 parallels each other. For the on-shell scattering operator $S(k)$, $k > 0$ in $L^2(S^{n-1})$, $n = 2,3$ one obtains

$$S(k) = 1 - 2\pi i \, T(k), \quad k > 0, \quad k^2 \notin \xi \tag{3.19}$$

where

$$(T(k)\phi)(\omega) = e^{-i\pi(n+1)/4}(2\pi)^{-(n+1)/2} k^{(n-1)/2} \int_{S^{n-1}} d\omega' \, f(k,\omega,\omega')\phi(\omega'),$$
$$k > 0, \; k^2 \notin \xi, \quad \omega \in S^{n-1}, \; \phi \in L^2(S^{n-1}), \; n = 2,3. \tag{3.20}$$

Here $f(k,\omega,\omega')$ denotes the on-shell scattering amplitude given by ([22], [23], [33] - [35])

$$f(k,\omega,\omega') = 2^{-1}(2\pi)^{-(n-1)/2} k^{(n-3)/2} e^{i\pi(n+1)/4} (v\, e^{ik\omega\cdot x}, (1+uG_k v)^{-1} u e^{ik\omega'\cdot x'}),$$
$$k > 0, \; k^2 \notin \xi, \quad n = 2,3 \tag{3.21}$$

and the exceptional set ξ reads

$$\xi = \{k^2 \geq 0 / uG_k v\, g = -g \text{ for some } g \in L^2(\mathbb{R}^n), \; n = 2,3, \; k \geq 0\}. \tag{3.22}$$

Under the hypotheses (3.5) and (3.6) $S(k)$ then has a meromorphic continuation into $\tilde{\Omega}_n = \{k \in \Omega_n / |\mathrm{Im} k| < a\}$, $n = 2,3$ (cf. [28], [33] - [35], [43], [62], [63], [71], [78] and [87]). In addition, it has been shown in [49] (based on results of [11]) that nonreal poles of $S(k)$ coincide with bound states or resonances of H for $n = 3$. For $n = 2$ such a result is contained in [50] under assumption on V stronger than (3.5).

For a recent result including exponentially decreasing interactions plus dilation analytic potentials for $n = 3$ we refer to [13].

Since by assumptions (3.5) and (3.6) the exceptional set ξ is discrete ([81], [74]) and nonzero points in ξ can be shown to be absent under reasonable additional conditions on V (cf. e.g. [31]), it remains to discuss $(1+uG_k v)^{-1}$ in a neighbourhood of $k = 0$. We confine ourselves to $n = 3$ (see [18], [64] for $n = 2$) and assume V to be real-valued. It has been shown in [3] (cf. also [15], [51], [64], [68]) that under hypothesis (3.6)

$$(1+uG_k v)^{-1} = \sum_{n=-M}^{\infty} t_n k^n , \quad M = 0,1, \text{ or } 2, \quad 0 < |k| \text{ small enough.} \tag{3.23}$$

If $M = 1$, then H has a zero-energy resonance (necessarily simple) whereas if $M = 2$, then H has zero-energy bound states or zero-energy bound states plus an additonal zero-energy resonance. In any case $S(k)$ never has a pole at $k = 0$

$$S(k) \underset{k\to 0}{=} O(1) , \quad M = 0,1,2 \tag{3.24}$$

(cf. [3] for the detailed behaviour of $S(k)$ near $k = 0$).

Finally, assume that the interaction V depends on an additional parameter $\lambda \in \Lambda$ (as discussed in Section II) denoted by V_λ. Moreover, suppose that conditions (3.4) - (3.6) are fulfilled for some fixed $a > 0$ for all $\lambda \in \Lambda$ and that u_λ and v_λ represent an appropriate splitting ov V_λ into

$$V_\lambda = u_\lambda v_\lambda$$

(e.g. if $V_\lambda = \lambda V$ one can choose $u_\lambda = \lambda u$, $v_\lambda = v$ etc.). Then, if $u_\lambda G_k v_\lambda$ is analytic with respect to $(k,\lambda) \in \Omega_n \times \Lambda$, $1 \le n \le 3$, Theorem 2.9 applies by identifying

$$E_\lambda^* = v_\lambda , \quad F_\lambda = u_\lambda , \quad K_\lambda(k) = u_\lambda G_k v_\lambda , \quad H_\lambda = -\Delta \dotplus V_\lambda .$$

Perturbation theory of eigenvalues and resonances in this spirit (including embedded eigenvalues and eigenvalues absorbed into the continuous spectrum) have been treated in [7], [8], [41], [47], [48], [61], [72], [73] and [83]. By a suitable choice of variables (modified) Fredholm determinants can also be used to describe situations where no Puiseux expansions in $(\lambda-\lambda_0)$ (i.e. analytic expansions in $(\lambda-\lambda_0)^{1/m}$) exist. E.g. in [38] the absorption of the ground state of two-dimensional systems into the continuous spectrum has been given in terms of analytic expansions in the variables $e^{-\text{const.}(\lambda-\lambda_0)^{-1}}$, $(\lambda-\lambda_0)^{-2} e^{-\text{const.}(\lambda-\lambda_0)^{-1}}$, etc.

We also would like to remark that the above approach has been extensively used to characterize bound states and resonances of Schrödinger operators on the half-line $(0,\infty)$ (see e.g. [10], [21], [24], [52], [53], [56], [65], [66] and [77] and the references therein).
For extensions of this formalism to the three particle case for $n = 3$ we refer to [12] and [67].

IV. Nonlocal Short-Range Interactions

The purpose of this section is to sketch how the results of Section III extend to a certain class of nonlocal interactions. It suffices to treat the three-dimensional case (we freely use the notation employed in Section III for $n = 3$).

Let W be a self-adjoint trace class operator in $L^2(\mathbb{R}^3)$, $W \in \mathcal{B}_1(L^2(\mathbb{R}^3))$. Furthermore assume that W can be written as the product of two Hilbert-Schmidt operators $W_1, W_2 \in \mathcal{B}_2(L^2(\mathbb{R}^3))$.

$$W = W_1 W_2 \tag{4.1}$$

such that the kernels $W_j(x,y)$ of W_j, $j = 1,2$ satisfy

$$e^{a|x|}\hat{u}_j \,,\quad e^{a|x|}\hat{v}_j \in L^1(\mathbb{R}^3) \cap L^2(\mathbb{R}^3) \quad \text{for some } a > 0,\ j = 1,2 \tag{4.2}$$

where

$$\hat{u}_j(x) = \left(\int d^3y \, |W_j(x,y)|^2\right)^{1/2},\quad \hat{v}_j(y) = \left(\int d^3x \, |W_j(x,y)|^2\right)^{1/2},\ j = 1,2. \tag{4.3}$$

The Hamiltonian H in $L^2(\mathbb{R}^3)$ is then defined by

$$H = -\Delta + W \,,\quad \mathcal{D}(H) = H^{2,2}(\mathbb{R}^3) \tag{4.4}$$

and, by identifying

$$E^* = W_1 \,,\quad F = W_2 \,,\quad K(k) = W_2 G_k W_1 \,, \tag{4.5}$$

Lemma 2.1 applies ([19]). In addition by conditions (4.2) one infers that $W_2 G_k W_1$ has a $\mathcal{B}_2(L^2(\mathbb{R}^3))$-valued analytic continuation into Ω_3. Thus Lemma 2.3 applies as well. From now on one can follow the case $n = 3$ of Section III step by step ([14], [19], [69]).

V. Point Interactions in Dimension 1,2 and 3

In this section we discuss certain models which explicitly allow the calculation of bound states and resonances. These models arise e.g. from those of Sections III and IV by introducing a particular limiting procedure in which the range of the interaction tends to zero ([2], [3]).

We start with one dimension. One of the simplest ways to introduce the point interaction Hamiltonian $-\Delta_{\alpha,y}$ in $L^2(\mathbb{R})$ which describes a δ-interaction of strength $\alpha \in \mathbb{C}$ centered at the point $y \in \mathbb{R}$ proceeds via the definition of its resolvent ([8], [9], [36], [88])

$$(-\Delta_{\alpha,y} - k^2)^{-1} = G_k - 2k\alpha(i\alpha+2k)^{-1} (\bar{g}_k,\cdot)g_k , \qquad (5.1)$$

$$\alpha \in \mathbb{C} , \quad \operatorname{Im} k > 0, \quad k \neq -i\alpha/2$$

where

$$g_k(x) = (i/2k)\, e^{ik|x-y|} .$$

Elements g of the domain of $-\Delta_{\alpha,y}$ then fulfill boundary conditions of the type

$$g(y_-) = g(y_+), \quad g'(y_+) - g'(y_-) = \alpha g(y), \; \alpha \in \mathbb{C} . \qquad (5.2)$$

Bound states and resonances (both simple by Eq. (5.1)) are then defined to be the poles of the kernel $(-\Delta_{\alpha,y} - k^2)^{-1}(x,x')$ in the upper - respectively in the lower k-plane. In particular, in the self-adjoint case $\alpha \in \mathbb{R}$ one obtains a simple bound state at $k_0 = -i\alpha/2$ for $\alpha < 0$ (with energy $E_0 = k_0^2 = -\alpha^2/4$ and corresponding eigenfunction $\psi_0(x) = \text{const.}\, e^{\alpha|x-y|/2}$) and a simple resonance at $k_0 = -i\alpha/2$ for $\alpha > 0$.

The corresponding on-shell scattering matrix $S_{\alpha,y}(k)$ then can be meromorphically continued to all of $\mathbb{C}$ such that poles of $S_{\alpha,y}(k)$ coincide with bound states or resonances of $-\Delta_{\alpha,y}$. This facts can be directly read off from ([9])

$$S_{\alpha,y=0}(k) = (i\alpha + 2k)^{-1} \begin{pmatrix} 2k & -i\alpha \\ -i\alpha & 2k \end{pmatrix} , \; \alpha \in \mathbb{C} , \quad k \geq 0. \qquad (5.3)$$

Next let us introduce the following family of scaled short-range Hamiltonians $H_{\varepsilon,y}$

$$H_{\varepsilon,y} = H_o \dotplus \lambda(\varepsilon)\varepsilon^{-2} V((.-y)/\varepsilon), \ \varepsilon > 0 \tag{5.4}$$

where $e^{2a|x|} V \in L^1(\mathbb{R})$ for all $a > 0$ and λ is analytic around $\varepsilon = 0$ with $\lambda(0) = 0$. Then it can be shown that H_ε converges in norm resolvent sense to $-\Delta_{\alpha y}$ as $\varepsilon \to 0_+$ ([8], [9])

$$\underset{\varepsilon\to 0_+}{\text{n-lim}} \ (H_{\varepsilon,y} - k^2)^{-1} = (-\Delta_{\alpha,y} - k^2)^{-1}, \quad k^2 \notin \sigma(-\Delta_{\alpha,y}) \tag{5.5}$$

where α is given by

$$\alpha = \lambda'(0) \int_{\mathbb{R}} dx\, V(x). \tag{5.6}$$

In fact, using Fredholm determinants, one can show that bound states and resonances of $H_{\varepsilon,y}$ are analytic in ε near $\varepsilon = 0$ and converge to those of $-\Delta_{\alpha,y}$ in the limit $\varepsilon \to 0_+$ ([8], [9]). More precisely, assume that <u>$\alpha = \lambda'(0) \int_{\mathbb{R}} dx\, V(x) \neq 0$</u>: Then, for ε small enough, $H_{\varepsilon,y}$ has precisely one simple bound state (for $\operatorname{Re}\alpha < 0$) or a simple resonance (for $\operatorname{Re}\alpha \geq 0$) k_ε analytic near $\varepsilon = 0$

$$k_\varepsilon = -(i/2)\lambda'(0) \int_{\mathbb{R}} dx\, V(x) - (i/4)\lambda''(0)\varepsilon \int_{\mathbb{R}} dx\, V(x)$$
$$- (i/4)\lambda'(0)^2 \varepsilon \int_{\mathbb{R}^2} dx\, dx'\, V(x)|x-x'|V(x') + O(\varepsilon^2). \tag{5.7}$$

<u>If $\alpha = \lambda'(0) \int_{\mathbb{R}} dx\, V(x) = 0$:</u> Then, for ε small enough, $H_{\varepsilon,y}$ either has a simple, negative eigenvalue $E_\varepsilon = k_\varepsilon^2 < 0$ if $\operatorname{Im}k_\varepsilon > 0$ which is absorbed into the continuous spectrum as $\varepsilon \to 0_+$ or, if $\operatorname{Im}k_\varepsilon < 0$, $H_{\varepsilon,y}$ has a simple resonance. In any case k_ε is unique and analytic near $\varepsilon = 0$

$$k_\varepsilon = -(i/4)\lambda''(0)\varepsilon \int_{\mathbb{R}} dx\, V(x) - (i/4)\lambda'(0)^2\varepsilon \int_{\mathbb{R}} dx\, dx'\, V(x)|x-x'|V(x')$$
$$+ O(\varepsilon^2). \tag{5.8}$$

For further details and generalizations to N-center point interactions we refer to [8], [9], [36] and the references therein.

In two-dimensions the corresponding model can be defined by ([9], [26], [36])

$$(-\Delta_{\alpha,y} - k^2)^{-1} = G_k - 2\pi[\alpha + \Psi(1) - \ln(k/2i)]^{-1} (\bar{g}_k, .) g_k ,$$

$$\alpha \in \mathbb{C} , \quad \mathrm{Im} k > 0, \quad k \neq 2i \exp[\alpha + \Psi(1)] \tag{5.9}$$

where

$$g_k(x) = -(i/4) H_0^{(1)}(k|x-y|), \quad x \neq y$$

and $\Psi(z)$ denotes the psi function [1]. Again $-\Delta_{\alpha,y}$ describes a point interaction centered at the point $y \in \mathbb{R}^2$.

Being a zero-range interaction, $-\Delta_{\alpha,y}$ differs from $H_0 = -\Delta$ only in the s-wave. In fact the corresponding s-wave boundary condition reads

$$-\alpha g_0 + g_1 = 0, \quad \alpha \in \mathbb{C} \tag{5.10}$$

where ([76])

$$g_0 = \lim_{r \to 0_+} [r^{1/2} \ln r]^{-1} g(r), \quad g_1 = \lim_{r \to 0_+} r^{-1/2}[g(r) - g_0 r^{1/2} \ln r].$$

Spectral properties now directly follow from the pole structure of Eq. (5.9). In particular in the self-adjoint case $\alpha \in \mathbb{R}$ one obtains precisely one simple bound state at $k_0 = 2i \exp[\alpha + \Psi(1)]$ and no resonance in $\mathrm{Im} k < 0$, $-\pi < \arg k < 0$.

The on-shell scattering matrix $S_{\alpha,y}(k)$, $k > 0$ in $L^2(S^1)$ associated with $-\Delta_{\alpha,y}$ is given by ([9])

$$S_{\alpha,y=0}(k) = 1 - i\pi[\alpha + \Psi(1) - \ln(k/2i)]^{-1} (Y_0, .) Y_0 , \tag{5.11}$$

$$k > 0, \quad Y_0(\omega) = (2\pi)^{-1/2}.$$

Hence $S_{\alpha,y}$ has a meromorphic continuation into the cut plane $\{k \in \mathbb{C} \setminus \{0\} / -\pi < \arg k < \pi\}$ such that the pole of $S_{\alpha,y}(k)$ coincides with the cound state resp. resonance of $-\Delta_{\alpha,y}$.

Recently, among other things, approximations by means of separable scaled short-range interactions have been discussed in [26].

Finally we turn to $n = 3$. Again we introduce the point interaction Hamiltonian $-\Delta_{\alpha,y}$ in $L^2(\mathbb{R}^3)$, centered at the point $y \in \mathbb{R}^3$, in terms of its resolvent ([3], [9], [30], [36], [88])

$$(-\Delta_{\alpha,y}-k^2)^{-1} = G_k - [(ik/4\pi) - \alpha]^{-1} (\bar{g}_k,\cdot)g_k \,, \tag{5.12}$$
$$\alpha \in \mathbb{C} \,, \quad \mathrm{Im}k > 0 \,, \quad k \neq -i4\pi\alpha$$

where

$$g_k(x) = (4\pi|x-y|)^{-1} e^{ik|x-y|}, \quad x \neq y \,.$$

The corresponding s-wave boundary condition (again the interaction is of zero range and hence an s-wave interaction) then reads

$$-4\pi\alpha\, g(0_+) + g'(0_+) = 0 \,, \; \alpha \in \mathbb{C} \,. \tag{5.13}$$

Since the kernel $(-\Delta_{\alpha,y}-k^2)^{-1}(x,x')$, $x \neq x'$ extends meromorphically to all of $\mathbb{C}$ bound states and resonances can be determined directly from Eq. (5.12). E.g. in the self-adjoint case $\alpha \in \mathbb{R}$ one obtains precisely one simple, negative bound state at $k_0 = -i\,4\pi\alpha$ for $\alpha < 0$ ($E_0 = k_0^2 = -(4\pi\alpha)^2$, $\psi_0(x) = \text{const.}\, e^{4\pi\alpha|x-y|}/|x-y|$, $x \neq y$ the corresponding eigenfunction) and a simple resonance at $k_0 = -i\,4\pi\alpha$ for $\alpha \geq 0$.

Similar to the one-dimensional case the on-shell scattering operator $S_{\alpha,y}(k)$, $k > 0$ in $L^2(S^2)$ associated with $-\Delta_{\alpha,y}$ can be meromorphically continued to all of $\mathbb{C}$ as can be seen from ([20])

$$S_{\alpha,y=0}(k) = 1 - 2ik\,(ik - 4\pi\alpha)^{-1}\,(Y_{00},\cdot)Y_{00} \,, \tag{5.14}$$
$$\alpha \in \mathbb{C},\; k \geq 0, \quad Y_{00}(\omega) = (4\pi)^{-1/2} \,.$$

The pole of $S_{\alpha,y}$ obviously coincides with the bound state or resonance of $-\Delta_{\alpha,y}$ as long as $\alpha \neq 0$.

In contrast to the one- and two-dimensional case, norm resolvent convergence of scaled short-range Hamiltonians $H_{\varepsilon,y}$ of the type

$$H_{\varepsilon,y} = -\Delta \dot{+} \lambda(\varepsilon)\varepsilon^{-2}\, V((\cdot-y)/\varepsilon), \; \varepsilon > 0 \tag{5.15}$$

(with $e^{2a|x|}\, V \in R$ for all $a > 0$ and λ analytic near $\varepsilon = 0$, $\lambda(0) = 1$) to the point interaction $-\Delta_{\alpha,y}$ (with α finite) as $\varepsilon \to 0_+$ crucially depends on zero-energy properties of $-\Delta \dot{+} V(\cdot-y)$ ([2]). In fact, it has been proven in [3] that if and only if $-\Delta \dot{+} V(\cdot-y)$ has a zero-energy resonance then $H_{\varepsilon,y}$ converges in norm resolvent sense to $-\Delta_{\alpha,y}$ with α finite whereas otherwise $H_{\varepsilon,y}$ converges to $-\Delta$ as $\varepsilon \to 0_+$. In particular, one can use modified Fredholm determinants to study the corresponding convergence of bound states and resonances of $H_{\varepsilon,y}$ to those of $-\Delta_{\alpha,y}$ ([3], [9]). Since the effective range parameter ([5]) of H_ε is of order ε these convergence results confirm the fact that $-\Delta_{\alpha,y}$ describes a zero-range interaction.

For further results and references to other literature see e.g. [7], [9], [32], [36], [39], [40] and [88]. Generalizations to nonlocal point interactions in $n = 1,2$ and 3 are studied in [26].

Finally we would like to mention that the approach of Section II also directly applies to systems where the point interaction Hamiltonian $-\Delta_{\alpha,y}$ is perturbed by a potential V_λ ([7])

$$H_\lambda = -\Delta_{\alpha,y} \dot{+} V_\lambda \quad , \quad \lambda \in \Lambda \subset \mathbb{C} \tag{5.16}$$

where for all $\lambda \in \Lambda$, V_λ satisfies conditions (3.4) - (3.6) for all $a > 0$ and, for $n = 2,3$ $u_\lambda g_k \in L^2(\mathbb{R}^n)$ for all $(k,\lambda) \in \{k \in \mathbb{C}\backslash\{0\}/ -\pi < \arg k < \pi\} \times \Lambda$ for $n = 2$ resp. for all $(k,\lambda) \in \mathbb{C} \times \Lambda$ for $n = 3$. The resolvent equation

$$(H_\lambda - k^2)^{-1} = (-\Delta_{\alpha,y} - k^2)^{-1}$$

$$-(-\Delta_{\alpha,y}-k^2)^{-1}v_\lambda[1+u_\lambda(-\Delta_{\alpha,y}-k^2)^{-1}v_\lambda]^{-1}u_\lambda(-\Delta_{\alpha,y}-k^2)^{-1} \tag{5.17}$$

then shows that poles due to $-\Delta_{\alpha,y}$ explicitly cancel in Eq. (5.17) (cf. Lemmas 2.5-2.7). Hence bound states and resonances of H_λ are exactly determined by eigenvalues -1 of $u_\lambda(-\Delta_{\alpha,y}-k^2)^{-1}v_\lambda$ which has a $\mathcal{B}_2(L^2(\mathbb{R}^n))$-valued meromorphic continuation to all of $\mathbb{C}$ for $n = 1,3$ and to the cut plane $\{k \in \mathbb{C}\backslash\{0\}/ -\pi < \arg k < \pi\}$ for $n = 2$. As a consequence, the whole machinery of Section II is applicable in this case.

VI. Coulomb-Type Interactions

We briefly sketch how one can generalize the results of Section III to long-range interactions. It suffices to treat the case $n = 3$.

Let H^C denote the Coulomb Hamiltonian in $L^2(\mathbb{R}^3)$

$$H^C = -\Delta + \gamma|x|^{-1} \ , \ \gamma \in \mathbb{R}, \quad \mathcal{D}(H^C) = H^{2,2}(\mathbb{R}^3) \tag{6.1}$$

with resolvent $G_{\gamma,k}$

$$G_{\gamma,k} = (H^C - k^2)^{-1}, \quad \mathrm{Im}k > 0, \quad k \neq -i\gamma/2n, \quad n \in \mathbb{N}. \tag{6.2}$$

Then the kernel $G_{\gamma,k}(x,x')$ of $G_{\gamma,k}$ is explicitly given by ([42])

$$G_{\gamma,k}(x,x') = (4\pi|x-x'|)^{-1}\Gamma(1+i\gamma/2k)\left[\left(\frac{d}{d\alpha} - \frac{d}{d\beta}\right)M_{\frac{-i\gamma}{2k};\frac{1}{2}}(\alpha)W_{\frac{-i\gamma}{2k};\frac{1}{2}}(\beta)\right]\Bigg|_{\substack{\alpha=-ikx_- \\ \beta=-ikx_+}}$$

$$\mathrm{Im}k > 0, \quad k \neq -i\gamma/2n, \quad n \in \mathbb{N}, \quad x_\pm = |x|+|x'|\pm|x-x'| \tag{6.3}$$

where $M_{\mu;\nu}(z)$, $W_{\mu;\nu}(z)$ denote the Whittaker functions [1]. H^C now plays the role of the "unperturbed" Hamiltonian H_o of Section II. In particular, if V fulfills condition (3.6) (respectively if W obeys assumption (4.2)) the Hamiltonian H in $L^2(\mathbb{R}^3)$ is defined by the form sum

$$H = H^C + V \tag{6.4}$$

(respectively by $H = H^C + W$, $\mathcal{D}(H) = H^{2,2}(\mathbb{R}^3)$). Moreover, from the fact that $uG_{\gamma,k}v$ can be written as ([17])

$$uG_{\gamma,k}v = \{\gamma\Psi(1+i\gamma/2k)+ik+\gamma\ln(2k/i|\gamma|)\}\, A(k^2) + B(k^2), \tag{6.5}$$

$$\mathrm{Im}k > 0, \quad k \neq -i\gamma/2n, \quad n \in \mathbb{N}$$

where A and B are analytic with respect to k^2 as long as $\mathrm{Im}k > -a$, one infers that $uG_{\gamma,k}v$ has a $\mathcal{B}_2(L^2(\mathbb{R}^3))$-valued analytic continuation into Ω_γ

$$\Omega_\gamma = \{k \in \mathbb{C}\backslash\{0\}/\mathrm{Im}k > -a,\ k \neq -i\gamma/2n,\ n \in \mathbb{N},\ -\pi < \arg k < \pi\}. \tag{6.6}$$

(The same result holds if V is replaced by W and hence $uG_{\gamma,k}v$ by $W_2G_{\gamma,k}W_1$). Thus one can follow the case $n = 3$ in Section III step by step.

VII. Coulomb Plus Point Interactions

Finally, we shortly treat an exactly solvable model for charged particles under the influence of an additional zero-range interaction in three dimensions. This model can be obtained from the Hamiltonians of Section VI if the range of the short-range part tends to zero in an appropriate way ([4]).

The resolvent for the Coulomb plus point interaction Hamiltonians $H^C_{\alpha,y}$ in $L^2(\mathbb{R}^3)$ with the point interaction centered at $y \in \mathbb{R}^3$ reads

$$(H^C_{\alpha,y} - k^2)^{-1} = G_{\gamma,k,y}$$

$$- 4\pi[\gamma\Psi(1+i\gamma/2k)+ik+\gamma \ln(2k/i|\gamma|) + \gamma(2C-1)-4\gamma\alpha]^{-1}(\bar{g}_{\gamma,k},.)g_{\gamma,k}, \quad (7.1)$$

$$\alpha \in \mathbb{C}, \quad \mathrm{Im}k > 0, \quad k^2 \notin \sigma(H^C_{\alpha,y})$$

where C denotes Euler's constant [1] and

$$g_{\gamma,k}(x) = (4\pi|x-y|)^{-1}\Gamma(1+i\gamma/2k)W_{\frac{-i\gamma}{2k};\frac{1}{2}}(-2ik|x-y|), \quad x \neq y.$$

Again $H^C_{\alpha,y}$ differs from H^C_y only in the s-wave where the corresponding boundary conditions is given by ([76])

$$-4\pi\alpha\, g_0 + g_1 = 0, \quad \alpha \in \mathbb{C}, \quad (7.2)$$

where

$$g_0 = g(0_+), \quad g_1 = \lim_{r\to 0_+} r^{-1}\{g(r) - g_0[1 + \gamma r \ln(|\gamma|r)]\}.$$

Bound states and resonances are now obtained directly from the pole structure of Eq. (7.1) as in Section V. (Note that the pure Coulomb poles at $k_n = -i\gamma/2n$, $n \in \mathbb{N}$ indeed cancel in Eq. (7.1).) In the self-adjoint case $\alpha \in \mathbb{R}$ one obtains precisely one simple, negative bound state of $H^C_{\alpha,y}$ if $\alpha < \gamma(2C-1)/4\pi$ and no bound states at all if $\alpha \geq \gamma(2C-1)/4\pi$ in the repulsive Coulomb case $\gamma \geq 0$. For attraction, $\gamma < 0$, $H^C_{\alpha,y}$ has always infinitely many simple and negative eigenvalues accumulating at zero.

The on-shell scattering operator $S^C_{\alpha,y}(k)$, $k > 0$ in $L^2(S^2)$ associated with this model is given by

$$S^C_{\alpha,y=0}(k) = S^C(k) - 2ik\, e^{-\pi\gamma/2k}\, \Gamma(1+i\gamma/2k)^2 \,\cdot$$

$$\cdot\, [\gamma\Psi(1+i\gamma/2k)+ik+\gamma\ln(2k/i|\gamma|)+\gamma(2C-1)-4\pi\alpha]^{-1}(Y_{00},.)Y_{00}, \quad (7.3)$$

$$\alpha \in \mathbb{C}, \quad k > 0$$

where

$$S^C = \frac{(\frac{1}{2} + (L^2 + \frac{1}{4})^{1/2} + \frac{i\gamma}{2k})}{(\frac{1}{2} + (L^2 + \frac{1}{4})^{1/2} - \frac{i\gamma}{2k})} , \quad k > 0 \tag{7.4}$$

(L^2 being the square of the angular momentum operator) denotes the pure Coulomb on-shell scattering operator.

Approximations of $H^C_{\alpha,y}$ by means of scaled Coulomb plus short-range Hamiltonians $H_{\gamma,\varepsilon,y}$

$$H_{\gamma,\varepsilon,y} = -\Delta + \gamma|x-y|^{-1} \dot{+} \lambda(\varepsilon,\gamma\varepsilon\ell n\varepsilon)\varepsilon^{-2} V((\cdot-y)/\varepsilon), \quad \varepsilon > 0 \tag{7.5}$$

with $\lambda(\cdot,\cdot)$ analytic near $(0,0)$, $\lambda(0,0) = 1$ and V obeying condition (3.6) again crucially depend on zero-energy properties of $-\Delta \dot{+} V(\cdot-y)$. For a detailed treatment of these questions and applications in the context of nucleon-nucleon scattering and level shifts in mesic atoms see [4], [6] and [9].

Finally, perturbation theory around $H^C_{\alpha,y}$ for Hamiltonians H_λ

$$H_\lambda = H^C_{\alpha,y} \dot{+} V_\lambda$$

can be developed as indicated at the end of Section V.

Acknowledgements

It is a pleasure to thank S. Albeverio, D. Bollé, H. Holden, R. Høegh-Krohn, W. Kirsch and L. Streit for joint collaborations which led to most of the results presented above. I am indebted to Prof. L. Streit for the warm hospitality extended to me at the Zentrum für interdisziplinäre Forschung der Universität Bielefeld, FRG. Financial support by the Alexander von Humboldt Stiftung is gratefully acknowledged.

References

[1] M. Abramowitz and I.A. Stegun, "Handbook of Mathematical Functions" (Dover, New York, 1972)

[2] S. Albeverio and R. Høegh-Krohn, J. Operator Theory 6, 313 (1981)

[3] S. Albeverio, F. Gesztesy and R. Høegh-Krohn, Ann. Inst. H. Poincaré A37, 1 (1982)

[4] S. Albeverio, F. Gesztesy, R. Høegh-Krohn and L. Streit, Ann. Inst. H. Poincaré A 38, 263 (1983)

[5] S. Albeverio, D. Bollé, F. Gesztesy, R. Høegh-Krohn and L. Streit, Ann. Phys. 148, 308 (1983)

[6] S. Albeverio, L.S. Ferreira, F. Gesztesy, R. Høegh-Krohn and L. Streit, Phys. Rev. C29, 680 (1984)

[7] S. Albeverio and R. Høegh-Krohn, "Perturbation theory of resonances in quantum mechanics", J. Math. Anal. Appl. 100 (1984), in print

[8] S. Albeverio, F. Gesztesy, R. Høegh-Krohn and W. Kirsch, "On point interactions in one dimension", J. Operator Theory, 12, 101 (1984)

[9] S. Albeverio, F. Gesztesy, H. Holden and R. Høegh-Krohn, "Solvable Models in Quantum Mechanics", book in preparation

[10] A.A. Arsenev, Theoret. Math. Phys. 15, 505 (1973)

[11] D. Babbitt and E. Balslev, J. Math. Anal. Appl. 54, 316 (1976)

[12] E. Balslev, "Resonances in three-body scattering theory", Aarhus Preprint Series 1981/82, No. 3

[13] E. Balslev, "Local spectral deformation techniques for Schrödinger operators", Preprint 1982

[14] H. Baumgärtel, M. Demuth and M. Wollenberg, Math. Nachr. 86, 167 (1978)

[15] D. Bollé and S.F.J. Wilk, J. Math. Phys. 24, 1555 (1983)

[16] D. Bollé, F. Gesztesy and S.F.J. Wilk, "A complete treatment of low-energy scattering in one dimension", J. Operator Theory, in print

[17] D. Bollé, F. Gesztesy and W. Schweiger, "Scattering theory for long-range systems at threshold", Univ. Bielefeld, ZiF-Preprint 1984, No. 76

[18] D. Bollé, C. Daneels, F. Gesztesy and S.F.J. Wilk, in preparation

[19] D. Bollé, F. Gesztesy, C. Nessmann and L. Streit, in preparation

[20] E. Brüning and F. Gesztesy, J. Math. Phys. 24, 1516 (1983)

[21] K. Chadan and P.C. Sabatier, "Inverse Problems in Quantum Scattering Theory" (Springer, New York 1977)

[22] M. Cheney, J. Math. Phys. 25, 95 (1984)

[23] M. Cheney, "Two-dimensional scattering: The number of bound states from scattering data", Stanford preprint

[24] M. Ciafaloni and P. Menotti, Nuovo Cimento 35, 160 (1965)

[25] J.M. Combes, Int. J. Quantum Chem. 14, 353 (1978)

[26] L. Dabrowski and H. Grosse, "On nonlocal point interactions in one, two and three dimensions", Preprint, Univ. Vienna, UWThPh-1983-15

[27] E.B. Davies, Lett. Math. Phys. 1, 31 (1975)

[28] C.L. Dolph, J.B. McLeod and D. Thoe, J. Math. Anal. Appl. 16, 311 (1966)

[29] N. Dunford and J.T. Schwartz, "Linear Operators II. Spectral Theory" (Interscience, New York, 1963)

[30] G. Flamand, "Mathematical theory of non-relativistic two- and three-particle systems with point interactions", Cargèse Lectures, F. Lurcat (ed.) (Gordon and Breach, New York, 1967)

[31] R. Froese, I. Herbst, M. Hoffmann-Ostenhof and T. Hoffmann-Ostenhof, J. Anal. Math. 41, 272 (1982)

[32] F. Gesztesy and L. Pittner, Rep. Math. Phys. 19, 143 (1984)

[33] A. Grossmann and T.T. Wu, J. Math. Phys. 2, 710 (1961)

[34] A. Grossmann, J. Math. Phys. 2, 714 (1961)

[35] A. Grossmann and T.T. Wu, J. Math. Phys. 3, 684 (1962)

[36] A. Grossmann, R. Høegh-Krohn and M. Mebkhout, J. Math. Phys. 21, 2377 (1980)

[37] E.M. Harrell, Commun. Math. Phys. 86, 221 (1982)

[38] H. Holden, "On coupling constant thresholds in two dimensions", Univ. Bielefeld, ZiF-Preprint 1984, No. 44

[39] H. Holden, R. Høegh-Krohn and S. Johannesen, Adv. Appl. Math. 4, 402 (1983)

[40] H. Holden, R. Høegh-Krohn and M. Mebkhout, "The short-range expansion for multiple well scattering theory", Preprint, CNRS-Marseille, CPT-84/PE.1590

[41] L.P. Horwitz and I.M. Siegal, Helv. Phys. Acta 51, 685 (1978)

[42] L. Hostler, J. Math. Phys. 8, 642 (1967)

[43] J.S. Howland, Proc. Am. Math. Soc. 21, 381 (1969)

[44] J.S. Howland, Arch. Rat. Mech. Anal. 39, 323 (1970)

[45] J.S. Howland, Proc. Am. Math. Soc. 28, 177 (1971)

[46] J.S. Howland, J. Math. Anal. Appl. 36, 12 (1971)

[47] J.S. Howland, Pac. J. Math. 55,157 (1974)

[48] J.S. Howland, J. Math. Anal. Appl. 50, 415 (1975)

[49] A. Jensen, J. Math. Anal. Appl. 59, 505 (1977)

[50] A. Jensen, Ann. Inst. H. Poincaré A33,209 (1980)

[51] A. Jensen and T. Kato, Duke Math. J. 46, 583 (1979)

[52] R. Jost, Helv. Phys. Acta 20, 256 (1947)

[53] R. Jost and A. Pais, Phys. Rev. 82, 840 (1951)

[54] T. Kato, "Perturbation Theory for Linear Operators", 2nd ed. (Springer, New York, 1980)

[55] T. Kato, Math. Ann. 162, 258 (1966)

[56] N.N. Khuri, Phys. Rev. 107, 1148 (1957)

[57] M. Klaus, Ann. Phys. 108, 288 (1977)

[58] M. Klaus, Helv. Phys. Acta 55, 49 (1982)

[59] M. Klaus and B. Simon, Ann. Phys. 130, 251 (1980)

[60] R. Konno and S.T. Kuroda, J. Fac. Sci. Univ. Tokyo I13, 55 (1966)

[61] S.N. Lakaev, Theoret. Math. Phys. 44, 810 (1980)

[62] J.B. McLeod, Quart. J. Math. Oxford 18, 219 (1967)

[63] K. Meetz, J. Math. Phys. 3, 690 (1962)

[64] M. Murata, J. Func. Anal. 49, 10 (1982)

[65] R.G. Newton, "Scattering Theory of Waves and Particles", 2nd ed. (Springer, New York, 1982)

[66] R.G. Newton, J. Math. Phys. 13, 880 (1972)

[67] R.G. Newton, Czech. J. Phys. B24, 1195 (1974)

[68] R.G. Newton, J. Math. Phys. 18, 1348 (1977)

[69] R.G. Newton, J. Math. Phys. 18, 1582 (1977)

[70] H.M. Nussenzveig, Nucl. Phys. 11, 499 (1959)

[71] J. Nuttall, J. Math. Phys. 8, 873 (1967)

[72] A.G. Ramm, J. Math. Anal. Appl. 88, 1 (1982)

[73] J. Rauch, J. Func. Anal. 35, 304 (1980)

[74] M. Reed and B. Simon, "Methods of Modern Mathematical Physics III: Scattering Theory" (Academic, New York, 1979)

[75] M. Reed and B. Simon, "Methods of Modern Mathematical Physics IV: Analysis of Operators" (Academic, New York 1978)

[76] F. Rellich, Math. Z. 49, 702 (1943/44)

[77] H. Rollnik, Z. Physik 145, 639 (1956)

[78] N. Shenk and D. Thoe, Rocky Mountain J. Math. 1. 89 (1971)

[79] H.K.H. Siedentop, Phys. Lett. 99A, 65 (1983)

[80] H.K.H. Siedentop, "On the width of resonances", Caltech-Preprint 1984

[81] B. Simon, "Qauntum Mechanics for Hamiltonians defined as Quadratic Forms" (Princeton Univ. Press 1971)

[82] B. Simon, Ann. Math. 97, 247 (1974)

[83] B. Simon, Ann. Phys. 97, 279 (1976)

[84] B. Simon, Adv. Math. 24, 244 (1977)

[85] B. Simon, Int. J. Quantum Chem. 14, 529 (1978)

[86] B. Simon, "Trace Ideals and their Applications", London Math. Soc. Lecture Notes Series 35 (Cambridge Univ. Press 1979)

[87] S. Steinberg, Arch. Rat. Mech. Anal. 38, 278 (1970)

[88] J. Zorbas, J. Math. Phys. 21, 840 (1980)

The resonance expansion for the Green's function of the Schrödinger and wave equations

by

S. Albeverio[#,*] and R. Høegh-Krohn[##,**]

Zentrum für interdisziplinäre Forschung
Universität Bielefeld

ABSTRACT

We give a survey of some recent mathematical work on resonances, in particular on perturbation series, low energy expansions and on resonances for point interactions. Expansions of the kernels of $e^{-it\sqrt{H_+}}$ and e^{-itH} in terms of resonances are also given (where H_+ is the positive part of the Hamiltonian).

Mathematisches Institut, Ruhr-Universität, D-4630 Bochum 1 (Fed. Rep. Germany)

* Centre de Physique Théorique, CNRS, Université d'Aix-Marseille II

Université de Provence and Centre de Physique Théorique, CNRS, Marseille

** Matematisk Institutt, Universitetet i Oslo

1. Introduction

The problem of the definition and the study of resonances in quantum mechanics has attracted quite a lot of attention in physics and mathematics, in recent years, and several approaches have been taken. The whole present conference has been on this topics and we are happy to point at the collection of all essays for surveys of several areas of investigation on this topics. In the present lecture we shall discuss resonances for operators of the form $-\Delta + V$ with V a local potential (in some of the applications, of compact support).We shall discuss certain particular questions and we refer to other contributions in this volume for complements. Even with such restrictions there is a large literature on the subject. G. Gamov discussed in 1928 resonant states in relation to complex poles of the scattering matrix in the study of α decay (complex frequencies had actually appeared in the physical literature much earlier, e.g. in work by J.J. Thomson in 1884).

Through work by Siegert (1939), Humblet and Rosenfeld (1961) and many others a nuclear reaction theory involving resonance reactions in nuclei was developed. In particular a relation between analytic properties of the S-matrix in the complex plane and the space-time behaviour of wave packets was obtained. In the radial symmetric case the connection between resonances and poles of the scattering matrix was investigated extensively, starting from fundamental work of R. Jost [1], see also e.g. the references in [2] and [3].
The relation between a rapid variation of the phase shift with the probability of finding particles inside the region of interaction was also investigated, see e.g. [2].

The association of resonances with poles of the resolvent taken between suitable states has been discussed starting with classical work by Titchmarsh see e.g. [2], [4-9]. Resonances can be produced by perturbing eigenvalues embedded in the continuum, this is a well studied subject, see e.g. [8], [10], [14].

For work using dilation techniques see e.g. [11] and the contributions by Balslev, Graffi and others at this meeting. For work relating resonances to the local decay of solutions of Schrödinger equation and the corresponding wave (acoustic) equation see e.g. [15] - [17]. and references therein. Resonances also enter discussions of asymptotic completeness (see e.g. [11], [18], and low energy expansions of the S-matrix and the resolvent [19] - [22]. Let us shortly describe the content of the different sections of this lecture. Sections 2-4 are essentially based on [24].
In section 2 we give the basic definition and study the basic quantities related to resonances.
We touch upon results related to [8], [23], [27].
We recall explicit formulae for the residues of the S-matrix and the resolvent kernel at a simple resonance, the latter being entirely analogue to those for the eigenvalues

[24].

In particular we discuss a suitable normalization for resonance functions in the case of potentials of compact support. There is actually a large physical literature about normalization and orthogonalization of resonance functions (Gamow vectors), see e.g. [9], [28] - [36]. For the case of dilation analytic potentials see [18].

We shall recover the basic results, not assuming analyticity but rather compact support for the potential V.

In section 3 we study perturbation of resonances. In fact we look at a family of Schrödinger operators of the form $-\Delta + V_\lambda$, with V_λ real analytic in λ.
We study in particular, following [24], the resonances on the imaginary axis as functions of λ, giving in particular perturbation formulae for them and their corresponding resonance functions.
In section 4 we study resonances for point interactions and the related short range expansions for Schrödinger operators with local potentials.
Hamiltonians with point interactions have received a lot of attention in recent years, see e.g. [19]-[21],[38],[41],[67],[68], and references therein.Application to solid state physics [20],[55],[58], nuclear physics [19],[20],[61],[62], electromagnetism [20],[54],polymer physics [40],[49] and quantum field theory[40],[49]have been given. For simple point interactions the resonances can be explicitly computed and some result are given here. Due to the fact that the Hamiltonian with point interactions can be well approximated (norm resolvent sense) by Schrödinger operators with scaled potentials, the explicit results on their resonances yield also results on the resonances of the latter.

In section 5 we discuss some forms of "completeness" for resonance functions. In the physics literature there have been extensive discussions of completeness properties of resonance functions, see e.g. [33], [42-45].
Under a reasonable assumption on the resolvent kernel (which we here verify e.g. for the case of point interactions) we prove formulae expressing $\exp(-it\sqrt{H_+})(x,y)$, H_+ being the positive part of $H = -\Delta + V$, in terms of resonances and their resonance functions (for large times and fixed x,y). We also derive a corresponding formula for $\exp(-itH_+)(x,y)$ in terms of such quantities and a rest integration over a continuum, the latter contribution being then evaluated asymptotically for $t \to +\infty$, using low energy expansions of [19].

We conclude this lecture with two remarks, one on the acoustic equation approach to quantum mechanical resonances ([15] - [17]) and one on the relation between scattering theory for automorphic forms, resonances and the Riemann zeta function ([16], [46]).

2. Definition and basic properties of resonances

Let V be a real valued measurable function on $\mathbb{R}^3$ satisfying, for some $\gamma > 0$, $|V|_\gamma^2 \equiv (4\pi)^{-2} \iint |V(x)V(x)| \ \{\exp[2\gamma(|x| + |y|)]\} / |x-y|^2 \ dxdy < \infty$. It is then easy to show, see e.g. [11] , pp. 167-170, that the form $-(\varphi,\Delta\varphi) + (\varphi,V\varphi)$ with (,) the $L^2(\mathbb{R}^3,dx)$ scalar product, is closed densely defined (with domain $D((-\Delta)^{1/2})$, lower bounded and defines uniquely a lower bounded self-adjoint operator $H = -\Delta + V$ in $L^2(\mathbb{R}^d,dx)$. Let $v \equiv |V|^{1/2}$, $u(x) \equiv v(x)\text{sign } V(x)$ (i.e. $u(x) = v(x)$ if $V(x) \geq 0$ and $u(x) = -v(x)$ if $V(x) \leq 0$). Set, for $\text{Im } k^2 \notin [0,\infty)$ $G_k \equiv (-\Delta - k^2)^{-1}$. The kernel $G_k(x,y)$ of G_k is equal to $G_k(x-y) = \frac{e^{ik(x-y)}}{4\pi(x-y)}$ for $\text{Im } k \geq 0$. One observes that $G_k(x-y)$ extends, for $x \neq y$, to an analytic function of k in the whole complex plane. As a function of k^2, $G_k(x-y)$ is analytic with a cut along the positive real axis. For $\text{Im } k > 0$ ("the physical half plane") we have

$$(H-k^2)^{-1} = G_k - G_k v(1 + uG_k v)^{-1} \ uG_k . \qquad (2.1)$$

For $\text{Im } k > -\gamma$ we have that $uG_k v$ is Hilbert-Schmidt, hence the resolvent kernel $(H-k^2)^{-1}(x,y)$ is, for almost all $x \neq y$, an analytic function of k in $\text{Im } k > -\gamma$ with poles on a discrete subset of $\text{Im } k > -\gamma$ consisting of k such that $uG_k v$ has the eigenvalue -1, as an operator in $L^2(\mathbb{R}^3,dx)$. If V has in addition compact support then $(H-k^2)^{-1}(x,y)$ for $x \neq y$ is analytic in the whole k-plane, except for isolated poles.

In the general case for $\text{Im } k > 0$ the poles are all simple and situated on the imaginary axis $\text{Re } k = 0$ and they coincide with the negative eigenvalues of H, of finite multiplicities. The poles of the resolvent kernel $(H-k^2)^{-1}(x,y)$ for $x \neq y$ in $\text{Im } k < 0$, ("the unphysical halfplane") are called <u>resonances</u> of H.

Because of $G_k(x,y) = \overline{G}_{-\bar{k}}(x,y)$ we have that the resonance of H lie symmetrically with respect to the negative imaginary axis. As function of k^2, $(H-k^2)(x,y)$ is analytic in the cut plane $k^2 \notin [0,\infty)$ (cut along the positive real axis), with poles in $k^2 < 0$, corresponding to eigenvalues, and resonances at points k^2 with $\text{Im } k < 0$. The resonances coincide, by the above definition, with the zeros in $-\gamma < \text{Im } k < 0$ of the modified Fredholm determinant

$$\det_2(1+uG_k v) \equiv \exp \sum_{n=2} \frac{1}{n} \operatorname{tr}((G_k^{1/2} V G_k^{1/2})^n) \equiv d(k,V).$$

The eigenvalues are the solutions of $d(k,v) = 0$ in $\text{Im } k > 0$.
The determinant d is a joint analytic function of V and k in $\text{Im } k > -\gamma$. The only possible singularities of its zeros $k_n(V)$ as functions of V in the complex Banach space R_γ with norm $|\ |_\gamma$ are branch points of finite order. The latter form a subset of R_γ of a t most codimension 1 (the condition for having a branch point at V_o being precisely $\frac{\partial}{\partial k} d(V_o,k(V_o)) = 0$ at $k = k_n$). V_o is a branch point of order ℓ for $k_n(V)$ if $\frac{\partial^j}{\partial k^j} d(V_o,k(V_o)) = 0$ at $k = k_n$, for $j=1,\dots,\ell-1$, while $\frac{\partial^\ell}{\partial k^\ell} d(V_o,k(V_o)) \neq 0$

at $k = k_n$. In this case there are resonances

$k_n(V),\dots,k_{n+\ell-1}(V)$ s.t. $k_n(V_o) = \dots = k_{n+\ell-1}(V_o)$.

Remark. As remarked above the poles of $(H-k^2)^{-1}(x,y)$, $x \neq y$ are simple if k^2 is an eigenvalue i.e. $\operatorname{Im} k > 0$. The poles are however not necessarily simple in $\operatorname{Im} k \leq 0$, in particular for the resonances, as discussed in [47], [26], [38]. Moreover as discussed in [48], [12], [38] the multiplicity of the eigenvalue -1 of $uG_k v$ at $k=k_n$ coincides with the multiplicity of the zero of d at k_n iff $(1+uG_k v)^{-1}$ has a pole of 1. oder at k_n.

It is possible to put in relation the eigenvalues and resonances of H with poles of the T-matrix. Let in fact $T(k) \equiv V - V(H-k^2)^{-1}V$, for k^2 not in the spectrum of H. (T is the off-shell T matrix). We note that the scattering matrix (S-matrix) is determined by T, since it is defind as the operator with kernel
$S(p,q) = \delta(p-q) - 2\pi i\delta(p^2-q^2)\ (2\pi)^{-3} \iint e^{-i(px-py)}T(p)(x,y)\,dxdy,\ p,q \in \mathbb{R}^3$.

One has that the kernel $T(k)(p,q)$ of $T(k)$ is equal to
$(-i(2\pi)^{-2}(e^{ipx}v,\ (1+uG_k v)^{-1}e^{iqy}u)$, and hence the eigenvalues of it are the poles of $T(k)$ in $\operatorname{Im} k > 0$, while the resonances are the poles of $T(k)$ in $\operatorname{Im} k < 0$. (see [22], [56], [38]).

For the detailed study of the situation of the on-shell scattering amplitude, meromorphic in $|\operatorname{Im} k| < \gamma$ with non real poles coinciding with bound states or resonances of H see [38] .
Let k_o be a resonance such that $uG_{k_o}v$ has the eigenvalue -1, with eigenfunction φ_o.

Let $\psi_o \equiv G_{k_o}v\varphi_o$, then one has $(-\Delta+V-k_o^2)\psi_o = 0$

in the sense of distributions. ψ_o is called the resonance function (or the Gamow vector) corresponding to the resonance k_o. It satisfies $\psi_o, \nabla\psi_o \in L^2_{loc}(\mathbb{R})$, $\psi_o \notin L^2(\mathbb{R}^3)$. As shown in [24] one has, if φ_o is the only eigenfunction of $uG_{k_o}v$ to the eigenvalue -1, (i.e. if the resonance is simple):

$$\lim_{k\to k_o} (k-k_o)T(k)(p,q) = \tag{2.2}$$

$$\left(\iint e^{-i(px-qy)}\ \bar{\tilde{\varphi}}_o(x)u(x)\ \varphi_o(y)v(y)\,dxdy\right)$$

$$\left(\iint \bar{\tilde{\varphi}}_o(x)u(x)G'_{k_o}(x-y)\ \varphi_o(y)v(y)\,dxdy\right)^{-1},$$

with $\tilde{\varphi}_o$ the solution of $vG_{k_o}u\,\tilde{\varphi}_o = -\tilde{\varphi}_o$, i.e. $\tilde{\varphi}_o = \operatorname{sign} V\varphi_o$.

Set

$$<\psi_o,\psi_o> \equiv \int (-\Delta-k_o^2)\psi_o G^2_{k_o}(-\Delta-k_o^2)\psi_o\,dxdy, \tag{2.3}$$

with $G_k^2(x-y) = \frac{i}{8\pi k} e^{ik|x-y|} = \frac{1}{2k} \frac{\partial}{\partial k} G_k(x-y)$ (2.4)

(defined by analytic contimation of $G_k \cdot G_k$ from $\mathrm{Im}\, k > 0$).

Using $(-\Delta - k_o^2)\psi_o = v\varphi_o$ we get

$$\langle\psi_o,\psi_o\rangle = (2k_o)^{-1} \iint \bar{\varphi}_o(x)\ u(x)\ G'_{k_o}(x-y)\varphi_o(y)v(y)dxdy, \quad (2.5)$$

and thus from (2.2), using also $v\varphi_o = -V\psi_o$:

$$\lim_{k\to k_o} (k-k_o)T(k)(p,q) = -i(2\pi)^{-2}(2k_o)^{-1} \quad (2.6)$$

$$\langle\psi_o,\psi_o\rangle^{-1} \iint e^{-i(px-qy)} V(x)\psi_o(x)V(y)\psi_o(y)dxdy.$$

Moreover

$$\lim_{k\to k_o} (k-k_o)(H-k^2)^{-1}(x,y) = (2k_o)^{-1} \langle\psi_o,\psi_o\rangle^{-1}\psi_o(x)\psi_o(y). \quad (2.7)$$

Remark: For eigenvalues we have $\mathrm{Im}\, k_o > 0$, $G_{k_o}^2 = G_{k_o} \cdot G_{k_o}$, the resonance functions become real eigenfunction and the corresponding formula (2.6) holds with $\langle\psi_o,\psi_o\rangle = (\bar{\psi}_o,\psi_o)$, $(\ ,\ )$ being the L^2-inner product.

Remark: Although ψ_o grows exponentially at infinity (see e.g. [47],[18]) $(-\Delta-k_o^2)\psi_o = v\varphi_o \in L^2(\mathbb{R}^d,dx)$ (and has even compact support if v has compact support) which makes it understandable, why, despite the exponential growth at infinity in $\mathrm{Im}\, k_o < 0$ of $G_{k_o}^2(x-y)$ (for $x \neq y$), we have $\langle\psi_o,\psi_o\rangle < \infty$.

Define for any $\varphi_i \in L^2(\mathbb{R}^3)$, $i=1,2$, $k_1 \neq k_2$:

$$\langle G_{k_1} v\varphi_1,\ G_{k_2} v\varphi_2\rangle_o \equiv$$

$$\equiv \frac{1}{k_1^2-k_2^2} \iint v(x)\varphi_1(x)\ [G_{k_1}(x,y)-G_{k_2}(x,y)]\ \ v(y)\varphi(y)dxdy \quad (2.8)$$

Define also

$$\langle G_k v\varphi, G_k v\varphi\rangle_o = \frac{1}{2k} \int v(x)\varphi(x) \frac{d}{dk} G_k(x,y)v(y)\varphi(y)dxdy \quad (2.9)$$

We remark that (2.9) follows from (2.8) by taking the limit $k_1 \to k_2 = k$. Then we have

Proposition 2.1. For φ_o such that $\psi_o \equiv G_{k_o} v\varphi_o$ is a resonance function to the resonance value k_o we have

$\langle G_{k_o} v\varphi_o,\ G_{k_o} u\varphi_o\rangle_o = \langle\psi_o,\psi_o\rangle$ with $\langle\psi_o,\psi_o\rangle$ defined by (2.3).

Moreover $\langle G_{k_o} v\varphi_o, G_{k_o} v\varphi_o\rangle$ can be computed by analytic continuation of the r.h.s. in (2.9) to $\mathrm{Im}\, k > 0$ and use of $\frac{1}{k^2-k'^2}(G_k - G_{k'}) = G_k * G_{k'}$ for $\mathrm{Im}\, k > 0$.

Proof: The proof uses the analyticity properties of the integrand and the fact that $v\varphi = (-\Delta - k^2)\psi$. □

Theorem 2.2. Let V be as before and assume in addition that V is C^1 with compact support. Let ψ_0,ψ_1 with $\psi_i = G_{k_i} v\varphi_i$, i=0,1 be resonance functions corresponding to two different resonance values $k_0 \neq k_1$. Then

$$\langle G_{k_o} v\varphi_o, G_{k_1} v\varphi_1\rangle_o = \langle\psi_o,\psi_1\rangle,$$

$$\text{with } \langle\psi_0,\psi_1\rangle \equiv \frac{1}{k_0^2-k_1^2}\iint (-\Delta_x-k_0^2)\psi_0(G_{k_0}-G_{k_1})(-\Delta_y-k_1^2)\psi_1 dxdy.$$

Moreover $\langle\psi_0,\psi_1\rangle = 0$.

Proof: The first equality is immediate from $v\varphi_i = (-\Delta-k_i^2)\psi_i$. By the definition of $\langle\psi_0,\psi_1\rangle$, use of $(-\Delta-k_2^2)\psi_i = -V\psi_i$, i=0,1 and integration by parts we have

$$\langle\psi_0,\psi_1\rangle = \frac{1}{k_0^2-k_1^2}\int \psi_0(x)(-V\psi_1)(x)dx - \frac{1}{k_0^2-k_1^2}\int (-V\psi_0)(x)\varphi_1(x)dx = 0. \quad \square$$

3. Perturbation of resonances

In this section we discuss shortly the perturbation theory of resonances. This has been done in some details in [24], where also further references were given, see also the contribution of Gesztesy to these Proceedings , and e.g. [26], [65].

Let us assume the potential V depends on an additional parameter λ in some open connected subset Λ of R containing 0 and s.t. $|V_\lambda|_\gamma < \infty$ for all $\lambda \in \Lambda$. If for $\lambda = 0$ the resonance k_0 has a branch part of order ℓ then there are ℓ branches $k_j(\lambda)$, j=1,...,ℓ given by $k_j(\lambda) = k_0 + k_{0,j}^{(1)} \omega^j (\lambda)^{1/\ell} + O(|\lambda|^{2/\ell})$, where ω is a primitive ℓ-th root of the unit.

If ℓ is odd it is possible to choose $k_j(\lambda)$ s.t. each of them has a well defined tangent at $\lambda = 0$ and the angles between the tangents of $k_j(\lambda)$ are multiples of $2\pi/\ell$. If, on the other hand, the order is even, then the tangents from the left and those from the right exist and the angles between them are multiples of π/ℓ. Both the left tangents and right tangents form angles which are multiples of $2\pi/\ell$ with each other.

If $k_1(\lambda)$ is a simple resonance on the imaginary axis, then, due to the symmetry of resonances with respect to that axis, the only way in which $k_1(\lambda)$ can come off the imaginary axis is by colliding with another resonance $k(\lambda)$ at $\lambda = \lambda_1$, on the same axis. If we have a branch point of even order the resonances leave the imaginary axis in the way described above, if the branch point is of odd order then $k_1(\lambda)$ cannot leave the imaginary axis. As proven in [24] we have that if $k_0(0)$ is a resonance of $-\Delta + V_{\lambda=0}$, then there is a resonance $k_0(\lambda)$ of $-\Delta + V_\lambda$ holomorphic at $\lambda = 0$ and such that $k_0'(\lambda=0) = \frac{1}{2k_o(0)} \langle\psi_o,\psi_o\rangle^{-1} \int\psi_0 V_0'\psi_0 dx$, with $\langle\psi_0,\psi_0\rangle$ described in sect. 2 (especially Pr. 2.1). This formula corresponds exactly to the one for eigenvalues (Im $k_o > 0$).

More generally one has the result (Th. 3.2 in [24]) that if $k_o(0)$ is a resonance of order n then there are n resonances $k_1(\lambda),\ldots,k_n(\lambda)$ which are continuous in a neighborhood of 0 and such that $k_j(0) = k_o$, $j=1,\ldots,n$.
In the matrix $\langle\psi_i,\psi_j\rangle$ of resonance function $\psi_j = G_{k_o} v_o \varphi_j, \varphi_j + u_o G_{k_o} v_o \varphi_j = 0$, $v_\lambda \equiv |V_\lambda|^{1/2}$, $u_\lambda \equiv v_\lambda$ sign V_λ, is non degenerate, then the functions $k_j(\lambda)$ $j=1,\ldots,n$ are all analytic in a neighborhood of $\lambda = 0$ and we may choose $\langle\psi_i,\psi_j\rangle = \delta_{ij}$ and with this choice we have

$$k_j'(0) = \frac{1}{2k_o} \int \psi_j V_o' \psi_j dx.$$

If the matrix $\langle\psi_i,\psi_j\rangle$ is degenerate and at least one of the integrals $\int\psi_j V_o'\psi_j dx$ is different from zero, then at least one of the $k_j(\lambda)$ has a branch point at $\lambda = 0$.
Moreover we proved in [24] the following:

Theorem 3.1 Let $|V|_\gamma < \infty$ and let $V_\lambda = \lambda V$, $\lambda \in \mathbb{R}$. Let $k_o(\lambda_o)$ be a simple resonance of $H_{\lambda_o} = -\Delta + \lambda_o V$. Let $k(\lambda)$, φ_λ satisfy $\varphi_\lambda + uG_{k(\lambda)} v\varphi_\lambda = 0$, then if $k'(\lambda_o) \neq 0$ we have $\varphi'_{\lambda_o} = -\lambda_o k'(\lambda_o)(1+\lambda_o uG_{k_o} v)^{-1} uG'_{k_o} v\varphi_o$ and if $k'(\lambda_o) = 0$ then

$$\varphi'_{\lambda_o} = \lambda_o^{-1} (1+\lambda_o uG_{k_o} v)^{-1} \varphi_o))$$

where $(1 + \lambda_o uG_{k_o} v)^{-1}$ is the inverse of $(1+\lambda_o uG_{k_o} v)$ on the orthogonal complement of $\tilde{\varphi}_o$ and zero on φ_o if $(\tilde{\varphi}_o,\varphi_o) \neq 0$. $(\tilde{\varphi}_o,\varphi_o) = 0$ implies that $k'(\lambda_o) = 0$.
Moreover

$$k''(\lambda_o) = -2\lambda_o^{-1}k'(\lambda_o) - k_o^{-1}k'(\lambda_o)$$

$$(\tilde{\varphi}_o, uG'_{k_o} v\varphi'_{\lambda_o}) + \lambda_o^{-2}k_o^{-1}(\tilde{\varphi}_o, \varphi'_{\lambda_o}) - \frac{1}{2} k_o^{-1}k'(\lambda_o)^2(\tilde{\varphi}_o, uG''_{k_o} v\varphi_o)$$

where φ_o is so normalized that $(\tilde{\varphi}_o, \varphi_o) = 1$.
Similar formulae can be obtained for the higher derivatives of φ_λ and for the eigenvalues $k(\lambda)$ at $\lambda = \lambda_o$.

An application: Let us consider the radial symmetric Schrödinger equation
$[-\Delta + V(r)]\psi = k_o^2 \psi$.

In the usual way one can split the Hilbert space $L^2(\mathbb{R}^3,dx)$ in a direct sum of subspaces L^2 corresponding to fixed angular momentum ℓ. If $\psi_o(|x|)$ is a resonance function of the radial equation (corresponding to $\ell = 0$) then $Y_\ell^m \psi_o(|x|)$, with Y_ℓ^m spherical harmonics ($m = -\ell,\ldots,+\ell$) are resonances of the Schrödinger equation for fixed ℓ. The resonance subspace spanned by $Y_\ell^m\psi_o(|x|)$ carries a representation of the rotation group SO(3).

Replace now V by the family of potentials $V_\lambda(x) = V(|x|) + \lambda x_3 V'(|x|)$, where x_3 is the 3-d component of x. The resonances corresponding to each ℓ split then in $2\ell + 1$ different bands $k_m(\lambda)$, with $m = -\ell,\ldots,+\ell$, $k_m(0) = k_o$, with resonance functions

$\psi_m(x) = Y_\ell^m \psi_o(|x|)$ and, to 1. order in an analytic series:

$$k_m(\lambda) = k_o + \lambda k_m'(0) + O(\lambda^2) = (2k_o)^{-1} \frac{\langle\tilde{\psi}_m, x_3 V' \psi_m\rangle}{\langle\tilde{\psi}_m, \psi_m\rangle} + O(\lambda^2)$$

Above formula for $k''(\lambda_o)$ gives the expression for $k_m(\lambda)$ to 2-order.

We also remark that $k_m'(0)$ is not real. By splitting the integration in a radial and angular part we get

$$k_m'(0) = \frac{m}{2k_o} \gamma ,$$

where γ is a complex number, m independent, just dependent on ψ_o and V (in fact γ is the value $k_m'(0)$ would have taken if the potential would have just been radial symmetric).

Thus we see that for resonances we have a very similar situation as for eigenvalues (e.g. "selection rules").

4. Resonances for point interactions and short range expansions

In recent years much work has been invested in the study of point interactions, for surveys see e.g. [41] , [21], [38] and for a general exposition see [20].
We shall here shortly mention those results which have a direct relevance to the study of resonances, see also the contribution by F. Gesztesy [38].
Let Y be a subset of $\mathbb{R}^3$. A system with point interactions at the point of Y is by definition described by an Hamiltonian of the form

$$H_\lambda = - \Delta + \sum_{y \in Y} \lambda_y \delta_y(x) \tag{4.1}$$

with δ_y the Dirac measure at y, λ_y a constant (it turns out that at least as long as Y is discrete, λ_y has to be chosen infinitesimal negative to get $H_\lambda \neq -\Delta$).
Systems of above form have been discussed in the physical literature since the thirties (nuclear physics, many body theory, solid state physics, electromagnetism).
The most direct mathematical definition of H as a self-adjoint lower bounded operator in $L^2(\mathbb{R}^3, dx)$ is the one using methods of non standard analysis ([39], [40] , [49]). Other definitions use the theory of Dirichlet forms ([50]) or approximations by suitable regularized Hamiltonians (see the references in [41],[21]). In fact in recent years a systematic study of the limit of regularized (scaled) Hamiltonians of the form

$$H_\varepsilon = - \Delta + \sum_{y \in Y} \frac{\lambda_y(\varepsilon)}{\varepsilon^2} V_y\left(\frac{1}{\varepsilon} (x-y)\right), \tag{4.2}$$

which $|V_y|_\gamma < \infty$ has been undertaken. For reviews up to the fall of '83 see [41], [21] see also for a systematic account and recent work [20].

It turns out that any realization of the above Hamiltonian involving only "local" point interactions (for this concept see [51]) is given by an element of the family H_{α_Y}, $\alpha_y \equiv (\alpha_y, y \in Y), \alpha_y \in \mathbb{R} \cup \{+\infty\}$, ("the (renormalized) strength" of the point interaction). H_{α_Y} is the self-adjoint lower bounded operator in $L^2(\mathbb{R}^3, dx)$ whose resolvent is given for $\text{Im } k^2 \neq 0$ by

$$(H_{\alpha_Y} - k^2)^{-1} = G_k - \sum_{y,y' \in Y} G_k(\cdot - y) A_{y,y'} G_k(\cdot - y'),$$

with $A_{y,y'}$ the kernel of the operator A in $\ell^2(Y)$ defined as the inverse $A = B^{-1}$ of B, with B the operator in $\ell^2(Y)$ with kernel $B_{y,y'} = (\frac{ik}{4\pi} - \alpha_y)\, \delta_{y,y'} + \tilde{G}_k(y-y')$, with $\delta_{y,y'}$ the Kronecker symbol and $\tilde{G}_k(x) \equiv (1-\delta_{x,0}) G_k(x)$.

The eigenvalues and the resonances of H_{α_Y} are determined by the singularities in k of the kernel $A_{y,y'}$.

In particular for $|Y| = 1$ we have that A is the number $(\frac{ik}{4\pi} - \alpha)^{-1}$ and there is a simple resonance for $\alpha \geq 0$ at $k = -4\pi i \alpha$ and a simple eigenvalue of energy $k^2 = -16\pi^2\alpha^2$, at $k = -4\pi i\alpha$ for $\alpha < 0$. The corresponding resonance function resp. eigenfunction is $\psi_\alpha(x) = |x|^{-1} e^{4\pi\alpha|x|}$. α is connected with the scattering length a by $4\pi\alpha = -1/a$. The spectrum is $\{-16\pi^2\alpha^2\} \cup [0,\infty)$ for $\alpha < 0$ resp. $[0,\infty)$ for $\alpha \geq 0$, with $[0,\infty)$ the purely absolutely continuous spectrum.

For $|Y| < \infty$ we have that the eigenvalues resp. resonances of H_{α_Y} are the values k with $\text{Im } k > 0$ resp. $\text{Im } k \leq 0$ s.t. the determinant of the matrix $(\frac{ik}{4\pi} - \alpha_y)\delta_{y,y'} + \tilde{G}_k(y-y')$ is zero. For $2 \leq |Y| < \infty$ one has infinitely many resonances.

For $|Y| = 2$, $|y-y'| = L$ all eigenvalues and resonances are given as the values of k with $\text{Im } k > 0$ resp. $\text{Im } k \leq 0$ and $(ikL - 4\pi\alpha_{y_1} L)(ikL - 4\pi\alpha_{y_2} L) = e^{2ikL}$.

The solutions of this equation have been discussed in [24], [52],[53] . For L = 0.95 fm [53] find e.g. a lifetime of 2.10^{-23} sec. for the first resonance.

For $|Y| = 3$, $Y = \{y_1, y_2, y_3\}$, y_i at the vertices of an equilateral triangle of side L, the equation giving the eigenvalues and resonances is, with $z = kL = iy$: $(y-\gamma_1)(y-\gamma_2)(y-\gamma_3) - e^{-2y}(3y-\gamma_1 - \gamma_2 - \gamma_3) - 2e^{-3y} = 0$, with $\gamma_i = -4\pi\alpha_{y_i} L$.

In this case one has 0,1,2 or 3 eigenvalues and the energy spectrum is $(-\infty, E_0)$ for an $E_0 < 0$ (calculated in [53] to be at -2.23 MeV for L = 0,88 fm). There are 2 series of resonances. Another case discussed in [53] is y_i equidistant on a straight line, in which case we can have one or two eigenvalues.

For a computation of resonances in the case $|Y| < \infty$ with Y consisting of equidistant points on a circle (circular antenna) see [54] and [20] . In the case $|Y| = \infty$ the computation of eigenvalues and resonances is possible in the presence of some additional symmetry. Cases discussed in [55], [20] include

a) $Y = \bigcup_{n \in \mathbb{Z}} (0,0,an)$ for some $a \in \mathbb{R}_+$, $\alpha_y = \alpha \ \forall \ y \in Y$. In this case

$H_{\alpha_Y} = \oint_{\hat{Y}} H_\alpha(k)dk$, where the integral is over the dual $\hat{Y} = \mathbb{R}/\{\frac{2\pi}{a} n, n \in \mathbb{Z}\}$ of Y.

The reduced Hamiltonian $H_\alpha(k)$ has absolutely continuous spectrum on $[k^2,\infty)$ and has at most a simple eigenvalue $E_0(k) < k^2$.

If $-\frac{1}{a^2}[\ln z_0(k)]^2 \geq k^2$, with $z^2-(2\cos(ak) + e^{-4\pi a\alpha}) z + 1 = 0$,

then $-\frac{1}{a^2}[\ln z_0(k)]^2$ is a resonance (embedded in the continuum) and the other resonances are given by

$$E_n(k) = \frac{1}{a^2}[\pm i \ln z_0(k) + 2\pi n]^2, \ n = 1,2,\ldots$$

In [55],[20] results have also been given for $Y = \bigcup_{n \in \mathbb{Z}} (X+(0,0,an))$, with X an arbitrary finite subset of $\mathbb{R}^3$.

b) $Y = \bigcup_{n_i \in \mathbb{Z}} \{x \in \mathbb{R}^3 \mid x = n_1a_1+n_2a_2\})$,

some $a_i \in \mathbb{R}^2$, $\alpha_y = \alpha \ \forall \ y \in Y$. The resonance are in this case the solutions k with $\operatorname{Im} k < 0$ of

$$\alpha = (2\pi)^{-3} \lim_{\kappa \to \infty} \frac{1}{2} \sum_{\substack{|\gamma+q| \leq \kappa \\ \gamma \in \Gamma}} \left[\frac{|B|}{\sqrt{|\gamma+q|^2-E}} - 4\pi\kappa \right]$$

with Γ the orthogonal lattice to Λ, $|B|$ the volume of the Brillonin zone.

In [55],[20] results have been obtained also for Y replaces by Y+X, X an arbitrary finite subset of $\mathbb{R}^3$.

By the convergence results mentioned at the beginning of this section we can use the exact results on eigenvalues, resonances (and scattering quantities, which we did not discuss here, but for which we refer to the surveys [41],[21],[38]) to get information on the Hamiltonians H_ε of the type (4.2).

In particular, it has been shown in [56] , [19] , [57] , [58] , [38] that for V_y Rollnik of compact support, $\lambda_y(\varepsilon)$ differentiable with $\lambda_y(0) = 1$ and $|Y| < \infty$, or if $|Y| = \infty$ only finitely many of the λ_y different, then H_ε converges in norm resolvent sense as $\varepsilon \downarrow 0$ to H_{α_Y} with $\alpha_y = +\infty$ if -1 is not an eigenvalue of uG_0v, $u \equiv (\text{sign } V) \ |V|^{1/2}$, $v \equiv |V|^{1/2}$ or -1 is a simple or non simple eigenvalue but with all resonance functions in $L^2(\mathbb{R}^3)$,

$\alpha_y = \lambda_y'(0) \ (\text{sign } V_y\varphi_y,\varphi_y)/ \ |(v_y,\varphi_y)|^2$ (with $u_yG_0v_y\varphi_y = -\varphi_y, \varphi_y \in L^2(\mathbb{R}^3)$),

$(\ , \ )$ the $L^2(\mathbb{R}^3)$-scalar product in case $\psi_y \equiv G_0v\varphi_y \notin L^2(\mathbb{R}^3)$, $\alpha_y = \lambda_y'(0)/ \sum_j |(v,\varphi_y^i)|^2/ \ (\varphi_y^i, \text{ sign } V \varphi_y^j)$ in case $u_yG_0v_y\varphi_y^i = -\varphi_y^i, \varphi_y^i \in L^2(\mathbb{R}^3)$,

$\psi_y^i \equiv G_o v_y \varphi_y^i \notin L^2(\mathbb{R}^3)$ for at least one i.

This norm resolvent convergence implies of course a strong control on the eigenvalues and resonances of H_ε and, perhaps sometimes of more physical interest, of $-\Delta + \sum_y \lambda_y(\varepsilon)V_y$ as $\varepsilon \downarrow 0$. In fact $-\Delta + \sum_y \lambda_y(\varepsilon) V_y$ and H_ε are unitary equivalent under the scaling $(U_\varepsilon f)(x) \equiv \varepsilon^{-3/2} f(x/\varepsilon)$: $U_\varepsilon(-\Delta + \sum \lambda_y(\varepsilon)V_y)U_\varepsilon^{-1} = \varepsilon^2 H_\varepsilon$.

E.g. for $Y = \{0\}$ if -1 is not an eigenvalue of $uG_o v$, then $k(\lambda(\varepsilon)) = \varepsilon k_\varepsilon(\lambda(\varepsilon))$, with $k(\lambda(\varepsilon))$ the quantity giving eigenvalue and resonances of $-\Delta + \lambda(\varepsilon)V$ and $k_\varepsilon(\lambda(\varepsilon))$ the one of $-\Delta + \lambda(\varepsilon)\varepsilon^{-2}V(x/\varepsilon)$, is analytic in ε up to branching points and holomorphic in λ with $k_\varepsilon(\lambda(\varepsilon)) = \varepsilon^{-1} k(\lambda(0)) + O(1)$. In case -1 is a simple eigenvalue of $uG_o v$ with resonance function $\psi = G_o v\varphi$, $uG_o v\varphi = -\varphi$ not in $L^2(\mathbb{R}^3)$, then the eigenvalues and resonances of H_ε converge to the ones of $H_{\alpha_Y} = H_{\alpha_{\{0\}}}$ with $\alpha = \lambda'(0)\ (\text{sign } V\varphi,\varphi)/|(v,\varphi)|^2$.

Also results about expansions in ε around the limit ("short range expansions") have been obtained, see [19], [57], [58], [38].

Yet another problem which can be handled by the same technique is the one of discussing the asymptotics in $\varepsilon \downarrow 0$ of multiple-well Hamiltonians of the form

$$\hat{H}_\varepsilon = -\Delta + \sum_{y \in Y} V(x - \frac{1}{\varepsilon} y),$$

with V bounded and of compact support.

In fact one has $\hat{H}_\varepsilon = U_\varepsilon^{-1} \varepsilon^2 H_\varepsilon U_\varepsilon$.

For $|Y| = 2$ the resonance are given in the limit $\varepsilon \downarrow 0$ by ones discussed above for H_{α_Y}, in this case [24], [52], [20].

Let us also mention that, as discussed in the contribution by Gesztesy, also results on the case with $Y = \{0\}$ and where the "free part" $-\Delta$ is replaced by the Coulomb Hamiltonian $-\Delta + \frac{\gamma}{|x|}$ has been extensively studied, see [60] - [62], [20]. Moreover the case of point impurities in a crystal with point interaction has also been discussed quite extensively [59], [20].

The case of Hamiltonians of the form

$$H = H_{\alpha_Y} + \lambda V, \quad |Y| < \infty$$

with $V \in L^1(\mathbb{R}^3)$ of compact support has also been considered [24], [59], [20].

The eigenvalue and resonances of H are given by the zeros of the modified Fredholm determinant $d(\lambda,k)$ of

$$1 + \lambda u F_k v, \quad v \equiv |V|^{1/2}, \quad u \equiv (\text{sign } V)v,$$

$$F_k \equiv (H_{\alpha_Y} - k^2)^{-1}, \text{ in } \operatorname{Im} k > 0 \text{ resp } \operatorname{Im} k \leq 0.$$

$d(\lambda,k)$ is analytic in $\lambda \in \mathbb{C}$ and meromorphic in $k \in \mathbb{C}$. In particular if k_o is a simple eigenvalue or resonance of H_{α_Y} then there exists an eigenvalue resp. resonance of H s.t. $k(0) = k_o$ and $k(\lambda)$ is analytic in λ in a neighbourhood of $\lambda = 0$.

We have $k'(0) = \frac{1}{2k_o} \langle\psi ,\psi\rangle^{-1} \int \psi(x)V(x)\psi(x)dx$,

with $\langle\psi,\psi\rangle = \frac{i}{8\pi k_o} \sum_{i,j} \lambda_i\lambda_j e^{ik_o|x_i-x_j|}$,

$\psi(x) = \sum_{y\in Y} \lambda_y G_{k_o}(x-y)$ being the resonance function to k_o.

For further discussion of Hamiltonians H of above form see [59], [20].

5. Formulae for $e^{-it\sqrt{H_+}}(x,y)$ and $e^{-itH}(x,y)$ in terms of resonances

Let $H = -\Delta + V$ with V real and s.t. H is self-adjoint. Let $H = \int_{\sigma(H)} \lambda dE(\lambda)$ be the spectral decomposition of H.

Let $H_- \equiv \int_{\sigma(H)\cap(-\infty,0]} \lambda dE(\lambda)$, $H_+ \equiv \int_{\sigma(H)\cap(0,\infty)} \lambda dE(\lambda)$.

We assume that V is such that $\sigma(H) \cap (-\infty,0)$ consists of finitely many eigenvalues (this is so e.g. if V is as in the preceding sections). Since $\sigma(H)\cap(-\infty,0)$ consists of finitely many negative eigenvalues $k_n^2 < 0$, with corresponding orthonormalized eigenfunctions $\psi_n \in L^2(\mathbb{R}^3)$, we can write for any $f \in C_o^\infty(\mathbb{R}_+)$, by the spectral calculus $f(\sqrt{H_-}) = \sum_n f(k_n)$ and $f(H_-) = \sum_n f(k_n^2)$. Do we have similar formulae for $f(\sqrt{H_+})$ and $f(H_+)$ in terms of resonances for some set of functions f?

This has been discussed in the physical literature e.g. in [32] - [37]. We shall see below that for special functions f it is possible indeed to obtained representations of above form, at least when V has compact support.

Assume that V is such that $(H-k^2)^{-1}(x,y)$ is analytic in the whole k plane, for $x \neq y$. This is the case if e.g. V is as in Sect. 2, with compact support.

Let first $f(\lambda) = e^{-it\lambda}$, $\lambda \in \mathbb{R}_+$. In this case $f(\sqrt{H_+})$, well defined by the spectral decomposition, is a bounded unitary operator. We have, by the functional calculus, in the sense of locally integrable functions for $x \neq y$

$$e^{-it\sqrt{H_+}}(x,y) = -(2i\pi)^{-1} \int_\Gamma e^{-itk}(H-k^2)^{-1}(x,y)2kdk,$$

where Γ runs along the positively oriented real axis (we assume throughout this section that H does not have the eigenvalue zero, otherwise we should have included the spectral point zero into the definition of H_+).

Denote by C_R^- the arc of circumference C_R of radius R, centered at the origin, lying in the complex lower half plane Im $k<0$. We give C_R^- the positive orientation from $(+R,0)$ to $(-R,0)$.

We shall assume from now on that V is such that one has the following estimate on the resolvent kernel:

$$|(H-k^2)^{-1}(x,y)| \le C(x,y)\ e^{\alpha(x,y)|k|} \tag{5.1}$$

for all $x \neq y$, some locally integrable function C and some function α of x,y, for all Im $k<0$ s.t. $|k|>k_o$, for some $k_o(x,y)$.
We shall denote by $\mathcal{V}$ the class of potential V fulfilling these conditions.
Let us first remark that $\mathcal{V}$ contains the trivial potential $V = 0$. In fact for $V = 0$ we have $(H-k^2)^{-1}(x,y) = (-\Delta-k^2)^{-1}(x,y) = G_k(x,y) = \frac{e^{ik|x-y|}}{4\pi|x-y|}$ and we see from the analyticity and an immediate estimate that for $x \neq y$

$$\left|\int_{C_R^-} e^{-itk}G_k(x,y)2kdk\right| \le \frac{1}{4\pi|x-y|} \int_{C_R^-} e^{-t|\mathrm{Im}\ k||x-y|}2kdk \to 0$$

as $R\to\infty$, as long as $t>t_o(x,y)$, for some $t_o>0$.
We also remark that $\mathcal{V}$ contains all potential V(x) of the form $\sum_{y\in Y}\lambda_y\delta(x-y)$, with $|Y|<\infty$, and λ_y chosen as in Sect. 3, so that $-\Delta+V(x)$ is realized as H_{α_Y}.
In fact then

$$(H-k^2)^{-1}(x,z) = G_k(x,z) - \sum_{y,y'\in Y} G_k(x-y)\ A_{y,y'}\ G_k(z-y').$$

Since we already handled the term G_k it remains to control the integral

$$\int_{C_R^-} e^{-itk}\ G_k(x,y)G_k(z-y')2kdk$$

as $R\to\infty$. This however vanishes as $R\to\infty$, for $t>t_o(x,z,Y)$.
Finally we like to point out that $\mathcal{V}$ contains radial symmetric integrable potentials of compact support, in fact for such potentials, by a convergent partial wave expansion, use of the compact support of V and of the formulae for the Green's function in 1-dimension, we get the above estimate for $(H-k^2)^{-1}(x,y)$ from the corresponding estimates on the partial wave components $G_{\ell,k}(|x|,|y|)$ of $(-\Delta-k^2)^{-1}(x,y)$, which in turn follow essentially from asymptotic estimates on Bessel functions, see [2], [33]. Now let $V\in\mathcal{V}$.
Consider the integral $\int_{\Gamma_R} f(k,x,y)dk$, with

$$f(k,x,y) \equiv e^{-itk}(H-k^2)^{-1}(x,y)2kdk \tag{5.2}$$

and with Γ_R the segment $[-R,+R]$ of Γ.
By the analyticity of the integrand we have by Cauchy's theorem that $-(2\pi i)^{-1}$ times

$$\int_{\Gamma_R} f(k,x,y)dk + \int_{C_R^-} f(k,x,y)dk$$

is the total residuum at the poles enclosed by Γ_R and C_R^-, hence by the resonances in C_R. Since as $R \to \infty$ the contribution of the integral over C_R^- vanishes due to $V \in \mathcal{V}$ we have finally

$$-\frac{1}{2\pi i} \int_\Gamma e^{-itk}(H-k^2)^{-1}(x,y)\ 2kdk = \sum_n \frac{e^{-itk_n}}{\langle\psi_n,\psi_n\rangle} \psi_n(x)\psi_n(y), \tag{5.3}$$

where we used the formula (2.7) for the residuum.

We have thus proven the following resonance expansion for the acoustic Green's function:

<u>Theorem 5.1</u> Let V be any potential belonging to the above class $\mathcal{V}$. Then there exists $t_o(x,y,V)$ such that for all $t > t_o$

$$e^{-it\sqrt{H_+}}(x,y) = \sum_n \frac{e^{-itk_n}}{\langle\psi_n,\psi_n\rangle} \psi_n(x),\psi_n(y),$$

with k_n resp. ψ_n the resonance values resp. resonance functions of $H = -\Delta + V$.

<u>Remark:</u> As we showed before the theorem the class $\mathcal{V}$ includes $V = 0$, V a sum of finitely many point interactions, as well as integrable radial symmetric potentials V of compact support. In the case $V = 0$ the sum over resonances is zero (there are then no resonances!), but of course in this case also $e^{-it\sqrt{H_+}}(x,y) = e^{-it\sqrt{-\Delta}}(x,y)$ is zero, for $t > t_o(x,y) \equiv |x-y|$ (this uses in an essential way the fact that we are in 3-dimensions!).

Is it possible to get corresponding formulae for the kernel of e^{-itH_+} instead of $e^{-it\sqrt{H_+}}$? The next theorem gives an answer to this question.

<u>Theorem 5.2</u> Let V be as in Theorem 5.1. Then there exists $t_o(x,y,V)$ such that for all $t > t_o$ one has the following resonance expansion:

$$e^{-itH_+}(x,y) = \sum_n e^{-itk_n^2}\psi_n(x)\psi_n(y) + (2\pi it)^{-3/2} e^{\frac{i}{2t}|x-y|^2}$$
$$-\frac{1}{2\pi}\int_0^\infty [\hat{R}(e^{-i\frac{\pi}{4}}\sigma) - \hat{R}(-e^{-i\frac{\pi}{4}}\sigma)]\, e^{-t\sigma}d\sigma,$$

with $\hat{R}(k) \equiv (H-k^2)^{-1} - (-\Delta-k^2)^{-1}$. The sum is over the resonances k_n lying in the sector $-\frac{\pi}{4} < \arg k_n < 0$. We have normalized the resonance functions such that $\langle\psi_n,\psi_n\rangle = 1$.

<u>Proof:</u> We have

$$e^{-itH_+}(x,y) = -\frac{1}{2\pi i}\int_\Gamma e^{-itk^2} R(k)(x,y)2kdk, \tag{5.4}$$

with $R(k) \equiv (H-k^2)^{-1}$ and Γ as in Theor. 2.1.

We write (5.4) in the form

$$\frac{1}{2\pi i}\int_\Gamma e^{-itk^2}\hat{R}(k)(x,y)2kdk + \frac{1}{2\pi i}\int_\Gamma e^{-itk^2}G_k(x,y)2kdk \tag{5.5}$$

We observe that

$$\frac{1}{2\pi i}\int_\Gamma e^{-itk^2}G_k(x,y)2kdk = (2\pi it)^{-3/2}\, e^{\frac{i|x-y|^2}{2t}} \tag{5.6}$$

We shall now study the first term in (5.5).

We first remark that it is equal

$$\frac{1}{2\pi i}\int_0^\infty [\hat{R}(k)-\hat{R}(-k)]\,(x,y)\, e^{-itk^2}\, 2kdk \tag{5.7}$$

Let C_R^o be the circle of radius R, centered at the origin. Let γ_R be the segment in C_R^o of the half-line γ: Rek = - Im k, Im k < 0, run towards the origin. Let C_R be the circumference of radius R and center at the origin and let $C_{R,-}$ be the arc of C_R between (R,0) and γ_R, run clockwise. By the analyticity of the integrand and Cauchy's formula we have that (5.7) is equal to

$$-\frac{1}{2\pi i}\int_{C_{R,-}\cup\gamma_R} e^{-itk^2}\,[\hat{R}(k)(x,y)-\hat{R}(-k)(x,y)]\; 2kdk \tag{5.8}$$

$$-\sum_n e^{-itk_n^2}\psi_n(x)\psi_n(y),$$

since $\hat{R}(-k)$ has no poles and the ones of $\hat{R}(k)$ are those of R(k), in the relevant region, and where we used (2.7) for the residua at the poles k_n of R(k)(x,y). Using the assumption $V \in \mathcal{V}$ we can show that for $t > t_0(x,y,V)$ the contribution to the integral over $C_{R,-}$ vanishes, as $R\to\infty$. Thus we have, for such t:

$$e^{-itH_+}(x,y) = \sum_n e^{-itk_n^2}\psi_n(x)\psi_n(y) + (2\pi it)^{-3/2}\, e^{\frac{i}{2t}|x-y|^2}$$

$$-\frac{1}{2\pi i}\int_\gamma e^{-itk^2}\,[\hat{R}(k)(x,y)-\hat{R}(-k)(x,y)]\; 2kdk \tag{5.9}$$

If we introduce the new integration variable $\sigma \in [0,\infty)$ though $k = e^{-i\pi/4}\sqrt{\sigma}$,
we can write

$$\int_\gamma e^{-itk^2}\hat{R}(k)2kdk = -\, i\int_0^\infty e^{-\sigma t}\,\hat{R}(e^{-i\pi/4}\,\sigma)d\sigma \tag{5.10}$$

Inserting (5.10) into (5.9) we get the Theorem. □

Remark. Using the well known expression

$$e^{-itH_-}(x,y) = \sum_n e^{-itk_n^2}\psi_n(x)\psi_n(y) \tag{5.11}$$

with ψ_n the eigenfunction to the eigenvalues $E_n = k_n^2$ (lying on arg $k_n = \frac{\pi}{4}$), and the fact that $H = H_+ + H_-$ (direct sum) we obtain from Theor. 5.2 and (5.9) the following expression for the Green's function of Schrödinger's equation:

$$e^{-itH}(x,y) = \sum_{-\frac{\pi}{4}\leq \arg k_n < 0} e^{-itk_n^2}\,\psi_n(x)\psi_n(y) +$$

$$+ \sum_n e^{-itE_n}\,\psi_n(x)\psi_n(y) + (2\pi it)^{-3/2}\, e^{\frac{i}{2t}|x-y|^2} -$$

$$- \frac{1}{2\pi}\int_0^\infty [\hat{R}(e^{-i\frac{\pi}{4}}\sigma) - \hat{R}(-e^{-i\frac{\pi}{4}}\sigma)]\, e^{-t\sigma}d\sigma \qquad (5.12)$$

We shall now see how one can use the formulae in Theor. 5.2 to derive an asymptotic expansion of $e^{-itH_+}(x,y)$ for given $x \neq y$ as $t \to \infty$.

Let V be as in Theor. 5.2. Assume $V \in \mathcal{V}$ is either the potential created by finitely many point interactions, in which case a direct treatment can be given, using the explicit formulae for the resolvent, or V is an integrable Rollnik potential ($|V|_0 < \infty$) of compact support. In the latter case we have, for $x \neq y$, as proven in [19]

$$\hat{R}(k) = -G_k v(1+uG_k v)^{-1} uG_k(x,y) \qquad (5.13)$$

In [56], [19], [22] the expansion of $\hat{R}(k)$ as $k \to 0$ is discussed. We consider here two cases:

1) H has no zero energy resonance i.e. -1 is not an eigenvalue of uG_0v (the generic case).

2) H has a zero energy resonance i.e. -1 is a simple eigenvalue of uG_0v and the corresponding function is not in $L^2(\mathbb{R}^3)$.

In both cases we have, by [56], [19], [22], Laurent expansions in k for the kernel $\hat{R}(k)(x,y)$ (we have no problems with the existence of all kernels for $x \neq y$ by ellipticity, at least when V is Hölder continuous, see also [64], or V is a finite sum of point interactions, by direct calculation):

$$\hat{R}(k) = A_{-1}k^{-1} + A_0 + A_1 k + O(k^2) \qquad (5.14)$$

as $k \to 0$.

Using (5.14) with $\int_0^\infty e^{-\sigma t}\,\sigma^{\alpha/2}d\sigma = t^{-(1+\frac{\alpha}{2})}\,\Gamma(1+\frac{\alpha}{2})$ together with Tauberian theorems we get then the asymptotics for $t \to \infty$ of the integrals in Theorem 5.2. We observe also that the terms containing even powers in k do not contribute to the integrals. Hence we get:

$$- \frac{1}{2\pi}\int_0^\infty [\hat{R}(e^{-i\frac{\pi}{4}}\sigma) - \hat{R}(-e^{-i\frac{\pi}{4}}\sigma)]\, e^{-t\sigma}\,d\sigma =$$

$$= - \frac{e^{i\frac{\pi}{4}}}{\sqrt{\pi}}\, A_{-1}^{(j)}\, t^{-\frac{1}{2}} - \frac{e^{-i\frac{\pi}{4}}}{2\sqrt{\pi}}\, A_1^{(j)} t^{-\frac{3}{2}} + O(t^{-2}),$$

the index $j = 1,2$ standing for the cases 1, 2 above.

In the case 1 we have $A_{-1}^{(1)} = 0$ and

$A_1^{(1)} = -G_1v(1+uG_ov)^{-1}uG_o - G_ov(1+uG_ov)^{-1}uG_1 + G_ov(1+uG_ov)^{-1}uG_1v(1+uG_ov)^{-1}uG_o$, with

$G_1 \equiv ik/4\pi$.

In the case 2 we have $A_{-1}^{(2)}(x,y) = \psi(x)\psi(y)$ with ψ the zero energy resonance function. $A_1^{(2)}$ is given by a more complicated expression, which can be derived from [19].

Thus we have the following

Theorem 5.3. Let V be as in Theorems 5,1, 5.2 and assume that V is Rollnik, Hölder continuous or a sum of finitely many point interactions. Then we have for $x \neq y$ the asymptotic expansion

a) if $-\Delta + V$ has no zero energy resonance then we have the asymptotic expansion as $t \to \infty$

$$e^{-itH_+}(x,y) = \sum_{\substack{n \\ 0 > \arg k_n > -\frac{\pi}{4}}} e^{-itk_n^2} \psi_n(x)\psi_n(y) +$$

$$+ (2\pi it)^{-3/2} e^{\frac{i}{2t}|x-y|^2} + \tilde{A}_1(x,y)t^{-\frac{3}{2}} + O(t^{-5/2}) \text{ with } \tilde{A}_1 \equiv -\frac{e^{-i\frac{\pi}{4}}}{2\sqrt{\pi}} A_1^{(1)}$$

b) if $-\Delta + V$ has a simple zero energy resonance with resonance function $\psi = G_ou\varphi$ not in $L^2(\mathbb{R}^3)$

$$e^{-itH_+}(x,y) = \sum_n e^{-itk_n^2} \psi_n(x)\psi_n(y) + (2\pi it)^{-3/2} e^{\frac{i}{2t}|x-y|^2} + \tilde{A}_{-1} t^{-1/2} +$$

$$+ \tilde{A}_1' t^{-3/2} + O(t^{-5/2}), \text{ with } \tilde{A}_{-1}(x,y) = (x)\ (y)\ \frac{e^{i\frac{\pi}{4}}}{\sqrt{\pi}}$$

$$\tilde{A}_1' \equiv -\frac{e^{-i\frac{\pi}{4}}}{2\sqrt{\pi}} A_1^{(2)} .$$

Remark. Asymptotic expansions for e^{-itH} for $t \to \infty$ have also been derived in [22], however without extraction of the free term $e^{it\Delta}$ and the resonance term.

We close with two short remarks, indicating other areas where an investigation of resonances is possible using related techniques as in quantum mechanics.

Remark 1. Lax and Phillips [15], [16] have developed an extensive study of the acoustic equation corresponding to the Schrödinger equation:

$$-\frac{\partial^2}{\partial t^2} u = -\Delta u + Vu.$$

The scattering matrix S_a for this equation is related to the one, S, for the Schrödinger equation by $S(k^2) = S_a(k)$.

For results using the acoustic equation approach see e.g. [15 - 17].

Remark 2. There is a very interesting connection between the problems of existence of resonances for a scattering problem in hyperbolic space and the set of zeros of Riemann ζ-function.

Let us consider the Poincaré-plane π i.e. the complex upper plane $\{z \in \mathbb{C} \mid y \equiv \text{Im } z > 0\}$ with Riemann metric $y^{-2}(dx^2+dy^2)$, $x \equiv \text{Re} z$.
The Laplace-Beltrami operator on π is
$\Delta \equiv y^2 \left(\frac{\partial^2}{\partial x^2} + \frac{\partial^2}{\partial y^2} \right)$. Let $\Gamma = SL(2,\mathbb{Z})$ be the discrete subgroup of $SL(2,\mathbb{R})$ consisting of matrices $\begin{pmatrix} a & b \\ c & d \end{pmatrix}$, $a,b,c,d \in \mathbb{Z}$, $ad - bc = 1$. A function f on π is automorphic if $f(\gamma z) = f(\gamma)$ $\forall \gamma \in \Gamma$, $\forall z \in \pi$. Δ is realized as a self-adjoint operator in $L^2(F,\lambda)$, with F the fundamental domain for π with respect to Γ, λ the $SL(2,\mathbb{R})$ - invariant measure. The continuous spectrum of $-\Delta$ is $(-\frac{1}{4}, \infty)$, with multiplicity 1 [16], [63].

As pointed out by Gelfand, Lax-Phillips and others, see [16], [63], the poles of the scattering matrix correspond to the zeros of the Riemann function
$\zeta(s) \equiv \sum_{n=1}^{\infty} n^{-s}$.
Riemann's hypothesis is equivalent with the assertion that a certain function $S''(z)$, essentially the scattering matrix, has no poles (no resonances) in the half plane $\text{Im } z < 1/4$.

ACKNOWLEDGEMENTS

We are very grateful to Fritz Gesztesy, Helge Holden, Werner Kirsch, Mohammed Mebkhout, Ludwig Streit for the joy of collaboration and most stimulating discussions and corrections. The second author would like to thank the organizers for the kind invitation to give a lecture. The first author is grateful to the Centre de Physique Théorique, CNRS, the Université d'Aix-Marseille II and the University of Oslo for the hospitality, as well as to the Norwegian Research Council for Science and the Humanities for financial support. Both authors have the pleasure to thank Prof. Dr. Ludwig Streit for the hospitality at ZiF, University of Bielefeld, at various stages, during Project No 2, which greatly stimulated our work. They also gratefully acknowledge the skilful typing by Mrs. Richter.

References

[1] R. Jost, Über die falschen Nullstellen der Eigenwerte der S-Matrix, Helv. Phys. Acta 20, 250-266 (1947)

[2] R. Newton, Scattering theory of waves and particles, McGraw Hill, 2nd ed. (1982)

[3] W.O. Amrein, J.M. Jauch, H.B. Sinha, Scattering theory in quantum mechanics, Reading, Benjamin (1977)

[4] A. Grossmann, Nested Hilbert spaces in quantum mechanics I, J. Math. Phys. 5, 1025-1037 (1964)

[5] L.P. Horwitz, E. Katznelson, A partial inner product space of analytic functions for resonances, J. Math. Phys. 24, 848-859 (1983)

[6] L.P. Horwitz, J.P. Marchard, The decay scattering system, Rocky Mount. J. Math. 1, 225-253 (1971)

[7] L.P. Horwitz, I.M. Sigal, On a mathematical model for non stationary physical systems, Helv. Phys. Acta 51, 685-715 (1978)

[8] J. Rauch, Perturbation theory for eigenvalues and resonances for Schrödinger Hamiltonians, J. Funct. Anal. 35, 304-315 (1980)

[9] G. Parravicini, V. Gorini, E.C.G. Sudarshan, Resonances, scattering theory, and rigged Hilbert spaces, J. Math. Phys. 21, 2208-2226 (1980)

[10] S. Albeverio, On bound states in the continuum of N-body systems and the virial theorem, Ann. Phys. 71, 167-276 (1972)

[11] M. Reed, B. Simon, Methods of Modern Mathematical Physics, III. Scattering Theory; IV, Analysis Operators, Academic Press, New York (1978).

[12] J.S. Howland, The Livsic matrix in perturbation theory, J. Math. Anal. Appl. 50, 415-437 (1975)

[13] H. Baumgärtel, M. Wollenberg, Mathematical Scattering Theory, Birkhäuser, Basel (1983)

[14] M. Demuth, On the perturbation theory of instable isolated eigenvalues, Math. Nachr. 64, 345-356 (1974)

[15] P.D. Lax, R.S. Phillips, Scattering theory, Acad. Press, New York (1967)

[16] P.D. Lax, R.S. Phillips, Scattering theory for automorphic forms, Princeton Univ. Press,Princeton(1976);Bull. Am. Math. Soc. 2, 265-295 (1980)

[17] C. Bardos, J.C. Guillot, J. Ralston, La relation de Poisson pour l'équation des ondes dans un ouvert non borné. Application à la théorie de la diffusion, Commun. Part. Diff. Eqts. 7, 905- 958 (1982)

[18] E. Balslev, these Proc.

[19] S. Albeverio, F. Gesztesy, R. Høegh-Krohn, The low energy expansion in non relativistic scattering theory, Ann. Inst. H. Poincaré A37, 1-28 (1982)

[20] S. Albeverio, F. Gesztesy, R. Høegh-Krohn, H. Holden, Solvable models of mathematical physics, book in preparation.

[21] S. Albeverio, F. Gesztesy, R. Høegh-Krohn, H. Holden, Some exactly solvable models in quantum mechanics and the low energy expansion, to appear Proc. Leipzig Conf. Operator algebras '83, Teubner

[22] A. Jensen, T. Kato, Spectral properties of the Schrödinger operators and time-decay of the wave functions, Duke Math. J. 46, 583-611 (1979)

[23] R. Newton, Non central potentials: The generalized Levinson theorem and the structure of the spectrums, J. Math. Phys. 18, 1348-1357 (1977)

[24] S. Albeverio, R. Høegh-Krohn, Perturbation of resonances in quantum mechanics, J. Math. Anal. Appl. 101, 491-513 (1984)

[25] B. Simon, On the absorption of eigenvalues by the continuum spectrum in regular perturbation problems, J. Funct. Anal. 25, 338-344 (1977)

[26] A.G. Ramm, Perturbation of resonances, J. Math. Anal. Appl. 88, 1-7 (1982)

[27] M. Klaus, B. Simon, Coupling constant thresholds in non relativistic quantum mechanics I, Ann. Phys. 130, 251-281 (1980)

[28] A. Baz, Ya. Zeldovich, A. Perelomov, Scattering, Reactions and Decay in Nonrelativistic Quantum Mechanics, Isr. Progr. Scient. Transl. Jerusalem 1969

[29] J.S. Bell, Electromagnetic properties of unstable particles, Nuovo Cim 24, 452-460 (1962)

[30] N. Hokkyo, A remark on the norm of the unstable state, Prop. Theor. Phys. 33, 1116-1128 (1965)

[31] A. Böhm, Resonance poles and Gamov vectors in the rigged Hilbert space formulation of quantum mechanics, J. Math. Phys. 22, 2813-2823 (1981)

[32] T. Berggren, A note on function space spanned by complex energy eigenfunctions, Phys. Letts. B 38, 61-63 (1972)

[33] W.J. Romo, A study of the completeness properties of resonant states, J. Math. Phys. 21, 311-326 (1980)

[34] B. Gyarmati, T. Vertse, On the normalization of Gamov functions, Nucl. Phys. A 160, 523-528 (1971)

[35] R. Moore, E. Gerjuoy, Properties of resonance wave functions, Phys. Rev. A7, 1288-1303 (1973)

[36] G. Garcia-Calderon, R. Peierls, Resonant states and their uses, Nucl. Phys. A 265, 443-460 (1976)

[37] A. Böhm, Quantum mechanics, Springer, New York (1979)

[38] F. Gesztesy, these Proceedings

[39] S. Albeverio, J.E. Fenstad, R. Høegh-Krohn, Singular perturbations and non standard analysis, Trans. Ann. Math. Soc. 252, 275-295 (1979)

[40] S. Albeverio, J.E. Fenstad, R. Høegh-Krohn, T. Lindstrom, Non standard methods in stochastic analysis and mathematical physics, Acad. Press

[41] S. Albeverio, R. Høegh-Krohn, Schrödinger operators with point interactions and short range expansions, Physica 124A, 11-28 (1984)

[42] T. Berggren, Inner product for resonant states and shell-model application, Nucl. Phys. A116, 618-636 (1968)

[43] T. Berggren, On the case of resonant states in eigenfunction expansions of scattering and reaction amplitudes, Nucl. Phys. A109, 265-287 (1968)

[44] T. Berggren, On resonance contributions to sum rules in nuclear physics, Phys. Lett. B44, 23-25 (1973)

[45] B. Berrondo, G. Garcia-Calderon, An eigenfunction expansion involving resonant states, Lett. Nuovo Cim 20, 34-38 (1977)

[46] Y. Colin de Verdière, Pseudo-Laplaciens I, II, Ann. Inst. Fourier 32,3,275-286 (1982); 33,2, 87-113 (1983)

[47] H.M. Nussenzveig, The poles of the S-matrix of a rectangular potential well or barrier, Nucl. Phys. 11, 499-521 (1959)

[48] R.G. Newton, Non central potentials, J. Math. Phys. 18, 1348-1357 (1977); Czech. J. Phys. B24, 1195-1204 (1974)

[49] S. Albeverio, J.F. Fenstad, R. Høegh-Krohn, W. Karwowski, T. Lindstrom, Perturbation of the Laplacian supported by zero measure sets, ZiF, Preprint 1984, to appear in Phys. Letts. A.

[50] S. Albeverio, R. Høegh-Krohn, L. Streit, Energy forms, Hamiltonians and distorted Brownian paths, J. Math. Phys. 18, 907-917 (1977)

[51] L. Dabrovsky, H. Grosse, On non local point interactions in one, two and three dimensions, J. Math. Phys.

[52] R. Høegh-Krohn, M. Mebkhout, The 1/r expansion for the critical multiple well problem, Commun. Math. Phys. 91, 65-73 (1983)

[53] H. Holden, R. Høegh-Krohn, T. Wahl, Some explicit results on point interactions, Oslo Preprint (1983)

[54] A. Grossman, T.T. Wu, A class of potentials with extremely narrow resonances I. Case with discrete rotational symmetry, CNRS-CPT Marseille Preprint, 1981.

[55] A. Grossman, R. Høegh-Krohn, M. Mebkhout, The one particle theory of periodic point interactions, Commun. Math. Phys. 77, 87-110 (1980)

[56] S. Albeverio, R. Høegh-Krohn, Point interactions as limits of short range interactions, J. Operator Th. 6, 313-339 (1981)

[57] H. Holden, R. Høegh-Krohn, S. Johannesen, The short range expansion, Adv. Appl. Math. 4, 402-421 (1983)

[58] H. Holden, R. Høegh-Krohn, S. Johannesen, The short range expansion in solid state, Ann. I. H. Poincare (1984)

[59] S. Albeverio, R. Høegh-Krohn, M. Mebkhout, Scattering by impurities in a solvable model of a 3-dimensional crystal, J. Math. Phys. 25, 1327-1334 (1984)

[60] S. Albeverio, F. Gesztesy, R. Høegh-Krohn, L. Streit, Charged particles with short range interactions, Ann. Inst. H. Poincaré A38, 303-333 (1983)

[61] S. Albeverio, D. Bollé, F. Gesztesy, R. Høegh-Krohn, L. Streit, Low energy parameters in non relativistic scattering theory, Ann. Phys. 140, 308-326 (1983)

[62] S. Albeverio, L.S. Ferreira, F. Gesztesy, R. Høegh-Krohn, L. Streit, Model dependence of Coulomb-corrected scattering lengths, Phys. Rev. C29, 680-683 (1984)

[63] S. Lang, $SL_2(\mathbb{R})$, Addison-Wesley (1975)

[64] B. Simon, Schrödinger semi groups, Bull. Am. Math. Soc. 7, 447-526 (1982)

[65] A.G. Ramm. Theory and Applications of Some New Classes of Integral Equations, Springer, New York (1980)

[66] O. Zohni, Generalized completeness relations in the theory of resonant scattering, J. Math. Phys. 23, 798-802 (1982)

[67] S. Fassari, On the Schrödinger equation with periodic point interactions in the 3-dimensional case, J. Math. Phys. 1984

[68] J. Persson, The wave equation with measures as potentials and related topics, to appear in Rend. Sem. Mat. Univ. Politec. Torino

[69] T.A. Osborn, R. Wong, Time Decay and Spectral Kernel Asymptotics, University of Maryland Preprint April 1984

WAVE FUNCTIONS ON SUBGROUPS OF THE GROUP OF AFFINE CANONICAL TRANSFORMATIONS

Alex Grossmann

and

Thierry Paul*

Centre de Physique Théorique, Section 2
CNRS - Luminy, Case 907
F-13288 MARSEILLE, Cedex 9
France

Content:

1. Introduction

This talk describes a method for constructing quantum-mechanical wave-functions, that generalizes the transformation to a coherent space representation. The method is based on general facts about square integrable representations of (possibly non-unimodular) groups. In the examples that we study, the group of phase-space shifts, which is basic in the usual formulations, is replaced by a suitable two-parameter subgroup of the group of affine canonical transformations. One of the parameters represents dilations; we obtain a representation in which the study of complex dilations, of importance in the theory of resonances, promises to be quite easy.

---

* Allocataire D.G.R.S.T.

2. Square Integrable Representations and the Coherent State Philosophy

The space of wave functions that we shall consider can be obtained as special cases of a general construction which will be discussed in this section.

Let G be a locally compact group and U a continuous unitary representation of G in a Hilbert space $H(U)$. (In the sections that follow, G will be a two-parameter subgroup of the group of affine canonical transformations, and U the natural representation of G). Remember that every locally compact group has a left-invariant (Haar) measure $d\mu$: $d\mu(g_1 g) = d\mu(g)$ $(g_1 \in G)$. The group G also carries a right-invariant measure $d_R\mu(g)$: $d_R\mu(gg_1) = d_R\mu(g)$ $(g_1 \in G)$. We do not assume that $d\mu = d_R\mu$, (which is expressed by saying that G is unimodular), since this assumption is not satisfied in our most interesting examples.

The representation U is said to be square integrable if

(i) U is irreducible, and

(ii) there exists in $H(U)$ at least one non-zero vector φ such that

$$\int |(U(g)\varphi,\varphi)|^2 \, d\mu(g) < \infty \tag{2.1}$$

It follows from (ii) that $\int |(U(g)\varphi,\varphi)|^2 \, d_R\mu(g) = \int |(U(g)\varphi,\varphi)|^2 \, d\mu(g)$, and we define the positive number

$$c_\varphi = \frac{1}{\|\varphi\|^2} \int |(U(g)\varphi,\varphi)|^2 \, d\mu(g) = \frac{1}{\|\varphi\|^2} \int |(U(g)\varphi,\varphi)|^2 \, d_R\mu(g). \tag{2.2}$$

A nonzero $\varphi \in H(U)$ satisfying (2.1) will be called admissible.

If G is unimodular and U square integrable, then every nonzero $\varphi \in H(U)$ is admissible, and the number c_φ is proportional to $\|\varphi\|^2$. If G is not unimodular and U is square integrable, then the admissible vectors form a dense linear subspace of $H(U)$, characterized by the condition (2.1). We shall see concrete examples in (§ 5) and (§ 6).

As an example of unitary irreducible representation that is not square integrable, consider the one-dimensional representation $x \to e^{iax}$ $(a \in R)$ of the real line.

From now on we shall assume that U is square integrable. Then there is a natural "realization" of $H(U)$ as a closed subspace of the space $L^2(G,d\mu(g))$ of complex-valued functions on G, square integrable with respect to $d\mu(g)$. It can be described as follows:

Let φ be a fixed admissible vector in $H(U)$.

<u>Proposition</u>: To every $\psi \in H(U)$ associate the function $f(g) = (L_\varphi \psi)(g)$ defined by

$$f(g) = (c_\varphi)^{-1/2} (U(g)\varphi, \psi).$$

Then L_φ is an isometry from $H(U)$ into $L^2(G, d\mu(g))$, i.e. a unitary map between H and a closed subspace $L_\varphi H$ of $L^2(G; d\mu(g))$. The functions f in $L_\varphi H$ are characterized by the fact that they satisfy the "reproducing equation"

$$f(g_o) = p_\varphi(g^{-1} g_o)\ d\mu(g) \qquad (2.3)$$

where

$$p_\varphi(g) = (c_\varphi)^{-1} (U(g)\varphi, \varphi) \ . \qquad (2.4)$$

The correspondence $\psi \to L_\varphi \psi$ intertwines U and the left regular representation, i.e. one has

$$(L_\varphi U(g_o)\psi)(g) = (L_\varphi \psi)(g_o^{-1} g), \quad (g_o g \in G). \quad (2.5)$$

<u>Remark</u>: There is of course also the realization of $H(U)$ as a closed subspace of $L^2(G, d_R\mu(g))$ (right-invariant measure), established by the correspondence $\psi \to R_\varphi \psi$,

$$(R_\varphi \psi)(g) = (c_\varphi)^{-1/2} (\varphi, U(g)\psi).$$

All the above results can be put into the language of "overcomplete continuous bases". If we write

$$|g) = (c_\varphi)^{-1/2}\ U(g)\varphi \ ,$$

then the isometry statement in the above proposition is equivalent to the formula

$$1 = \int |g)\ d\mu(g)\, (g| \qquad (2.6)$$

which expresses the identity in $H(U)$ as a weakly convergent integral of dyadic operators. In this notation, one has

$$f(g) = (g|\psi)$$

and

$$\int \bar{f}_1(g) f_2(g)\, d\mu(g) = (\psi_1, \psi_2).$$

References Section 2

A.L. Carey: Bull. Austral. Math. Soc. 15, 1 (1976)

M. Duflo, C.C. Moore: J. Funct. Anal. 21, 209 (1976)

S. Gaal: Linear Analysis and Representation Theory, Springer 1973

A. Grossmann, J. Morlet: to appear in S.I.A.M., Mathematical Analysis

A. Grossmann, J. Morlet: "Decomposition of functions into wavelets of constant shape and related transforms", to appear in: "Mathematics + Physics, Lect. on Recent Results", World Sc. Publ. Singapore

G. Warner: Harmonic Analysis on semi-simple Lie groups. Vol. 1 and Vol. 2, Springer Verlag 1972

3. Two-Parameter Subgroups of Affine Canonical Transformations in Quantum Mechanics

In this section we consider quantum-mechanical systems in one degree of freedom. The Hilbert space of states of any such system carries a projective representation of the group of affine canonical transformations, which will be described below. We may ask whether this representation, or its restriction to suitable subgroups, is square integrable. If it is, then the space of states can be realized by square integrable functions on the subgroup, in accordance with the results described in the preceding section. Such a realization will always have the covariance property described by (2.5). A suitable choice of φ can bring about other properties, as we shall see.

Let X and P be the canonical variables satisfying $[X,P] = i\mathbf{1}$ and let $\mathbf{1}$ be the identity operator in the Hilbert space of states H. Consider the affine transformation

$$\begin{aligned} X' &= \alpha X + \beta P + x\mathbf{1} \\ P' &= \gamma X + \delta P + p\mathbf{1} \end{aligned} \tag{3.1}$$

with $\alpha, \beta, \gamma, \delta, x, p$ real, and with $\det \begin{vmatrix} \alpha & \beta \\ \gamma & \delta \end{vmatrix} = 1.$

Then X' and P' are also canonical variables, i.e. satisfy $[X',P'] = i\mathbf{1}$. We have assumed that canonical variables act irreducibly in H. The uniqueness theorem of von Neumann (which, for technical reasons, is better stated in terms of exponentiated canonical variables) then says that there exists in H a unitary operator $U = U(g)$, with

$$g = \left\{ \begin{pmatrix} x \\ p \end{pmatrix}, \begin{pmatrix} \alpha & \beta \\ \gamma & \delta \end{pmatrix} \right\}$$

such that

$$X' = U(g)\, X\, U(g)^{-1} \tag{3.2}$$

and

$$P' = U(g)\ P\ U(g)^{-1}. \qquad (3.3)$$

The equations (3.2) and (3.3) do not determine the phase of the operator $U(g)$. These operators form a representation up to a phase factor (projective representation) of the group of affine canonical transformations.

We recall now the one-parameter subgroups of this group:

a) Space shifts:

$$X \to X+x, \quad \text{i.e.} \quad g = \left\{ \binom{x}{0}, \begin{pmatrix} 1 & 0 \\ 0 & 1 \end{pmatrix} \right\}$$

b) Momentum shifts:

$$P \to P+p, \quad \text{i.e.} \quad g = \left\{ \binom{0}{p}, \begin{pmatrix} 1 & 0 \\ 0 & 1 \end{pmatrix} \right\}$$

c) Dilations:

$$\begin{matrix} X \to \alpha X \\ P \to \alpha^{-1} P \end{matrix} \quad \text{i.e.} \quad g = \left\{ \binom{0}{0}, \begin{pmatrix} \alpha & 0 \\ 0 & \alpha^{-1} \end{pmatrix} \right\} \quad (\alpha \neq 0)$$

d) Harmonic oscillator motion:

$$\begin{matrix} X \to X\cos\theta + P\sin\theta \\ P \to -X\sin\theta + P\cos\theta \end{matrix} \quad \text{i.e.} \quad g = \left\{ \binom{0}{0}, \begin{pmatrix} \cos\theta & \sin\theta \\ -\sin\theta & \cos\theta \end{pmatrix} \right\}$$

e) Free particle motion:

$$\begin{matrix} X \to X+\beta P \\ P \to P \end{matrix} \quad \text{i.e.} \quad g = \left\{ \binom{0}{0}, \begin{pmatrix} 1 & \beta \\ 0 & 1 \end{pmatrix} \right\} .$$

The restriction of U to any one of these subgroups is highly reducible. We shall see that the situation becomes different when we consider two-parameter subgroups.

We consider now two-parameter subgroups of the group of affine canonical transformations.

A) Phase space shifts:

$$\begin{matrix} X \to X+x \\ P \to P+p \end{matrix} \quad \text{i.e.} \quad g = \left\{ \binom{x}{p}, \begin{pmatrix} 1 & 0 \\ 0 & 1 \end{pmatrix} \right\} \qquad (3.4)$$

B) Shifts in one variable, and dilations:

$$\begin{matrix} X \to \alpha X + x \\ P \to \alpha^{-1} P \end{matrix} \quad \text{i.e.} \quad g = \left\{ \binom{x}{0}, \begin{pmatrix} \alpha & 0 \\ 0 & \alpha^{-1} \end{pmatrix} \right\} \quad (\alpha \neq 0) \qquad (3.5)$$

(nothing essential is changed if we interchange X and P)

C) Free motion and dilations:

$$\begin{matrix} X \to \alpha X + & \beta P \\ P \to & \alpha^{-1} P \end{matrix} \quad \text{i.e.} \quad g = \left\{ \begin{pmatrix} 0 \\ 0 \end{pmatrix}, \begin{pmatrix} \alpha & \beta \\ & \alpha^{-1} \end{pmatrix} \right\} \quad (\alpha < 0) \qquad (3.6)$$

(here, again, X and P may be interchanged).

In contrast to the one-parameter case, the restriction of U to these subgroups splits into at most two irreducible components, and these components are square integrable. Consequently they give rise to realizations of the space of states, which will now be examined in more detail.

References Section 3

A. Grossmann: "Geometry of real and complex canonical transformations in quantum mechanics", in: Group Theoretical Methods in Physics, Tübingen 1977, edited by P. Kramer and A. Rieckers, Springer 1978

I. Daubechies: J. Math. Phys. 21 (6), 1377 (1980).

4. First Example: Phase-Space Shifts:

If we choose the two-parameter subgroup (A) and apply the procedure described in Sec. 2, we obtain the well-known coherent state (or equivalently Bargmann) representation of quantum mechanics. The steps are as follows:

(i) The restriction of U to phase-space shifts consists of Weyl-operators. They give rise to an irreducible representation of the Weyl-Heisenberg group (with compact center). The Weyl-Heisenberg group is unimodular.

(ii) The representation just introduced is square integrable. By the unimodularity of the Weyl-Heisenberg group, all vectors are admissible. One has

$$c_\varphi = 2\pi(\varphi,\varphi).$$

(iii) The transformation considered in Sec. 2 is now $\psi \to \Psi$, with

$$\Psi(x,p) = \frac{1}{\sqrt{2\pi}\,\|\varphi\|}\,(W(x,p)\varphi,\psi),$$

where $W(x,p)$ is the Weyl operator. The general isometry statement of Sec. 2 becomes simply

$$\int \Psi_1(x,p)\Psi_2(x,p)\,dxdp = (\psi_1,\psi_2).$$

(iv) The reproducing equation (2.3), characterizing the range of the transform, is now

$$\Psi(x,p) = \iint \bar{G}(x,p;x',p')\Psi(x',p')\ dx'dp'$$

with

$$G(x,p;x',p') = \frac{1}{2\pi\|\varphi\|^2}\ (Wx',p')\varphi, W(x,p)\varphi) . \tag{4.1}$$

(v) All the above was true for any choice of the "reference vector" φ in the space H. If we now choose the φ that is annihilated by $X+iP$ (i.e. the Gaussian in the x-representation), then the reproducing equation shows that the range of the transform consists exactly of the functions that are square integrable with respect to the measure $dx\,dp$ and of the form

$$\Psi(x,p) = e^{-(x^2+p^2)/2}\ f(x-ip)$$

with f entire analytic.

References Section 4

V. Bargmann: Comm. Pure Appl. Math. 14, 187 (1961)
V. Bargmann: Comm. Pure Appl. Math. 20, 1 (1967).

5. Second Example: Shifts in One Variable, and Dilations

Consider now the two-parameter subgroup (B). If we write $\{x,\alpha\}$ to abbreviate (3.5), we see that the product law is

$$\{x_1,\alpha_1\}\{x_2,\alpha_2\} = \{\alpha_1 x_2 + x_1, \alpha_1\alpha_2\} . \tag{5.1}$$

This is the so-called "ax+b"-group. Here we allow α to be negative, and so our group has two connected components. In contrast to (A), this group is not unimodular. Its left-invariant measure is

$$d\mu(g) = \frac{dx d\alpha}{\alpha^2} \quad ; \tag{5.2}$$

the right-invariant measure is

$$d_R\mu(g) = \frac{dx d\alpha}{|\alpha|} . \tag{5.3}$$

The representation in the Hilbert space $H = L^2(R,dx')$ is

$$U(x,\alpha)\psi(x') = |\alpha|^{-1/2}\psi(\frac{x'-x}{\alpha}) . \tag{5.4}$$

This representation is irreducible. It is square integrable. In accordance with the general results of Sec. 2, not every vector in H is admissible. The admissibility condition is best expressed in terms of the Fourier transform

$$\psi(p) = (2\pi)^{-1/2} \int e^{-ipx} \psi(x)dx.$$

It is then

$$c_\varphi = 2\pi \int |\psi(p)|^2 \frac{dp}{|p|} < \infty . \tag{5.5}$$

The transformation L_φ associates to every $\psi \in H$ the function

$$f(x,\alpha) = (c_\varphi)^{-1/2} |\alpha|^{-1/2} \int \bar{\varphi}(\frac{x-x'}{\alpha}) \psi(x')dx'. \tag{5.6}$$

It gives rise to wave functions defined on the (α,x)-plane, cut along the axis $\alpha = 0$. The range of L_φ is again characterized by a reproducing equation. By isometry, the transform (5.6) is inverted, on its range, by its adjoint:

$$\psi(x) = (\varphi_\varphi)^{-1/2} \int\int f(x',\alpha)|\alpha|^{-1/2} \varphi(\frac{x-x'}{\alpha}) \frac{dx'd\alpha}{\alpha^2} .$$

References Section 5

A. Grossmann, J. Morlet: to appear in S.I.A.M. Mathematical Analysis

A. Grossmann, J. Morlet: "Decomposition of functions into wavelets of constant shape and related transforms", to appear in "Mathematics + Physics, Lectures on Recent Results", World Sc. Publ. Singapore

6. Third Example: The "ax+b" Group as a Subgroup of SL(2,R):

In this section, we shall consider the subgroup (C) of Sec. 3, and a transform associated with it. The group of free motion and dilations is isomorphic to the connected component of the identity of the group considered in Sec. 5. This can be seen from the re-parametrization

$$a = \alpha^2$$
$$b = \alpha\beta$$

and the representation

$$(U(a,b)\psi)(p) = a^{1/4} e^{-ibp^2/2} \psi(a^{1/2}p). \tag{6.1}$$

This representation has two irreducible components, corresponding to even and odd functions in $L^2(R)$. We shall have to use two "reference functions", one even and one odd.

The admissibility condition can be written as

$$\int |\varphi(p)|^2 \frac{dp}{p^2} < \infty . \tag{6.2}$$

We choose

$$\varphi_e(p) = p^2 e^{-p^2/2} \quad \text{for the even part} \tag{6.3}$$

$$\varphi_o(p) = p\, e^{-p^2/2} . \quad \text{for the odd part} \tag{6.4}$$

This gives the transforms

$$g_e(a,b) = \frac{a^{5/4}}{\sqrt{2}\pi^{3/4}} \int p^2\, e^{i[b+ia]\frac{p^2}{2}} \psi(p)dp \quad (\psi \text{ even}) \tag{6.5}$$

and

$$g_o(a,b) = \frac{a^{3/4}}{2\pi^{3/4}} \int p\, e^{i[b+ia]\frac{p^2}{2}} \psi(p)dp \quad (\psi \text{ odd}). \tag{6.6}$$

Both g_e and g_o are square integrable with respect to the measure $a^{-2}dadb$. They are a product of a power of a with an analytic function of $b+ia$ on the open upper half plane $a > 0$. The analytic factors are

$$f_e(z) = \frac{1}{\sqrt{2}\pi^{3/4}} \int p^2\, e^{izp^2/2} \psi(p)dp \qquad (\psi \text{ even}) \tag{6.7}$$

$$f_o(z) = \frac{\wedge}{2\pi^{3/4}} \int p\, e^{izp^2} \psi(p)dp \qquad (\psi \text{ odd})$$

$$\text{with} \quad z = b+ia.$$

The function f (resp. f_o) is analytic in the upper half-plane $a > 0$ and square integrable with respect to the measure $a^{1/2}$ dadb (resp. $a^{-1/2}$dadb).

Certain quantum-mechanical operators are quite transparent in this representation:

a) The kinetic energy operator, i.e. the operator of multiplication by $p^2/2$ on $L^2(R,dp)$. It becomes, in this space of analytic functions:

$$H_o = -i \frac{d}{dz} .$$

It is interesting to notice that this operator is self-adjoint because of the analyticity of f, and positive because f is defined on the half-plane and not on the whole plane.

The free evolution is then just the translation along the real axis:

$$(e^{-iH_0 t}f)(z) = f(z-t).$$

b) Dilations: they act on $L^2(R)$ by

$$(D^y \psi)(p) = y^{-1/2}\ \psi(p/y). \tag{6.8}$$

In the spaces just introduced, they act by

$$(\underline{D}^y f_e)(z) = y^{3/2}\ f_e(y^2 z) \qquad (y > 0) \tag{6.9}$$

and

$$(\underline{D}^y f_o)(z) = y^{5/2}\ f_o(y^2 z) \qquad (y > 0). \tag{6.10}$$

c) The metaplectic representation (projective prepresentation of homogeneous linear canonical transforamtions) takes a particularly simple form:

On f_o:

$$(U\left\{\begin{pmatrix} a & b \\ c & d \end{pmatrix}^{-1}\right\} f_o)(z) = (cz + d)^{-3/2}\ f(\frac{az+b}{cz+d}) \tag{6.11}$$

and on f_e:

$$(U\left\{\begin{pmatrix} a & b \\ c & d \end{pmatrix}^{-1}\right\} f_e)(z) = (cz+d)^{-5/2}\ f_e(\frac{az+b}{cz+d})$$

$$+ \frac{c}{2}\ (cz+d)^{-3/2} \int_0^{+\infty} f_e(\frac{az+b}{cz+d} + i\lambda)d\lambda\ . \tag{6.12}$$

<u>Concluding remarks</u>:

1) The extension to more than one degree of freedom is not automatic, since square integrability is a strong restriction on a representation. Many questions can be treated with the help of suitable tensor product decompositions.

2) There are obvious generalizations in which $L^2(R)$ is replaced by spaces of distributions or by functions that increase exponentially (like resonance functions). We obtain functions that are still analytic but no more square integrable with respect to the measure introduced here.

3) From the formulas (6.11), (6.12) we can see that some of the subgroups can be taken with complex values of the parameters. This is

in particular the case for dilations, on which work is in progress.

References Section 6

T. Paul: "Functions analytic on the half-plane as quantum mechanical states", preprint Bielefeld Project Nr. 2/Nr. 22, to appear in J. Math. Physics.

T. Paul: Thèse de 3° cycle Université "Pierre et Marie Curie" Paris VI.

RESONANCES IN NUCLEAR PHYSICS

C. Mahaux,

Institut de Physique B5, Université de Liège,
Sart Tilman, B-4000 Liège 1 (Belgium)

In nuclear physics, the main characteristic features of resonances is that their width is narrow and their spacing is small. This reflects the complicated nature of the resonance states and calls for statistical theories rather than for a dynamical description of individual resonances. A microscopic understanding of the resonance process is nevertheless required, in particular for the identification and the interpretation of observed nonstatistical features.

1. INTRODUCTION

The main purpose of the present survey is to describe the characteristic features of the resonances observed in the realm of nuclear physics and to outline some of the theoretical approaches to the interpretation of these properties.

The word *resonance* first appeared in the fifteenth century in connection with acoustical phenomena. The theory of resonance cavities has been developed by Helmholtz in the second half of the nineteenth century. Helmholtz showed that *a resonance occurs whenever the frequency of the wave enclosed in the cavity is nearly equal to the frequency of one of the normal modes of the cavity*. Here, "nearly" indicates that the resonance can be excited even if the frequency of the imprisoned wave is not exactly equal to that of a normal mode. In other words, the resonance has a finite *width*.

Let us consider the example of electromagnetic waves. The amount of electromagnetic energy enclosed in the cavity decreases because of *losses* in the walls since these are not made of perfect conductors. This decrease is approximately exponential and can be characterized by a lifetime τ . One has

$$\tau \, \Delta\omega \approx 1 \quad , \tag{1.1}$$

where $\Delta\omega$ is the value of the width in the frequency scale.

These basic features (resonance energy $\rightarrow$ normal mode frequency;

resonance width → loss of probability) remain valid at the microscopic level. Then eq. (1.1) is usually written in the form

$$\tau \, \Gamma = \hbar \quad , \tag{1.2}$$

where $\Gamma = \hbar \, \Delta\omega$ is the width of the resonance in the energy scale. In nuclear physics, the loss of probability corresponds to the fact that the resonance state emits particles (or photons). If the possibility of emitting particles is suppressed, the width Γ vanishes. The lifetime τ then becomes infinite : in this limit, the resonance state becomes a bound state. Hence, narrow resonances reflect the existence of *quasi-bound* states. In Feshbach's projection operator formalism [1,2] the bound states correspond to the discrete eigenstates of the operator QHQ , where H is the Hamiltonian and Q is a projection operator (onto closed channels) which suppresses the possibility of emitting particles.

In other formulations of resonance theory [3], one writes

$$H = H_o + v \quad , \tag{1.3}$$

where H_o has bound eigenstates embedded in its continuous spectrum (section 6). These bound eigenstates of H_o lead to resonances when the "residual interaction" v is taken into account.

In section 2 we show a few examples of measured resonances, while in section 3 we discuss the relationship between resonances and poles of the scattering matrix. It is quite exceptional when a resonance can be interpreted as being due to the potential scattering by the average nucleon-nucleus potential; this exceptional case will be illustrated in section 4. In sections 5 and 6 we outline two main methods for the theoretical description of resonances, namely the R-matrix and shell-model approaches, respectively. The statistical model is considered in section 7 and some deviations from the statistical assumptions are briefly discussed in section 8.

2. EMPIRICAL FEATURES

By definition, a resonance is associated with a narrow peak in a cross section. Figure 2.1 shows resonance peaks observed in the total cross section of neutrons by ^{238}U . Most of the peaks present a minimum on their low-energy shoulder. This minimum is characteristic of the interference between an s-wave resonance and an s-wave background. The

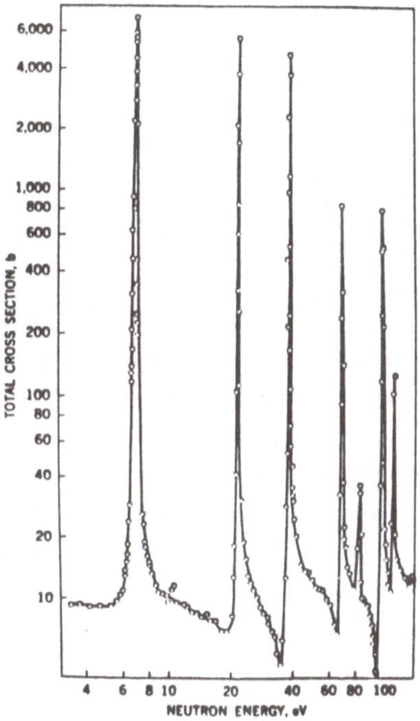

Fig. 2.1. Taken from ref. [4]. Dependence upon neutron energy (in eV) of the total cross section (in barns) of neutrons by ^{238}U . Note that the scales are logarithmic.

latter can approximately be associated with the elastic scattering by a hard sphere.

Figure 2.2 shows an isolated resonance which occurs at 1.1697 MeV in the elastic scattering cross section ^{20}Ne(p,p) . The possibility of measuring isolated resonances above 1 MeV is important because inelastic or reaction channels may then be open. Hence, one can then measure partial width amplitudes in different channels (sections 7.4 and 7.6).

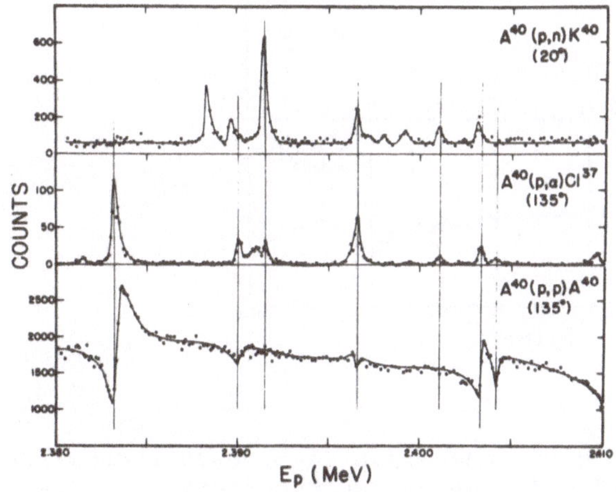

Fig. 2.2. Taken from ref. [5]. Dependence upon the proton energy (in MeV) of a resonance observed in the elastic differential cross section (arbitrary units) of protons by ^{20}Ne at 150°. The width of the measured peak (110 eV) is due to the energy resolution; the width of the resonance is 8 eV .

The empirical features of the resonance peaks lead to the following main observations.

(i) In the continuous spectrum of the Hamiltonian H there exist discrete "resonance" energies which obviously play a special role. This should be taken into account in any formulation of nuclear reaction theory at low energy.

(ii) The resonances occur in all open channels at the same excitation energy of the *compound nucleus*, i.e. of the system formed once the projectile has entered into the target. This property is illustrated in fig. 2.3, see also e.g. ref. [7]. It thus appears natural to associate nuclear resonances with the formation of excited metastable states of the compound nucleus.

In the compound nucleus model of N. Bohr [8], the projectile enters the target and then shares its energy among many of the target constituents, thus forming a metastable state whose properties (energy, decay modes, wave function) are nearly independent of the entrance channel which led to its formation. This picture is essentially a classical one and can therefore be criticized [9]. However, the assumption that the excitation energy of the compound nucleus is shared among *many* of its constituents is corroborated by the fact that the resonances are closely spaced; this is a signature of a process which involves many degrees of freedom.

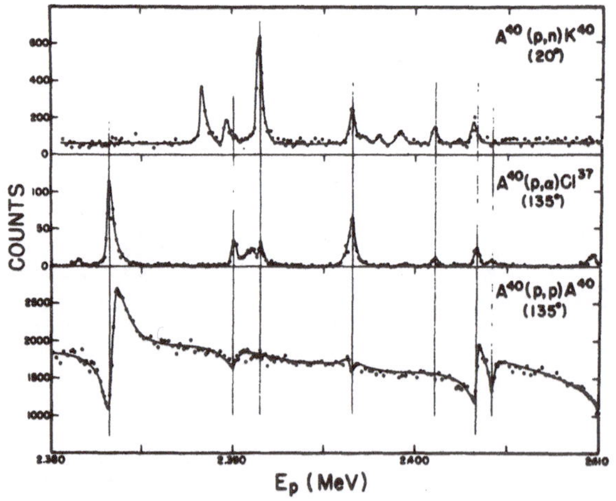

Fig. 2.3. Taken from ref. [6]. Dependence upon the proton energy (in MeV) of the differential cross sections of the reactions $^{40}A(p,n)$ (at 20°), $^{40}A(p,\alpha)$ (at 135°) and $^{40}A(p,p)$ (at 135°).

3. RESONANCES AND POLES OF THE SCATTERING MATRIX

3.1. Channels

The initial (or the final) state of a nuclear reaction is characterized by the nature (A_1,A_2) of the fragments, by their relative orbital angular momentum ℓ, by the sum of their spins (S), and by the total angular momentum (J) and its projection (M) on the quantization axis. All these quantum numbers are generically denoted by the channel index $c = \{A_1,A_2,\ell,S,J,M\}$.

3.2. Wave functions

The Schrödinger equation reads

$$H\,\Psi_E = E\,\Psi_E \quad , \tag{3.1}$$

where E is the total energy. It is important to realize that *if there exist Λ open channels at the energy E, there exist Λ linearly independent wave functions Ψ_E*. These can for instance be determined by specifying that only a given channel contains an incoming wave.

Let us denote by $\Psi_E^{(c)}$ the solution of eq. (3.1) which corresponds to the entrance channel c. For the sake of notational simplicity, we only consider s-wave neutron channels. The wave function $\Psi_E^{(c)}(1,\ldots,A)$ has the following asymptotic behaviour for $r_A \to \infty$

$$\Psi_E^{(c)} \sim \left(e^{-ik_c r_A}\,\delta_{cc'} - S_{cc'}(E)\,e^{ik_{c'} r_A}\right)\phi_{c'}/r_A \quad . \tag{3.2}$$

Here, $\phi_{c'}(1,\ldots,A-1)$ is essentially the wave function of the residual nucleus, $S_{cc'}$ is the (c,c') element of the *scattering matrix* and $k_{c'}$ is the wave number in the channel c'. If $\varepsilon_{c'}$ denotes the threshold energy of c', one has

$$k_{c'} = \left\{\frac{2m}{\hbar^2}\,(E - \varepsilon_{c'})\right\}^{\frac{1}{2}} \tag{3.3}$$

if c' is open (i.e. if $E > \varepsilon_{c'}$) and

$$k_{c'} = i\,\kappa_{c'} = i\left\{\frac{2m}{\hbar^2}\,(\varepsilon_{c'} - E)\right\}^{\frac{1}{2}} \tag{3.4}$$

if c' is closed (i.e. if $E < \varepsilon_{c'}$). The cross section is proportional to the quantity

$$\sigma_{cc'} = |S_{cc'} - \delta_{cc'}|^2 \quad . \tag{3.5}$$

3.3. Breit-Wigner formula

Let us consider the case of inelastic scattering $(c \neq c')$. In the vicinity of a resonance, the cross section is found to have the following "Breit-Wigner shape"

$$\sigma_{cc'}(E) = \frac{\Gamma_{\lambda c}\,\Gamma_{\lambda c'}}{(E - E_\lambda)^2 + \frac{1}{4}\Gamma_\lambda^2} \quad . \tag{3.6}$$

The fact that the resonance energy E_λ and the resonance width Γ_λ are independent of c and c' reflects the property that these quantities have the same value in all reactions in which the resonance is excited, see section 2.

The Breit-Wigner shape (3.6) reflects the following approximation for the scattering matrix

$$S_{cc'}(E) = e^{i(\xi_c+\xi_{c'})} \left\{B_{cc'} - i\,\frac{\gamma_{\lambda c}\,\gamma_{\lambda c'}}{E - \mathcal{E}_\lambda}\right\} \quad , \tag{3.7}$$

where $B_{cc'}$ is approximately independent of E while

$$\gamma_{\lambda c}^2 = \Gamma_{\lambda c} \quad , \tag{3.8}$$

$$\mathcal{E}_\lambda = E_\lambda - \frac{1}{2}\,i\,\Gamma_\lambda \quad . \tag{3.9}$$

The partial width amplitude $\gamma_{\lambda c}$ may be positive or negative.

3.4. Poles

Equations (3.7) and (3.9) show that when a resonance exists the scattering matrix has a pole at a complex energy $\mathcal{E}_\lambda$ which is located in the lower half of the complex E-plane. Near this pole, $S_{cc'}$ becomes very large compared to unity. At the pole, the analytic continuation

$$\psi_\lambda = \lim_{E\to\mathcal{E}_\lambda} \Psi_E^{(c)} \tag{3.10}$$

of the wave function $\Psi_E^{(c)}$ has the asymptotic behaviour (see eq. (3.2))

$$\Psi_\lambda \propto e^{ik_{\lambda c'} r_A}\,\phi_{c'}/r_A \tag{3.11}$$

in all open channels c' , with

$$k^2_{\lambda c} = \frac{2m}{\hbar^2} (E_\lambda - \varepsilon_{c'}) \quad . \qquad (3.12)$$

3.5. Gamow state

Equation (3.11) has the following important implication. *The* Λ *linearly independent solutions* $\Psi_E^{(c)}$ *become degenerate when* E *approaches the complex pole* E_λ . In a many open channels case, it is only at this complex energy that one can speak of "the" wave function without further ado. The complex resonance energy E_λ and the associated wave function Ψ_λ are therefore intrinsic properties of the full (projectile + target) system and appear to be good candidates for the resonance energy and wave function, respectively. This was pointed out by Breit [10] and developed by Humblet and Rosenfeld, see [11-13] and references contained therein.

The wave function Ψ_λ is the extension to the many channels case of the metastable state first introduced by Gamow [14] in his theory of α decay. It presents the difficulty that in all open channels the asymptotic value of Ψ_λ grows exponentially for $r_A \to \infty$. Hence, ψ_λ cannot be introduced in matrix elements without using an appropriate definition of the integral; the normalization, orthogonality and completeness properties of these Gamow states are therefore not straightforward nor physically transparent, see [15] and references contained therein. Finally, poles which lie far away from the real axis do not lead to any observable energy dependence of the cross section.

Difficulties are encountered when one continues $\Psi_E^{(c)}$ analytically in the complex E-plane. Indeed, $\Psi_E^{(c)}$ is truly a function of the wave numbers $k_{c'}$ in all channels c' (open and closed). It has branch points at each threshold energy $\varepsilon_{c'}$. Branch cuts must thus be defined. The way of drawing them is largely arbitrary. For a resonance pole which lied near a threshold $\varepsilon_{c'}$, it is relatedly somewhat arbitrary to consider the channel c' as being either open or closed.

3.6. Scattering wave functions near a resonance

Since real life takes place on the real energy axis, it is of interest to investigate to what extent the wave functions $\Psi_E^{(c)}$ (E real) differ from ψ_λ .

In [16] an exactly soluble many-channel model has been investigated. In keeping with section 1, it was found that narrow resonances are associated with the energies of the bound states which occur in the closed channel subspace when its coupling to the open channels is turned off. Near the resonance energy one has

$$\Psi_E^{(c)} \approx u_c(r_A, E)\, \phi_c + (2\pi)^{-\frac{1}{2}} \frac{\gamma_{\lambda c}}{E - \mathcal{E}_\lambda} \psi_\lambda \quad . \tag{3.13}$$

Here ψ_λ is the Gamow state while $u_c(r_A, E)$ is the potential scattering wave function in channel c in the absence of channel-channel coupling. In [17] a result equivalent to eq. (3.13) has been derived in the framework of R-matrix theory for the value taken by $\Psi_E^{(c)}$ inside the domain of interaction.

The probability of finding the system in the resonance state ψ_λ thus appears to be given by a Breit-Wigner law. This statement must be taken with caution since ψ_λ is not orthogonal to the configuration $u_c(r,E)\, \phi_c$.

The probability of finding the particle within the domain of interaction is dominated by the probability of finding it in the closed channel bound state configuration associated with the resonance. This corresponds to the following physical picture (see also section 6). The incoming particle excites the target into the configuration $\phi_{c'}$; thereby it looses energy and falls in a bound orbit $u_{\lambda c'}$ of the mean field created by $\phi_{c'}$. The system spends a long time in this bound configuration $u_{\lambda c'} \times \phi_{c'}$ before decaying.

4. SHAPE RESONANCES

In sections 1 and 3.6 we emphasized that in a many-body system most of the resonances are associated with closed channel configurations i.e. with bound configurations. In nuclear physics, only very few examples exist of resonances which can be interpreted as resonances in the average potential created by the target in its *ground state*, i.e. as shape (or potential scattering) resonances. In these exceptional cases, the projectile is trapped inside the target by a centrifugal of Coulomb surface barrier.

One of the very few known examples is shown in fig. 4.1. The location and the width of the resonance located at 1.00 MeV can be reproduced by a model in which a d-wave neutron is scattered by a standard average Woods-Saxon potential [19].

Note that this interpretation is not valid for any of the other resonances shown in fig. 4.1. These resonances can always be reproduced with a single-particle potential [1] but the latter then has a wild energy dependence and a complicated nonlocal dependence upon the coordinates.

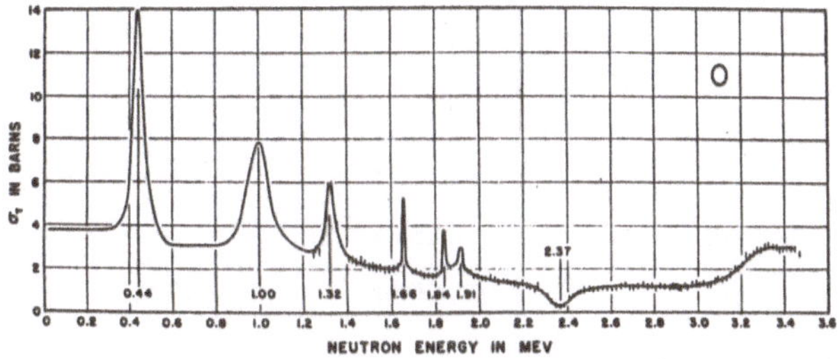

Fig. 4.1. Taken from [18]. Dependence upon neutron energy (in MeV) of the total cross section of neutron by oxygen.

5. R-MATRIX APPROACHES

The energy at which reactions occur necessarily lies in the continuous spectrum of the Hamiltonian. The observation of isolated resonances points to the interest of *defining* discrete resonance energies within this continuous spectrum.

The most obvious way of defining a discrete spectrum consists in enclosing the system inside a box centered on the target and in imposing that the eigenfunction of H have a prescribed logarithmic derivative on the surface of the box. This indeed defines an eigenvalue problem :

$$H_\lambda X_\lambda = e_\lambda X_\lambda \quad . \tag{5.1}$$

The eigenfunctions X_λ (1,...,A) are assumed to be of the form

$$X_\lambda = r_A^{-1} u_{\lambda c}(r_A) \phi_c \tag{5.2}$$

on the side $(r_A = a_c)$ of the box which corresponds to a channel c .

In the R-matrix approach of Wigner and Eisenbud 20 the eigenvalue problem is defined by imposing the boundary condition

$$\left. \frac{a_c u'_{\lambda c}}{u_{\lambda c}} \right|_{r_A = a_c} = B_c \quad , \tag{5.3}$$

where the real boundary condition parameters B_c can be chosen as one wishes. In the Kapur-Peierls approach [21], one takes the boundary condition

$$\left. \frac{a\, u'_{\lambda c}}{u_{\lambda c}} \right|_{r_A = a_c} = i\, k_c\, a_c \quad , \tag{5.4}$$

where k_c is the wave number in channel c .

The eigenfunctions X_λ form a complete set *inside* the box, where one can write

$$\Psi_E^{(c)} = \sum_{\lambda=1}^{\infty} A_\lambda^{(c)}\, X_\lambda \quad . \tag{5.5}$$

The R-matrix approach is of interest if at least one of the eigenvalues e_λ is close to an observed resonance energy E_λ . This depends upon the choice of the boundary condition parameters B_c and of the "channel radii" a_c . As expected from the previous sections, it turns out that is is mainly the choice of the boundary condition in the *closed* channels which matters [16].

If one wants to use the R-matrix approach to calculate the scattering matrix one must find a prescription to relate the value of $\Psi_E^{(c)}$ inside the box (see eq. (5.5)) to its value outside the box (see eq. (3.2)). The prescription is not unique because $\Psi_E^{(c)}$ (or its derivative) is not continuous on the box boundaries $r_A = a_c$ since in practice one must truncate the sum over λ on the right-hand side of eq. (5.5). Various prescriptions exist; they are critically reviewed in ref. [22].

6. SHELL-MODEL APPROACHES

In the shell model, the nucleons are assumed to move independently of one another in an average potential well. The corresponding single-particle Hamiltonian h_o has bound as well as scattering eigenstates, and correspondingly a discrete as well as a continuous spectrum. This is also true for the many-body shell-model Hamiltonian H_o

$$H_o(1,\ldots,A) = \sum_{j=1}^{A} h_o(j) \quad . \tag{6.1}$$

Let us call E_λ the discrete eigenvalues of H_o and E its continuous eigenvalues

$$H_0 \Phi_\lambda = E_\lambda \Phi_\lambda \quad , \tag{6.2}$$

$$H_0 \chi_{Ec} = E \chi_{Ec} \quad . \tag{6.3}$$

Here, the index c in χ_{Ec} means that when the radial coordinate r_A of one of the nucleons goes to infinity, the remaining (A-1) - nucleon system is in the state ϕ_c (see eq. (3.2)).

It is important to realize that in the many-body case the discrete energies E_λ can lie within the continuous spectrum of H_0 , i.e. that one may have $E_\lambda > \varepsilon_c$ where ε_c is the threshold energy for particle emission. Let us for instance consider the eigenstates of H_0 depicted in fig. 6.1. There, the drawing (a) represents a scattering eigenstate χ_E^c which is such that in the target ϕ_c (with energy ε_c) all orbits are occupied up to the Fermi momentum (see shaded area). The drawing (b) shows an eigenstate Φ_λ of H_0 which is bound since all the A nucleons are in bound orbitals. Note that one can write Φ_λ in the form

$$\Phi_\lambda = u_\lambda(r_A) \times \phi_{c'} \quad , \tag{6.4}$$

where $u_\lambda(r)$ is a bound eigenstate of h_0 with energy e_λ , while $\phi_{c'}$ is a bound eigenstate of $H_0(1,\ldots,A-1)$ with energy $\varepsilon_{c'}$. The value of E_λ can be larger than ε_c . Indeed, one can have

$$e_\lambda + \varepsilon_{c'} > \varepsilon_c \quad , \tag{6.5}$$

since $\varepsilon_{c'} > \varepsilon_c$ (note that $e_\lambda < 0$) . Then, Φ_λ is sometimes called a "bound state embedded in the continuum".

Finally, the drawing (c) of fig. 6.1 represents a scattering eigenstate $\chi_{Ec''}$ where one of the nucleons is in a scattering state while the other nucleons are in a (one particle-one hole) configuration $\phi_{c''}$ with energy $\varepsilon_{c''} > \varepsilon_{c'} > \varepsilon_c$.

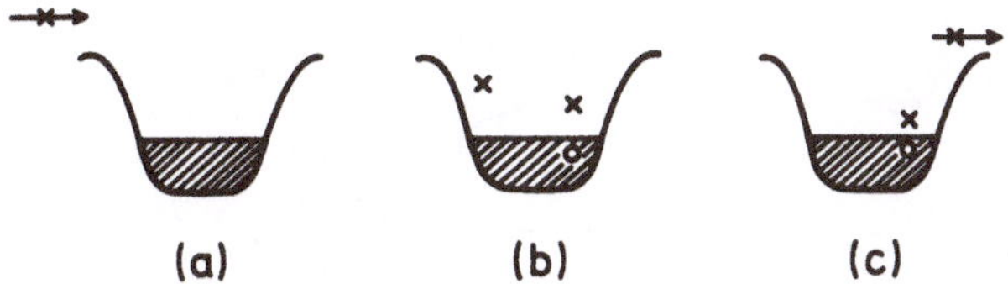

Fig. 6.1. Schematic representation of three eigenstates of the shell-model Hamiltonian H_0 . Those labelled (a) and (c) are associated with χ_{Ec} and $\chi_{Ec''}$, see eq. (6.3). The one labelled (b) is a bound eigenstate Φ_λ , see eq. (6.2).

In the shell-model approach, the resonance process is visualized in the following way. The initial configuration χ_{Ec} is depicted by the drawing (a). The incoming nucleon (cross) makes a collision with one of the target nucleons and falls in a bound single-particle state u_λ, while the target is excited into the state $\phi_{c'}$ with excitation energy $\varepsilon_{c'} - \varepsilon_c$. If the initial energy E is equal to the energy E_λ of the bound state of H_o, a resonance occurs since the energy is equal to that of normal mode i.e. of a bound state of the model Hamiltonian.

The probability amplitude for going from the scattering state χ_{Ec} to the bound configuration Φ_λ is proportional to the partial width amplitude

$$\gamma_{\lambda c} = \langle \Phi_\lambda | v | \chi_{Ec} \rangle \quad , \tag{6.6}$$

where v is the residual interaction, see eq. (1.3).

In the simple example of fig. 6.1, the excited configuration $\phi_{c'}$ is a one particle-one hole state. In practice, most of the target excitations which are important (in the sense that the matrix element (6.6) is large) are not of this type. Rather, they are collective excitations. This, however, does not modify the physical interpretation nor the basic features of the theoretical methods for solving the problem. The latter consists in diagonalizing the Hamiltonian $H = H_o + v$ in the configuration space spanned by the basis $\{\Phi_\lambda, \chi_{Ec'}\}$. They are surveyed in refs. [3,22]. One example is shown in fig. 6.2.

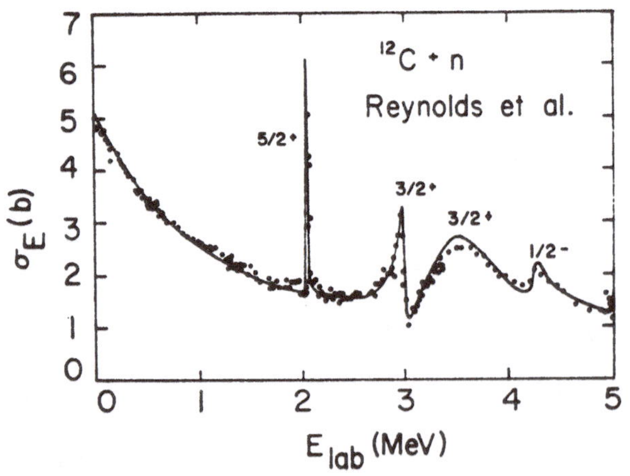

Fig. 6.2. Taken from ref. [23]. Measured (dots) and calculated (curve) elastic cross section $^{12}C(n,n)$. The ^{12}C levels ($\phi_{c'}$) introduced in the calculation are the 0^+ ground state, the 2^+ excited states at 4.43 and 3.18 MeV and the 0^+ excited states at 7.83 and 9.18 MeV.

The outcome of the calculation are usually the elements $S_{cc'}(E)$ of the scattering matrix rather than the resonance parameters. It is only for narrow and isolated resonances that eq. (3.7) is accurate and that these parameters are uniquely defined. Then, they can be determined by plotting the imaginary versus the real part of the calculated scattering matrix element (Argand diagram). One example is shown in fig. 6.3. There the angles θ and ϕ correspond to the phase of the background and resonance terms in eq. (3.7) when the latter is written in the form

$$S_{cc}(E) = d\,e^{i\phi} + \frac{\gamma^2_{\lambda c}\,e^{i\theta}}{E - E_\lambda + \frac{1}{2}\,i\,\Gamma_\lambda} . \qquad (6.7)$$

The property that the dots fall on a circle implies that all the quantities which appear in eqs. (3.7) or (6.7) are independent of energy; the radius of the circle is equal to $\gamma^2_{\lambda c}/\Gamma_\lambda$.

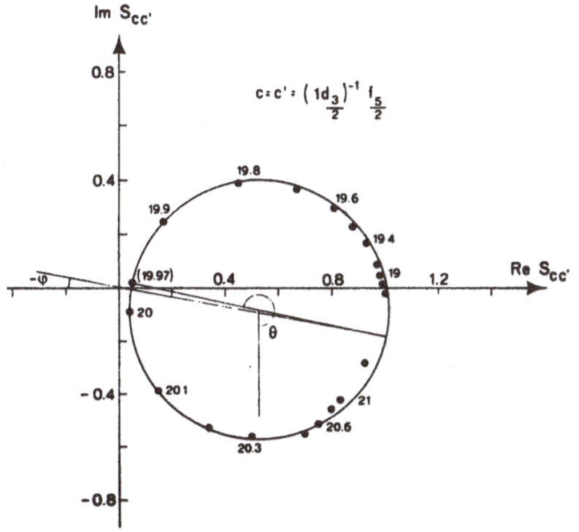

Fig. 6.3. Taken from ref. [24]. The dots represent the imaginary versus the real part of $S_{cc}(E)$ as calculated for the reaction $^{39}Ca(n,n)$ in the framework of the shell-model approach (coupled-channel method) for the f5/2 neutron channel and a 1^- resonance state in ^{40}Ca . The numbers attached to the dots denote the excitation energy in ^{40}Ca . The excitation energy of the resonance is equal to 19.97 MeV .

7. STATISTICAL MODELS

7.1. Introduction

With increasing excitation energy of the compound nucleus the spacing between the resonances decreases exponentially. At the same time the resonances become broader. They very soon can no longer be resolved experimentally. Then, one actually measures the energy average $\langle\sigma_{cc'}(E)\rangle$ of the cross section $\sigma_{cc'}(E)$ within an energy interval centered on E .

Equation (3.5) yields

$$<\sigma_{cc'}(E)> = \sigma^{DI}_{cc'}(E) + \sigma^{CN}_{cc'}(E) \quad , \tag{7.1}$$

where

$$\sigma^{DI}_{cc'}(E) = |<S_{cc'}(E)> - \delta_{cc'}|^2 \quad , \tag{7.2}$$

$$\sigma^{CN}_{cc'}(E) = <|S_{cc'}(E) - <S_{cc'}(E)>|^2> \quad . \tag{7.3}$$

The quantities (7.2) and (7.3) are respectively called the *direct interaction* and the *compound nucleus* contributions to the average cross section. Much effort has been devoted to the construction of approximation schemes for evaluating these two contributions, and to the study of the conditions of validity of these approximation schemes.

Since the energy averaging interval encompasses many resonances it is necessary to use many-level generalizations of the Breit-Wigner formula (3.7). The forms usually adopted are either the pole expansion

$$S_{cc'}(E) = S^{(o)}_{cc'} - i \sum_{\lambda} \frac{g_{\lambda c}\, g_{\lambda c'}}{E - e_{\lambda} + \frac{1}{2} i \Gamma_{\lambda}} \tag{7.4}$$

or the K-matrix expansion

$$S_{cc'}(E) = \left(\frac{1 + iK}{1 - iK}\right)_{cc'} \quad , \tag{7.5a}$$

where

$$K_{cc'} = K^{(o)}_{cc'} + \sum_{\lambda} \frac{\gamma_{\lambda c}\, \gamma_{\lambda c'}}{E_{\lambda} - E} \quad . \tag{7.5b}$$

The advantage of the K-matrix parametrization (7.5a,b) is that it automatically yields a unitary approximation for the S-matrix and that it only involves real quantities. Its disadvantage is that the matrix $(1 - iK)$ has to be inverted.

7.2. The optical model

The aim of the optical model is to reproduce the energy average $<S_{cc}>$ of the diagonal element of the scattering matrix. It is usually assumed that [25]

$$\langle S_{cc}(E)\rangle = S_{cc}(E + iI) \quad , \tag{7.6}$$

where I is the energy averaging interval. It has recently been shown [26] that care must be taken when using eq. (7.6) in conjunction with a parametrization of resonance data as done in refs. [27,28].

By definition, the optical-model potential is a single-particle potential whose scattering function is equal to $\langle S_{cc}\rangle$. Since $|\langle S_{cc}\rangle|^2$ is smaller than unity, the optical-model potential is not Hermitean. It is usually assumed that it can be written as a local complex operator.

Much effort has been devoted to the calculation of the optical-model potential. Most calculations of the imaginary part of the potential do not take into account the existence of resonances [29], see however ref. [30]. Hence, the microscopic understanding of the optical-model potential at low energy is yet not fully satisfactory. A related problem is the microscopic calculation of the *strength function*, i.e. of the low-energy limit of the quantity $\langle\gamma^2_{\lambda c}\rangle_\lambda/d$, where $\langle\ \rangle_\lambda$ refers to an average over the resonance index λ while d is the average spacing between the resonances.

The concept of an optical-model potential is also useful at negative energies, i.e. for the description of the properties of the single-particle states which are bound in the shell-model approximation [31-33]. These states can be observed in nucleon knock-out or pick-up reactions, e.g. $(e,e'p)$ or $(^3He,\alpha)$.

7.3. *Transmission coefficients*

When the off-diagonal elements of the direct interaction contribution (7.2) vanish, the compound nucleus part of the average cross section can be written in the following "Hauser-Feshbach" form [34]

$$\sigma^{CN}_{cc'} = T_c\, T_{c'}\, [\sum_{c''} T_{c''}]^{-1} \quad . \tag{7.7}$$

The quantity T_c is called the *transmission coefficient* in channel c . Recent progress in the evaluation of T_c has taken place since the review published in ref. [35], see e.g. [36,37]. The concepts of ergodicity [38] and of stochasticity [39] appear to be quite useful.

7.4. *Role of direct processes*

Engelbrecht and Weidenmüller [40] have provided the clue for extending the Hauser-Feshbach formula (7.7) to the case when direct reactions exist, i.e. when expression (7.2) does not vanish for $c \neq c'$.

From the general theory of direct reactions one expects that the partial width amplitudes $\gamma_{\lambda c}$ and $\gamma_{\lambda c'}$ are correlated, i.e. that $\langle \gamma_{\lambda c} \gamma_{\lambda c'} \rangle_\lambda$ does not vanish for $c \neq c'$ [41,42]. Here, $\langle \; \rangle_\lambda$ denotes an average over the resonance index λ. High resolution proton scattering data has recently enabled one to establish experimentally the existence of such correlations in the case of targets with mass number $A \approx 48$ [43].

7.5. *Ericson fluctuations*

In the region of overlapping resonances, the compound nucleus part (7.3) of the average cross section fluctuates with energy. Theoretical work is actively pursued to account for the main properties of these "Ericson fluctuations".

A related problem consists in the existence of *preequilibrium processes*. These involve interaction times which are intermediate between those which characterize direct reactions on the one hand and compound nucleus formation on the other hand, see [44,45] and references contained therein.

7.6. *Distribution of the partial width amplitudes*

One basic assumption of the statistical theory of nuclear reactions is that the quantities $\gamma_{\lambda c}$ (see eq. (7.5b)) have a Gaussian distribution with a mean value equal to zero. Recent experimental data have suggested that this assumption may be invalid [46]. It has since been argued, however, that the data are compatible with the Gaussian assumption when the experimental statistical errors are taken into account [47].

7.7. *Intermediate structure*

It happens that the energy average cross section displays a single resonance peak and that high resolution experiments show that this gross structure resonance is actually the result of a coherent contribution of many "fine structure" resonances of the actual (nonaverage) cross section. An example is shown in fig. 7.1. This phenomenon is called *intermediate structure* [48]. A review can be found in ref. [49].

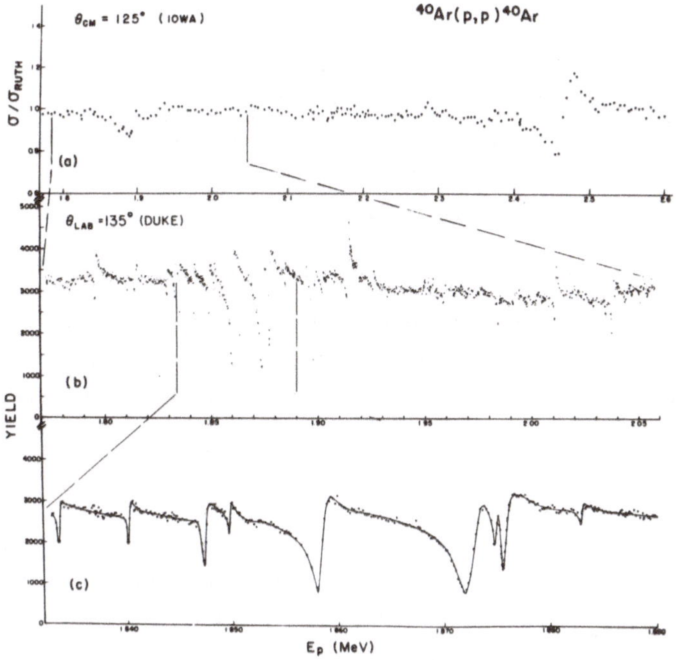

Fig. 7.1. Taken from ref. [5]. Differential cross section of $^{40}Ar(p,p)$ at $\theta = 125°$ (top) and $135°$ (middle and bottom), with successively expanded energy scales. The energy average data (top) shows a resonance at $E = 1.9$ MeV. The bottom drawing shows that this gross structure is actually composed of many fine structure peaks.

8. CONCLUSIONS

Millions of resonances have been detected and analyzed in the field of nuclear physics. The narrowness of the resonances, the smallness of their spacing and the strong character of the nuclear interaction imply that these resonances involve many degrees of freedom. This gives rise to a wide variety of phenomena. Correspondingly, the theory of nuclear reactions presents many facets. It makes use of some concepts and methods first developed in the field of classical physics or of atomic physics. Conversely, it has stimulated the development of techniques which have been applied to other fields of physics and chemistry. This provides a good example of interdisciplinary resonating stimulation.

REFERENCES

1. H. Feshbach, Ann.Phys. (N.Y.) 5 (1958) 357
2. H. Feshbach, Ann.Phys. (N.Y.) 19 (1962) 287
3. C. Mahaux and H.A. Weidenmüller, "Shell-Model Approach to Nuclear Reactions" (North-Holland Publ. Comp., Amsterdam, 1969)
4. Brookhaven National Laboratory Report BNL-325
5. E.G. Bilpuch, A.M. Lane, G.E. Mitchell and J.D. Moses, Phys.Reports 28C (1976) 145
6. E.G. Bilpuch, in "Isobaric Spin in Nuclear Physics", J.D. Fox and D. Robson, eds. (Academic Press, N.Y., 1966), p. 235
7. S.G. Kaufmann, E. Goldberg, L.J. Koester and F.P. Mooring, Phys.Rev. 88 (1952) 673
8. N. Bohr, Nature 137 (1937) 344
9. G. Breit, Revs.Mod.Phys. 36 (1964) 1071
10. G. Breit, Phys.Rev. 58 (1940) 506
11. J. Humblet and L. Rosenfeld , Nucl.Phys. 26 (1961) 529
12. J. Humblet, in "Fundamentals in Nuclear Theory", A. de-Shalit and C. Villi, eds. (IAEA, Vienna, 1967), p. 369
13. J. Humblet, P. Dyer and B. Zimmermann, Nucl.Phys. A271 (1976) 210; J. Humblet, Ann.Phys. (N.Y.) in press
14. G. Gamow, Z.Phys. 51 (1928) 204
15. E. Hernández and A. Mondragón, Phys.Rev. C29 (1984) 722
16. A. Lejeune and C. Mahaux, Nucl.Phys. A145 (1970) 613
17. J. Cugnon and C. Mahaux, Phys.Rev. C10 (1974) 4
18. C.K. Bockerman, D.W. Miller, R.K. Adair and H.H. Barschall, Phys. Rev. 84 (1951) 69
19. J.L. Fowler and H.O. Cohn, Phys.Rev. 109 (1958) 89
20. E.P. Wigner and L. Eisenbud, Phys.Rev. 72 (1947) 29
21. P.L. Kapur and R.E. Peierls, Proc.Roy.Soc.London, ser. A 166 (1938) 277
22. R.F. Barrett, B.A. Robson and W. Tobocman, Revs.Mod.Phys. 55 (1983) 155
23. R.J. Philpott and J. George, Nucl.Phys. A233 (1974) 164
24. C. Mahaux and A.M. Saruis, Nucl.Phys. A177 (1971) 103
25. G.E. Brown, Revs.Mod.Phys. 31 (1959) 893
26. C.H. Johnson, N.M. Larson, C. Mahaux and R.R. Winters, Phys.Rev. C27 (1983) 1913; C29 (1984) 1563
27. W.M. MacDonald, Nucl.Phys. A395 (1983) 221
28. W.M. MacDonald and M.C. Birse, Nucl.Phys. A403 (1983) 99
29. H. Dermawan, F. Osterfeld and V.A. Madsen, Phys.Rev. C29 (1984) 1075 and references contained therein

30. G. Bertsch, P.F. Bortignon and R.A. Broglia, Revs.Mod.Phys. 55 (1983) 287
31. C. Mahaux, in "Common Problems in Low- and Medium-Energy Nuclear Physics", B. Castel, B. Goulard and F.C. Khanna, eds. (Plenum Press, 1979), p. 265
32. C. Mahaux, Notas de Fisica 5 (1982) 147
33. C. Mahaux, in "Symmetries in Nuclear Structure", K. Abrahams, K. Allaart and A.E.L. Dieperink, eds. (Plenum Press, 1983)
34. W. Hauser and H. Feshbach, Phys. Rev. 87 (1952) 366
35. C. Mahaux and H.A. Weidenmüller, Ann.Rev.Nucl.Sci. 29 (1979) 1
36. H.M. Hofmann, T. Mertelmeier and H.A. Weidenmüller, Phys.Rev. C24 (1981) 1884
37. G. Lopez, P.A. Mello and T.H. Seligman, Z.Phys. A302 (1981) 351
38. W.A. Friedman and P.A. Mello, University of Wisconsin preprint (1984)
39. J.J.M. Verbaarschot, H.A. Weidenmüller and M.R. Zirnbauer, Phys. Rev.Lett. 52 (1984) 1597
40. C.A. Engelbrecht and H.A. Weidenmüller, Phys.Rev. C8 (1973) 859
41. C. Bloch, Nucl.Phys. 4 (1957) 503
42. J. Hüfner, C. Mahaux and H.A. Weidenmüller, Nucl.Phys. A105 (1967) 489
43. A.M. Lane, G.E. Mitchell, E.G. Bilpuch and J.F. Shriner Jr., Phys.Rev.Lett. 50 (1983) 321
44. H.A. Weidenmüller, in "Nuclear Data for Science and Technology", K.M. Böckhoff, ed. (Reidel Publ. Comp., Dordrecht, 1983), p. 495
45. K.W. McVoy and X.T. Tang, Phys.Reports 94C (1983) 139
46. J.F. Shriner Jr., E.G. Bilpuch, C.R. Westerfeldt and G.E. Mitchell, Z.Phys. A305 (1982) 307
47. H.M. Hofmann, T. Mertelmeier and H.A. Weidenmüller, Z.Phys. A311 (1983) 289
48. H. Feshbach, A.K. Kerman and R.H. Lemmer, Ann.Phys. (N.Y.) 41 (1967) 230
49. C. Mahaux, Ann.Rev.Nucl.Sci. 23 (1973) 193.

RESONANCES IN HEAVY-ION REACTIONS - STRUCTURAL vs DIFFRACTIONAL MODELS

N. Cindro and D. Počanić
Laboratory for Nuclear Spectroscopy, Rudjer Bošković Institute,
41001 Zagreb, Croatia, Yugoslavia

1. Introduction

The discovery of resonantlike processes in heavy-ion reactions (HIR) is about a quarter of a century old. It started with the discovery by Almqvist, Bromley and Kuehner[(1)] of correlated peaks in $^{12}C+^{12}C$ excitation functions (Fig. 1). Similar phenomena were later dis-

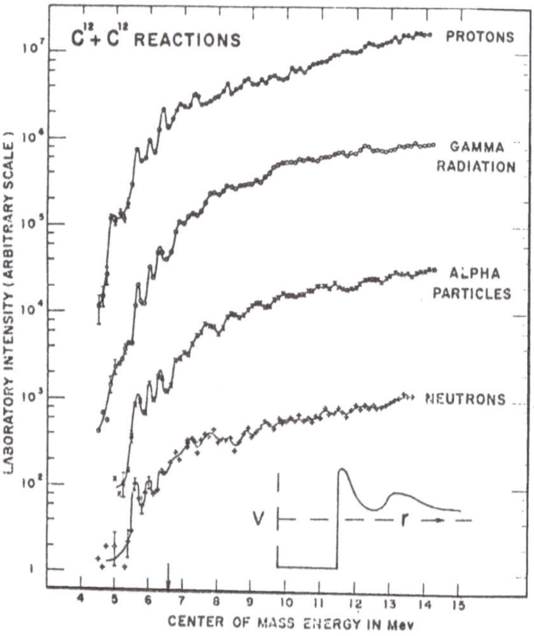

Fig. 1. Energy dependence of angle-integrated yields of neutrons, protons, alphas and gammas from the $^{12}C+^{12}C$ reaction. The astonishing feature is the presence of correlated peaks about 5.6, 6 and 6.6 MeV observed in all four excitation functions. This correlation was interpreted as due to a new class of eigenstates of ^{24}Mg at high excitation energies (Ref. 1).

covered in a wide range of nuclear systems ranging from the s-d shell (carbon+carbon) to the f-p shell (silicon+silicon).

The discovery was soon interpreted in terms of a new class of nuclear eigenstates. A characteristic feature of these states was their large partial widths for decaying into fragments of comparable mass, which led to picturing them as strongly deformed, fissionlike configurations. In this way the concept of nuclear molecular configurations made its entrance into nuclear physics, a concept that is presently used almost universally, but whose name implies much more than our knowledge of these states allows.

In this paper we present the subject of resonances in HIR, starting with a reminder of their main characteristics (section 2) and continuing with a description of the two classes of resonance models, the structural (section 3) and diffractional (section 4) models. However, before doing that, let us briefly discuss one important point: Why is the phenomenon of resonances in HIR so surprising and why does it contain so many seemingly paradoxical aspects? After all, resonances in atomic nuclei are a well-established phenomenon: low-energy neutron resonances, isobaric analog resonances, etc. have been present in nuclear physics for a long time and their nature understood in terms of long-lived, trapped configurations in nuclei, typical of a quantum-mechanical resonance. In fact, it is just this latter feature that appears to be missing from the description of resonances in HIR, as can be easily seen from the following argument.

A resonance in quantum mechanics is an unbound, but long-lived state, whose longevity stems from the fact that its decay is hindered by some physical reason. The lifetime of a resonance is characterized by a width Γ; this quantity is given by the so-called golden rule of quantum mechanics

$$\Gamma(E,J) = 2\pi |\langle\infty|V|0\rangle|^2 \cdot \rho(E,J) \quad . \tag{1}$$

The above expression states that the decay of a simple state $|0\rangle$ to more complex states $|\infty\rangle$ is governed by (a) the matrix element $|\langle\infty|V|0\rangle|^2$ describing the coupling of these states through the interaction V and (b) the density of available (final) complex states $\rho(E,J)$. Let us for the moment forget the matrix element, and concentrate on the level densities. Typical excitation energies where resonances in HIR are observed are of the order of 20-70 MeV, with corresponding level densities of 10^2-10^6 MeV^{-1} (see Table 1). It is thus obvious that we need extraordinary mechanisms to keep a resonant state from resolving into the complex noise of underlying nuclear levels. For isobaric analog resonances,

System	Excitation energy range (MeV)	Calculated level density (MeV^{-1})	Observed width (MeV)	Partial ("fission") width (MeV)
$^{12}C+^{12}C$	20-30	200	0.1-0.4	0.01-0.02
$^{12}C+^{16}O$	28-38	1000	0.5-1	
$^{12}C+^{28}Si$	34-47	70.000	0.5-1.5	
$^{28}Si+^{28}Si$	60-75	4.5×10^{6}	∿2	

Table 1

for instance, this hindering mechanism is the isospin conservation; states with T=1 are just absent below a certain energy. What is this mechanism for resonances in HIR? How can it be described? This is the main problem and the common difficulty of all structural resonance models.

2. Phenomenology of resonances in heavy-ion reactions

How are resonances in HIR observed? What are their main features? The conventional way to observe a resonance is through (a) an excitation function that shows correlated peaks at a given bombarding energy for several exit channels and (b) an angular distribution of emitted particles that, at this energy, shows a characteristic shape. This shape serves to identify the angular momentum associated with the resonance in question.

A good example of peaks correlated in several exit channels is given in Fig. 1 for $^{12}C+^{12}C$ decaying into three channels (p, n and α); note that the integration over angles and the integration over all states of residual nuclei (^{23}Na, ^{23}Mg and ^{20}Ne, respectively) was performed by the measuring method itself: the data were obtained by measuring characteristic gamma rays from the decay of the above nuclei. Fig. 2 shows measured total reaction cross sections of $^{12}C+^{12}C$ together with a simple calculation of the average cross section(2). As more and more data on total reaction cross sections exhibiting structure similar to that seen in Fig. 2 become available, attempts to interpret such data in terms of Ericson fluctuations gradually faded away.

A typical angular distribution of particles emitted at the resonant energy of 8.85 MeV(c.m.) from the $^{12}C(^{12}C,\alpha_0)^{20}Ne$ reaction is shown in Fig. 3. If all interacting particles have spins zero, as is the case for the above reaction, the angular distribution has a single squared Legendre polynomial $[P_L(\cos\theta)]^2$ shape. The order of the polynomial (L=6) gives the spin of the resonance. Essentially all the known HI resonance spins have been extracted in this way. The determination of spins for interacting systems of non-zero intrinsic spins is less

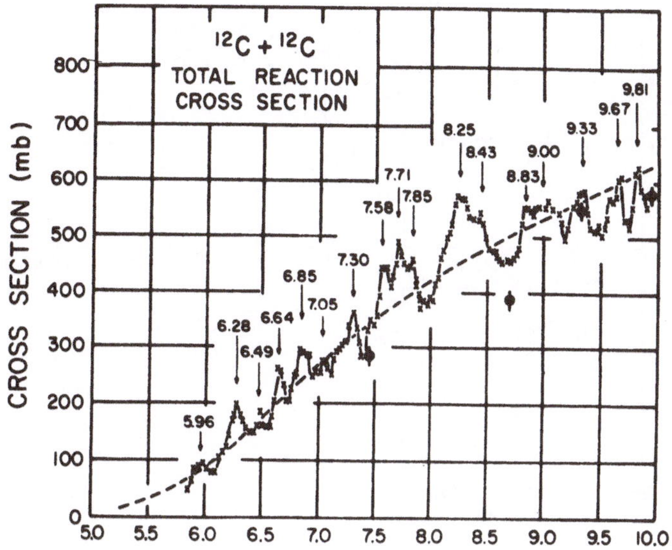

Fig. 2. $^{12}C+^{12}C$ total reaction cross sections measured from energies below the Coulomb barrier to about twice that energy, showing the observed resonances in that system. The dashed line is a simple calculation of the average value of the cross section (Ref. 2).

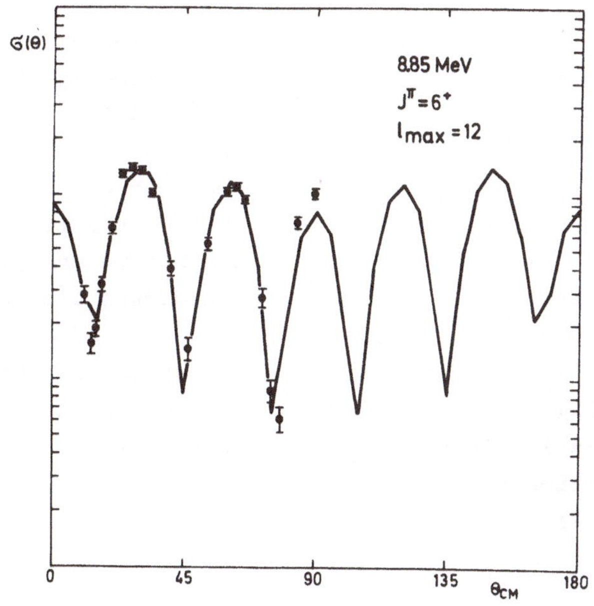

Fig. 3. Angular distribution and Legendre-polynomial fit of the $^{12}C(^{12}C,\alpha_0)^{20}Ne$ reaction at E_{CM} = 8.85 MeV (Ref. 3).

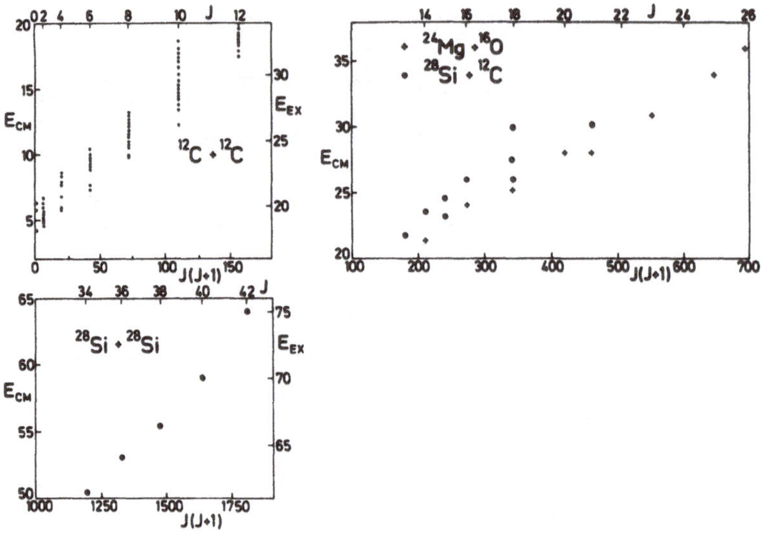

Fig. 4. Compendium of known resonances for $^{12}C+^{12}C$, $^{12}C+^{28}Si$ and $^{16}O+^{24}Mg$, and $^{28}Si+^{28}Si$. Data are shown in energy vs J(J+1) plots (J is the resonance spin). The scale on the vertical left is the center-of-mass energy E_{CM} of the colliding system; that on the right is the excitation energy of the corresponding compound system, $E_x=E_{CM}+E_B$, with E_B the binding energy of the two colliding nuclei in the composite system, taken with a positive sign (omitted for ^{40}Ca).

straightforward and subject to considerable ambiguities.

Fig. 4 summarizes data on resonances observed in three typical systems: $^{12}C+^{12}C$ (leading to ^{24}Mg), $^{12}C+^{28}Si$ and $^{16}O+^{24}Mg$ (leading to ^{40}Ca) and $^{28}Si+^{28}Si$ (leading to ^{56}Ni). These three systems span the mass region where resonances in HIR have been observed. Resonance data are plotted in an energy vs J(J+1) plot. Notice the tendency of resonances of a given spin J to group in rather narrow energy ranges as well as the fact that the centroids of such groups lie on approximately straight lines.

Experiments with nuclei leading to f-p shell composite systems, in particular $^{28}Si+^{28}Si$ [4], indicate that it may well be possible to observe resonances in HIR between very heavy ions. We shall return to this point in section 5.

An extended list of observed resonances in HIR can be found in Ref. 3.

3. Interpretation: the structural models

3.1. Quasimolecules. A structural model is a model that associates resonances in HIR with definite long-lived configurations in the composite system. To paraphrase an illustrious reference, such a model states that

Ab initio structure erat.

It can be safely said that, in spite of their apparent diversity, most structural models are based on the molecular (or quasimolecular, as some authors call it) picture of resonances. According to this picture, put forward by Bromley[1] and Vogt and McManus[5] almost a quarter of a century ago, the two colliding nuclei could, under certain conditions, form a diatomic configuration. The two "atoms" of the "molecule" stick together for some time, giving rise to the resonant state responsible for the structure in excitation functions.

The idea of molecular configurations has two solid sources in the experimental evidence. The first source is the fact already mentioned that partial decay widths of these resonant states for decaying into what one could call fission channels are relatively large. For symmetric systems such as $^{12}C+^{12}C$, the typical fission channel is, of course, the entrance channel. The corresponding observed partial widths Γ_C are of the order of 10-20 keV (Table 1). These widths are surprisingly large, in any case many (20-100) times larger than the average compound elastic widths of $^{12}C+^{12}C$ at ^{24}Mg excitation energies corresponding to resonances.

The second source of the molecular concept in HIR is the observation that resonances in given composite systems follow a rotationlike formula, i.e. that their median energies (for given angular-momentum values) can be accounted for by an expression of the type

$$E = E_o + \alpha \cdot J(J+1) \qquad (2)$$

typical of a rigid rotor. This statement is visualized in Fig. 4. A striking feature is that the "experimental" values of the constants E_o and α in expression (2) (i.e. values obtained by fitting experimental resonance data in an E vs J(J+1) plot) correspond to rotational molecular parameters. In fact, E_o is roughly equal to E_B+E_C, with E_B the binding energy of the projectile + target in the composite system and E_C the projectile + target Coulomb-barrier energy, while α can be expressed as $\hbar^2/2\mathcal{J}_{eff}$, with $\mathcal{J}_{eff}$ an effective moment of inertia deduced from the average slope of resonance data. Now, the values of $\mathcal{J}_{eff}$ obtained in such a way are very close to values of the moment of inertia

Table 2. Comparison of effective and calculated values of the bandhead energy E_o and the moment of inertia $\mathfrak{J}$ for several resonant systems.

Projectile + target system	Calculated $E_o = E_B+E_C$	"Experimental" value of E_o	Calculated moment of inertia (a)	"Experimental" value of $\mathfrak{J}_{eff}$
	MeV	MeV	10^{-42}MeV s^2	10^{-42}MeV s^2
$^{12}C+^{12}C$	20.6	19±1	3.1	2.6±0.2
$^{12}C+^{16}O$	25.4	25±1	4	3.3±0.4
$^{12}C+^{18}O$	32.1	31±1	4.4	4.4
$^{16}O+^{16}O$	27.5	28±1	5	5
$^{12}C+^{24}Mg$	28.4	30±1	5.6	5.7
$^{12}C+^{28}Si$	27.1	29±2	6.5	~7
$^{16}O+^{24}Mg$	31.6	29±2	7.1	~7
$^{16}O+^{28}Si$	28.8	29±1	8.1	8
$^{28}Si+^{28}Si$	39.2	40±2	12.8	12.7

(a) $\mathfrak{J}$ was calculated using

$$\mathfrak{J} = 0.01044\left[\tfrac{2}{5}(A_1^{5/3}+A_2^{5/3})+\frac{A_1A_2}{A_1+A_2}(A_1^{1/3}+A_2^{1/3})^2\right]r_o^2\cdot 10^{-42}\ \text{MeV s}^2,$$

$(r_o = 1.3)$.

of two osculating spheres having projectile and target masses, respectively, as seen in Table 2.

Thus, combining the two facts mentioned above - the large partial width for decaying into the entrance channel and the rotationlike energy-spin dependence - leads to the quasimolecular picture of resonances. In a more formal way, we should say that these facts point to the presence of a relatively simple configuration which has a large diatomic molecular component. This relative simplicity prevents the configuration from decaying into the (complex) neighbouring compound nuclear states; hence the relatively long lifetime of such a resonance.

In spite of such an inviting picture of the global characteristics of resonances in HIR, the molecular models of resonances leave quite a few open questions.

First and foremost, the obvious one: How come that only a limited number of systems display resonances and others do not? In fact, the above description fits all heavy-ion systems; yet, we observe resonances in $^{12}C+^{12}C$ but not in $^{10}B+^{14}N$. To answer this question, various models have proposed different mechanisms for producing conditions necessary for resonance observation. We shall briefly describe some of these models.

3.2. The double-excitation model. The first successful quantitative description of resonances in HIR was given independently by Imanishi[6] and by Greiner and Scheid[7], who introduced the idea of double excitations. The mechanism of double excitation is shown in Fig. 5. We consider two classes of states in the nucleus-nucleus potential: virtual states above the Coulomb-barrier top and quasibound states below the barrier top, but still having the possibility to escape through the barrier. A partial wave (ℓ=12 in the figure) resonates with the virtual state. Should the spreading into compound states with the same (high) angular momentum be small, the resonating partial wave will feel only a small imaginary potential. Hence this wave will be only weakly absorbed and will give rise to broad structure in the excitation function. This resonating partial wave, characterized by a given angular momentum J, can be visualized as representing the system orbiting in an ℓ=J orbit for a given (short) time.

Now let us turn to quasibound states. Because of friction, the system de-excites into a quasibound state, provided energy and angular-momentum matching conditions are met. The mechanism suggested for this

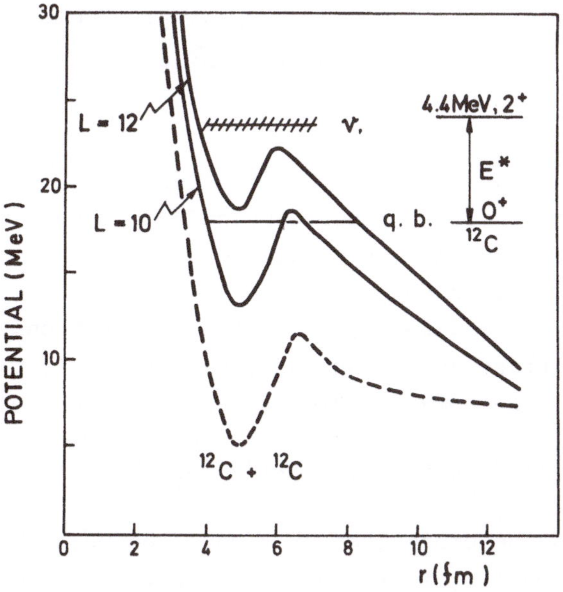

Fig. 5. The double-excitation mechanism: $^{12}C+^{12}C$. An elastic partial wave (ℓ=10) resonates with a virtual state (v). One of the ^{12}C nuclei is excited to its 2^+ state and the system drops to a quasibound state (qb) with matching energy and angular momentum (Ref. 3).

process by the double-excitation model is the excitation of the first collective level in one of the interacting nuclei; in Fig. 5 this is the 4.43 MeV 2^+ state of ^{12}C. In the excitation process, energy is lost and the system is trapped in the relatively long-lived quasibound state, thus giving rise to sharp resonant structure in the excitation function.

It is quite clear that the double-excitation process can occur only at well-defined incident energies and relative angular momenta; hence the periodicity of the phenomenon.

The double-excitation model has been successfully applied to the $^{12}C+^{12}C$ and $^{16}O+^{16}O$ systems. The model, however, leaves several questions open. Why, for instance, $^{14}N+^{14}N$ shows no trace of resonant behaviour, although strong, low-lying collective states exist in this nucleus? What is that produces structure in $^{12}C+^{12}C$ and somehow fails to produce it in $^{14}N+^{14}N$ or $^{13}C+^{13}C$?

3.3. Other molecular models. Models that were directly or indirectly based on the double-excitation model tried to answer these questions and to predict the presence of resonances in certain HI systems and their absence in others.

The orbiting-cluster model[8,9] is based on the resonance-window concept developed by Greiner and Scheid[7] and used by various authors[10]. In this model, the simple configuration that gives rise to molecular states is a diatomic molecular configuration which produces a rotational spectrum:

$$E = E_0 + \frac{\hbar^2}{2\mathcal{J}} J(J+1) \tag{3}$$

with $\mathcal{J}$ the moment of inertia of two osculating spheres (see Table 2). This expression yields the median energy of resonances in a given system. However, in order to account for the fragmentation of resonances, one assumes that in the collision, the two nuclei not only rotate, but also vibrate; the surface vibration is coupled to the rotational mode, giving rise to the well-known rotation-vibration spectrum[11]

$$E_{JKn_2n_0} = [J(J+1)-K^2]\frac{1}{2}\varepsilon + (\tfrac{1}{2}|K|+1+2n_2)E_\gamma + (n_0+\tfrac{1}{2})E_\beta \tag{4}$$

with $\frac{1}{2} = \hbar^2/2\mathcal{J}$ and K the projection of J onto the major body axis. The rotation-vibration coupling is responsible for the fragmentation of the rotational resonance. A comparison with experiment is shown in Fig. 6. In the calculation, the vibration deformation energies E_β and E_γ are free parameters and there is some freedom in the choice of n_2 and n_0. Nevertheless, one obtains fair agreement with experiment using values

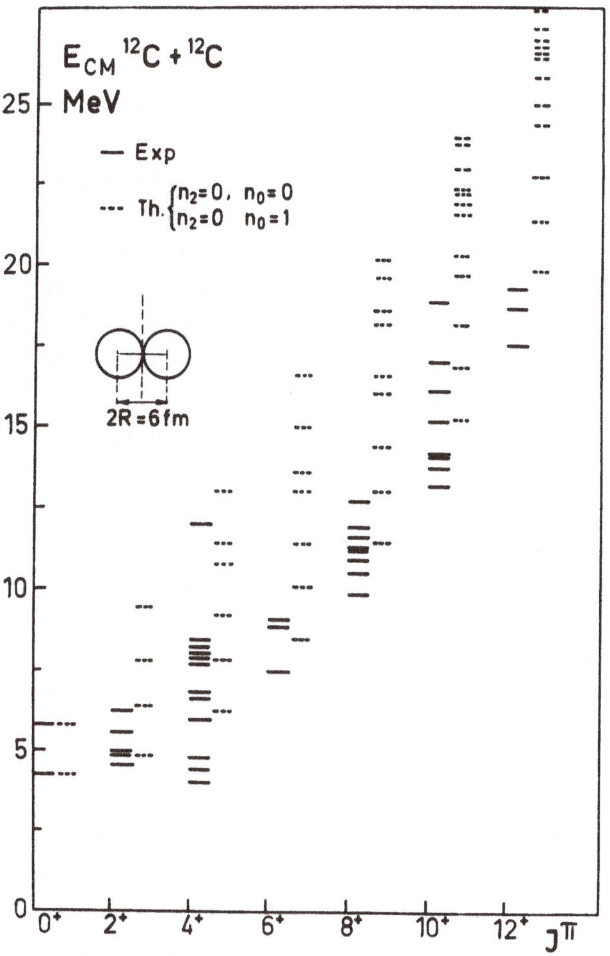

Fig. 6. A comparison of the experimental and calculated spectrum of resonances in $^{12}C+^{12}C$. Full lines represent experimentally observed resonances, dotted lines the result of a calculation using expression (3) with E_{β} = 1.65 MeV and E_{γ} = 3.4 MeV (Ref. 8).

of the above physical quantities well within the limits established in low-energy vibrational spectra.

The orbiting-cluster model thus provides a mechanism for generating resonances in HIR as well as for their fragmentation (i.e. the fact that all resonances of a given spin are packed in a relatively narrow energy range). For the basic question of the presence of resonances in some HI systems and their absence in others, the orbiting-cluster model relies on the resonance-window concept[7]; in order to observe resonances in HIR, their spreading width $\Gamma^{\downarrow}$ into complex (compound nuclear) states should be small. This width is related to the imaginary (absorptive) part W of the optical potential by the expression (see also eq. (1))

$$\Gamma^{\downarrow}(\text{elast} \rightarrow \text{comp}) = 2iW = 2\pi|\langle\text{comp}|V|\text{elast}\rangle|^2 \cdot \rho(E,J) \ . \qquad (5)$$

Hence, the requirement that $\Gamma^{\downarrow}$ be small for grazing angular momenta is equivalent to postulating a surface transparent optical potential for the interacting two-ion system. This is not unreasonable. If one wants the two osculating nuclei to form a quasimolecular configuration which will live for a certain time and not dissolve immediately into more complex configurations, there should be little absorption present.

The predicting power of the orbiting-cluster model relies on expression (5) and certain assumptions. These are that the square of the matrix element in (5) varies smoothly with E and J, while decreasing sharply with the mass of the composite system. It is not so easy to justify these assumptions except a posteriori; however, taking them for granted, the determinant factor for resonance observability is the level density of the composite system at high energy: resonances are likely to be observed where this quantity is small.

We have recently applied the orbiting-cluster model to 92 different ion pairs(12). The results are shown in Table 3 and Fig. 7. The agreement with known data on resonant systems is quite impressive.

If this model is applied to heavy nuclei - and it is not obvious that it can be - then we expect resonances for heavy-ion pairs having such large negative binding energies in composite systems, that they compensate for the Coulomb-barrier energies necessary to make contact. In this way, composite systems are left with excitations near zero and - provided the systems do not immediately fission - resonances may be observed. A system with such properties is, for instance, $^{48}Ca+^{208}Pb$ (binding energy in ^{256}No, E_B = -154 MeV, Coulomb-barrier energy $E_{CB} \gtrsim$ 160 MeV).

Other models have treated the problem of the presence or absence of resonances in HI collisions in a different way. We shall briefly mention two of them. Matsuse, Abe and Kondo(13) have extended the double-excitation model by including many of the degrees of freedom involved when one or both interacting nuclei are excited during the interaction. Again, this excitation reduces the kinetic energy of the system, leaving it trapped in a quasibound state. However, in addition to the elastic rotational band (no intrinsic excitation), one obtains several inelastic bands that are due to the coupling of the channel and orbital angular momenta. These additional bands may have crossing points with the elastic band, in which case strong coupling between

Table 3. Summary of the orbiting-cluster model calculations

Composite system	Entrance channel	E_B+E_C (MeV)	a_{FG} a) (MeV^{-1})	a a) (MeV^{-1})	Δ a) (MeV)	J_ℓ b) ($\hbar$)	$\rho_w(J_\ell)$ c) (MeV^{-1})	ρ_w^{max} c) (MeV^{-1})	$R^{max}_{\Gamma^\downarrow}$ d)	Resonance observation
^{22}Ne	$^{9}Be+^{13}C$	27.1	2.64	3.13	5.13	21.6	300	1920	14	Possible
^{22}Na	$^{10}B+^{12}C$	23.0	2.64	3.13	0	21.4	330	2590	18	Possible
^{23}Na	$^{11}B+^{13}C$	23.9	2.76	3.68	2.67	22.6	550	5790	28	
^{24}Mg	$^{10}B+^{14}N$	35.4	2.88	3.32	5.13	23.5	4300	3.5×10^4	113	Not observed
	$^{12}C+^{12}C$	20.7					39	312	1.0	Observed
^{25}Mg	$^{12}C+^{13}C$	23.0	3.00	3.85	2.46	24.7	530	6250	14	Possible
^{26}Mg	$^{12}C+^{14}C$	25.8	3.12	4.08	4.26	25.8	1100	1.5×10^4	22	
	$^{13}C+^{13}C$	29.1					3100	4.7×10^4	69	
^{28}Mg	$^{14}C+^{14}C$	27.5	3.36	4.00	4.13	27.9	1500	2.2×10^4	14	Possible
^{26}Al	$^{10}B+^{16}O$	26.9	3.12	3.65	0	25.6	4000	3.1×10^4	46	
	$^{12}C+^{14}N$	22.8					720	8080	12	Possible
^{27}Al	$^{12}C+^{15}N$	24.9	3.24	3.45	1.80	26.8	500	5080	5.0	Possible
	$^{13}C+^{14}N$	30.8					2600	3.0×10^4	30	
^{28}Si	$^{12}C+^{16}O$	25.4	3.36	3.05	3.89	27.6	130	1000	0.66	Observed
	$^{14}N+^{14}N$	41.7					8600	8.3×10^4	55	Not observed
^{29}Si	$^{12}C+^{17}O$	29.6	3.48	3.57	2.09	28.8	2600	2.6×10^4	12	Possible
	$^{13}C+^{16}O$	28.8					1700	2.0×10^4	9.0	Possible
	$^{14}N+^{15}N$	33.5					6300	8.1×10^4	36	
^{30}Si	$^{12}C+^{18}O$	32.1	3.60	3.81	3.76	30.0	6000	6.3×10^4	19	Observed
	$^{13}C+^{17}O$	35.2					1.2×10^4	1.6×10^5	48	
	$^{14}C+^{16}O$	31.1					3000	4.5×10^4	14	Observed
	$^{15}N+^{15}N$	33.1					5500	8.4×10^4	25	
^{32}S	$^{12}C+^{20}Ne$	29.3	3.84	3.39	3.29	31.6	1300	8880	1.2	Observed
	$^{16}O+^{16}O$	27.5					420	4870	0.66	Likely
^{34}S	$^{14}C+^{20}Ne$	36.0	4.08	4.12	3.49	34.1	3.3×10^4	4.8×10^5	30	
	$^{16}O+^{18}O$	35.2					1.9×10^4	3.7×10^5	23	
^{36}S	$^{18}O+^{18}O$	39.7	4.32	4.88	3.66	36.4	3.7×10^5	9.2×10^6	260	Not observed
^{34}Cl	$^{10}B+^{24}Mg$	32.9	4.08	3.72	0	33.6	5.6×10^4	1.8×10^5	11	Possible
^{36}Ar	$^{12}C+^{24}Mg$	28.4	4.32	4.03	3.48	35.6	5900	3.1×10^4	0.87	Observed
	$^{16}O+^{20}Ne$	31.7					5600	8.8×10^4	2.5	Likely
^{38}K	$^{10}B+^{28}Si$	31.0	4.56	4.29	0	37.6	2.6×10^5	4.9×10^5	6.3	Possible
	$^{14}N+^{24}Mg$	31.5					4.7×10^4	5.0×10^5	6.4	Possible
^{40}Ca	$^{12}C+^{28}Si$	27.0	4.80	5.00	3.87	39.4	4.2×10^4	1.4×10^5	0.84	Observed
	$^{16}O+^{24}Mg$	31.6					4.3×10^4	7.8×10^5	4.5	Observed
^{42}Ca	$^{14}C+^{28}Si$	33.5	5.04	6.58	3.47	42.0	5.2×10^6	6.8×10^7	180	Not observed
	$^{16}O+^{26}Mg$	32.8					1.8×10^6	4.8×10^7	126	
	$^{18}O+^{24}Mg$	39.0					1.1×10^7	5.7×10^8	1500	

Table 3 (continued)

Composite system	Entrance channel	E_B+E_C (MeV)	a_{FG} a) (MeV^{-1})	a a) (MeV^{-1})	Δ a) (MeV)	J_ℓ b) ($\hbar$)	$\rho_W(J_\ell)$ c) (MeV^{-1})	ρ_W^{max} c) (MeV^{-1})	$R^{max}_{\Gamma^+}$ d)	Resonance observation
^{43}Sc	$^{16}O+^{27}Al$	30.6	5.16	6.55	1.64	42.6	1.6×10^6	3.8×10^7	68	
^{44}Sc	$^{17}O+^{27}Al$	36.1	5.28	6.97	0	43.9	4.2×10^7	1.7×10^9	2000	
^{45}Sc	$^{18}O+^{27}Al$	39.2	5.40	7.59	1.44	45.1	2.2×10^8	1.3×10^{10}	1.1×10^4	
^{44}Ti	$^{12}C+^{32}S$	26.8	5.28	5.46	3.37	43.0	2.1×10^5	4.1×10^5	0.49	Observed
	$^{16}O+^{28}Si$	28.9					6.6×10^4	8.6×10^5	1.0	Observed
	$^{20}Ne+^{24}Mg$	35.2					2.7×10^5	9.4×10^6	11	Possible
^{46}Ti	$^{23}Na+^{23}Na$	43.6	5.52	6.93	3.17	45.7	9.2×10^7	7.8×10^9	4200	
^{47}V	$^{19}F+^{28}Si$	38.3	5.64	6.79	1.44	46.2	3.0×10^7	1.4×10^9	520	
	$^{23}Na+^{24}Mg$	38.6					2.1×10^7	1.6×10^9	570	
^{48}Cr	$^{16}O+^{32}S$	31.7	5.76	6.00	2.79	46.6	9.6×10^6	1.0×10^7	2.6	Likely
	$^{20}Ne+^{28}Si$	35.5					1.2×10^6	4.3×10^7	11	Possible
	$^{24}Mg+^{24}Mg$	36.7					1.3×10^6	6.7×10^7	16	Observed
^{50}Cr	$^{24}Mg+^{26}Mg$	41.6	6.00	6.54	2.89	49.3	2.0×10^7	1.4×10^9	160	
^{51}Mn	$^{23}Na+^{28}Si$	37.1	6.12	6.29	1.54	49.6	4.3×10^6	2.4×10^8	18	Possible
	$^{24}Mg+^{27}Al$	40.3					1.3×10^7	7.7×10^8	58	
^{52}Fe	$^{12}C+^{40}Ca$	31.8	6.24	6.02	3.08	49.9	1.4×10^6	1.1×10^7	0.55	Likely
	$^{24}Mg+^{28}Si$	37.7					1.7×10^6	8.4×10^7	4.2	Likely
^{54}Fe	$^{12}C+^{42}Ca$	35.9	6.48	6.13	2.84	52.7	1.1×10^8	7.6×10^7	1.7	Likely
	$^{14}C+^{40}Ca$	42.4					2.6×10^8	7.4×10^8	17	Possible
	$^{19}F+^{35}Cl$	48.3					2.2×10^8	5.0×10^9	110	
	$^{23}Na+^{31}P$	46.4					5.2×10^7	2.5×10^9	59	
^{55}Co	$^{24}Mg+^{31}P$	41.8	6.60	5.17	1.30	52.9	2.0×10^6	6.3×10^7	0.98	Likely
	$^{27}Al+^{28}Si$	41.7					1.6×10^6	6.1×10^7	0.94	Likely
^{56}Co	$^{14}N+^{42}Ca$	41.2	6.72	5.51	0	54.3	1.1×10^8	2.4×10^8	2.5	Likely
	$^{27}Al+^{29}Si$	43.2					8.1×10^6	3.8×10^8	4.0	Likely
^{57}Co	$^{18}O+^{39}K$	46.9	6.84	5.95	1.27	55.6	2.6×10^8	3.1×10^9	22	
^{56}Ni	$^{16}O+^{40}Ca$	37.9	6.72	4.67	2.50	53.0	9.0×10^5	3.2×10^6	0.035	Likely
	$^{24}Mg+^{32}S$	41.7					4.2×10^5	9.6×10^6	0.10	Likely
	$^{28}Si+^{28}Si$	39.2					1.5×10^5	4.3×10^6	0.045	Observed
^{58}Ni	$^{16}O+^{42}Ca$	40.3	6.96	5.44	2.47	55.9	1.6×10^7	6.0×10^7	0.28	Likely
	$^{18}O+^{40}Ca$	47.8					7.2×10^7	6.4×10^8	3.1	Likely
	$^{26}Mg+^{32}S$	45.4					6.7×10^6	2.7×10^8	1.3	Likely
	$^{27}Al+^{31}P$	46.4					8.6×10^6	3.7×10^8	1.8	Likely
	$^{28}Si+^{30}Si$	42.3					2.2×10^6	1.0×10^8	0.47	Likely
^{59}Cu	$^{27}Al+^{32}S$	42.7	7.08	5.58	1.27	56.0	5.7×10^6	2.4×10^8	0.77	Likely
	$^{28}Si+^{31}P$	40.2					2.4×10^6	1.0×10^8	0.33	Likely
^{60}Zn	$^{24}Mg+^{36}Ar$	40.6	7.20	5.81	2.33	56.0	5.2×10^6	1.5×10^8	0.33	Likely
	$^{28}Si+^{32}S$	38.3					1.5×10^6	6.8×10^7	0.14	Likely
^{61}Zn	$^{28}Si+^{33}S$	40.0	7.32	6.39	1.06	57.4	1.4×10^7	8.5×10^8	1.2	Likely
^{62}Zn	$^{28}Si+^{34}S$	41.0	7.44	6.94	2.35	58.8	3.8×10^7	2.8×10^9	2.7	Likely
^{64}Zn	$^{28}Si+^{36}S$	41.9	7.68	8.03	2.56	61.5	3.9×10^8	4.4×10^{10}	20	Possible

Table 3 (continued)

Composite system	Entrance channel	E_B+E_C (MeV)	a_{FG} a) (MeV^{-1})	a a) (MeV^{-1})	Δ a) (MeV)	J_ℓ b) (ħ)	$\rho_w(J_\ell)$ c) (MeV^{-1})	ρ_w^{max} c) (MeV^{-1})	$R_{\Gamma^+}^{max}$ d)	Resonance observation
^{63}Ga	$^{31}P+^{32}S$	39.6	7.56	7.35	1.29	58.8	6.8×10^7	6.3×10^9	4.1	Likely
^{64}Ge	$^{24}Mg+^{40}Ca$	39.0	7.68	7.81	2.65	58.7	1.9×10^8	8.0×10^9	3.5	Likely
	$^{28}Si+^{36}Ar$	37.6					5.4×10^7	4.3×10^9	1.9	Likely
	$^{32}S+^{32}S$	37.8					4.7×10^7	4.7×10^9	2.1	Likely
^{65}Ge	$^{32}S+^{33}S$	39.0	7.80	8.34	1.36	58.8	3.6×10^8	4.2×10^{10}	12	Possible
^{66}Ge	$^{32}S+^{34}S$	40.7	7.92	8.94	2.77	61.6	9.5×10^8	1.6×10^{11}	33	
^{68}Ge	$^{32}S+^{36}S$	44.7	8.16	9.94	2.86	64.3	2.7×10^{10}	7.2×10^{12}	660	
^{69}As	$^{32}S+^{37}Cl$	42.1	8.28	10.37	1.50	64.3	3.5×10^{10}	9.7×10^{12}	600	
	$^{34}S+^{35}Cl$	40.9					1.8×10^{10}	5.5×10^{12}	340	
^{68}Se	$^{28}Si+^{40}Ca$	35.9	8.16	7.99	2.84	61.1	4.1×10^7	2.5×10^9	0.23	Likely
	$^{32}S+^{36}Ar$	37.0					4.3×10^7	4.1×10^9	0.37	Likely
^{72}Se	$^{35}Cl+^{37}Cl$	45.3	8.64	11.33	2.93	66.9	3.3×10^{11}	1.4×10^{14}	2600	
	$^{32}S+^{40}Ar$	45.0					3.4×10^{11}	1.2×10^{14}	2300	
^{72}Kr	$^{32}S+^{40}Ca$	35.7	8.64	9.43	2.67	63.3	3.2×10^8	4.0×10^{10}	0.75	Likely
	$^{36}Ar+^{36}Ar$	36.6					4.0×10^8	6.1×10^{10}	1.2	Likely

a) a_{FG} = 0.12 A, Fermi-gas level-density parameter; a = $a_{observed}$ when available; otherwise a = a_{GC}; Δ = pairing energy correction; E_{eff} = E-Δ (a_{obs}, a_{GC} and Δ taken from A. Gilbert and A.G.W. Cameron, Can. J. Phys. 43 (1965) 1446.

b) J_ℓ = limiting angular momenta for compound nuclei, after S. Cohen, F. Plasil and W.J. Swiatecki, Ann. Phys. (NY) 82 (1974) 557.

c) CN level densities calculated using the Gilbert-Cameron parametrization along OCM rotational bands (eq. (2)):
ρ_w^{max} = maximum values along rotational bands,
$\rho_w(J_\ell)$ = values corresponding to the upper limit of the band, i.e. $J = J_\ell$.

d) $R_{\Gamma^+}^{max}$ = relative maximum spreading width. $R_{\Gamma^+}^{max} = \frac{(\Gamma^+)^{max}}{(\Gamma_{CC}^+)^{max}}$, for details see Ref. 12.

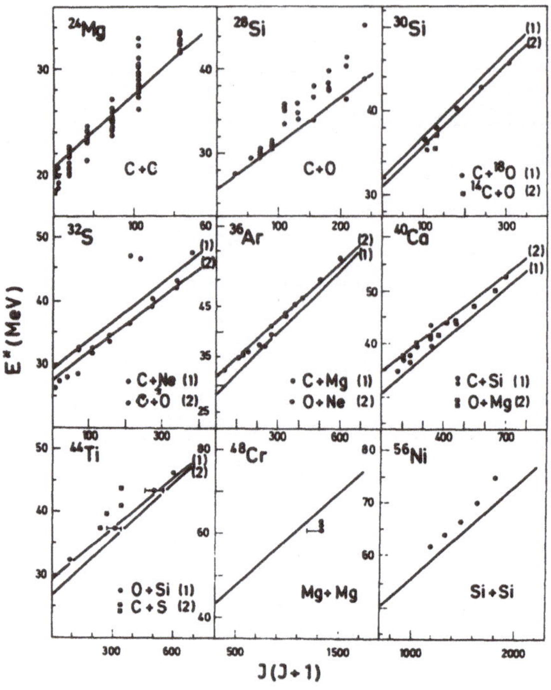

Fig. 7. Comparison of known data on resonances in heavy-ion reactions for nine systems with molecular bands predicted by the OCM. Full circles and squares represent isolated resonances with established spins. Open circles and squares represent resonant-like structure or strong indications for resonances with known dominant partial wave. The mass number is not specified for nuclei with N=Z (Ref. 12).

the elastic and inelastic channels occurs and considerable reaction strength may be routed through the inelastic channels; this, in turn, gives rise to resonant structure in the respective excitation functions (the band-crossing model of resonances).

Haas and Abe(14) have related the number of open channels in the collision available for direct reactions to surface transparency. The existence of a molecular-resonance region in their approach is seen as a consequence of the fact that in several heavy-ion systems the minimum of the number of open channels in the collision (region of surface transparency) coincides with the band-crossing zone. Then, owing to weak absorption, the two nuclei will rotate forming a molecular-type configuration.

Both these models (band-crossing and open-channel) have been successfully applied to lighter systems.

<u>3.4. Dynamic symmetries: algebraic vs geometrical approaches.</u> The analogy of the $^{12}C+^{12}C$ spectra with the spectra of diatomic molecules led Iachello to classify the $^{12}C+^{12}C$ resonances using a group-theoretical method(15). Seen in a broader context, Iachello's idea falls into the category of classifying nuclear and other spectra using spectrum-generating algebras, as shown in Fig. 8. SU(3) (colour) symmetry suf-

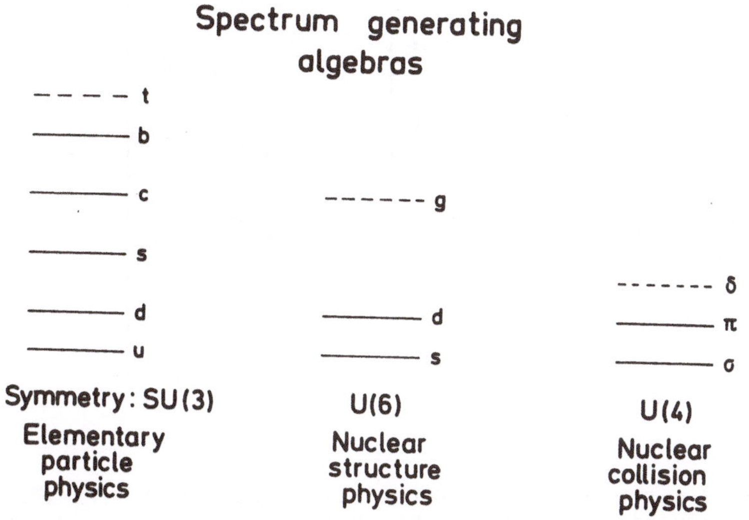

Fig. 8. Spectrum-generating algebras for elementary particle, nuclear structure and nuclear collision physics.

fices in ordering quantum-chromodynamic solutions, while for nuclear structure, the IBM model (symmetry: U(6)) has been successful in classifying nuclear bound states. Following these lines, Iachello assumed that in classifying the unbound states, the U(4) group could be sufficient. Erb and Bromley(16) applied the idea to spectra of $^{12}C+^{12}C$, relying on the fact that any molecular Hamiltonian having U(4) symmetry necessarily generates a vibration-rotation spectrum of the form

$$E(n_o,J) = \sum_{i,j} A_{ij}(n_o + \tfrac{1}{2})^i \, [J(J+1)]^j \quad . \qquad (6)$$

Empirically, this rule is well known in atomic physics under the name of Dunham expansion(17). Using only the first four terms of the expansion, i.e.

$$E(n_o,J) = -D + a(n_o + \tfrac{1}{2}) + b(n_o + \tfrac{1}{2})^2 + cJ(J + 1) \quad , \qquad (7)$$

Erb and Bromley were able to reproduce the energies of 28 known resonances in $^{12}C+^{12}C$ with an average rms deviation of 44 keV! Fig. 9b

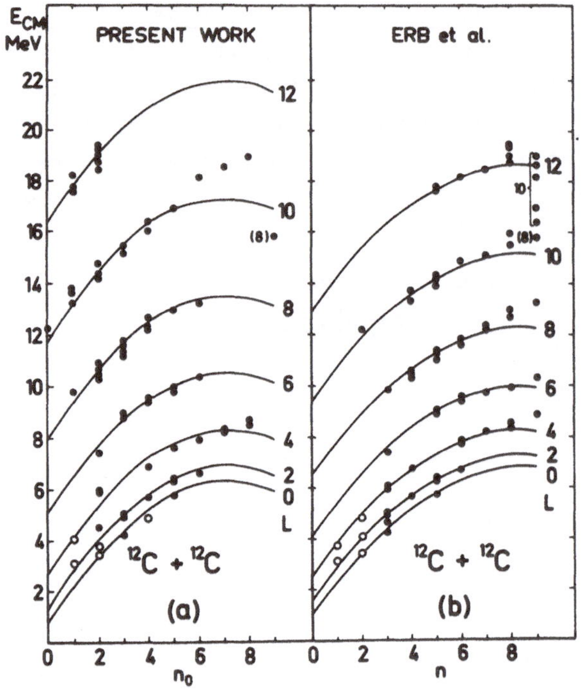

Fig. 9. $^{12}C+^{12}C$ resonances classified using expression (7) (Fig. 9a) and expression (9) (Fig. 9b). Data: full circles: resonances whose spin is known; open circles: resonances without spin assignment (Ref. 18).

shows the result of their classification for an enlarged set of data. The agreement is striking. Of course, three parameters (D, a and b) are free in the fit; also, the vibrational number n_0 is a sliding quantity.

The group-theoretical approach is certainly a new way of classifying resonant states. Whether it contains anything more than a classifying scheme remains to be seen. In this respect it is worth noting that an expression of the same type as expression (7) can be obtained from a standard geometrical approach by simply adding a quartic term to the bonding potential

$$V = V_0 + \frac{1}{2} C_2 \xi^2 + C_4 \xi^4 \ . \qquad (8)$$

The quantization of this quartic term in the potential (8) leads to a term in the energy spectrum proportional to $[(n_0 + \frac{1}{2})^2 + \text{const}]$ (the expectation value of a cubic term is zero). The energy spectrum of such a rotation-(anharmonic) vibration Hamiltonian has the form (assuming, for reasons of symmetry, only states with projections K=O)

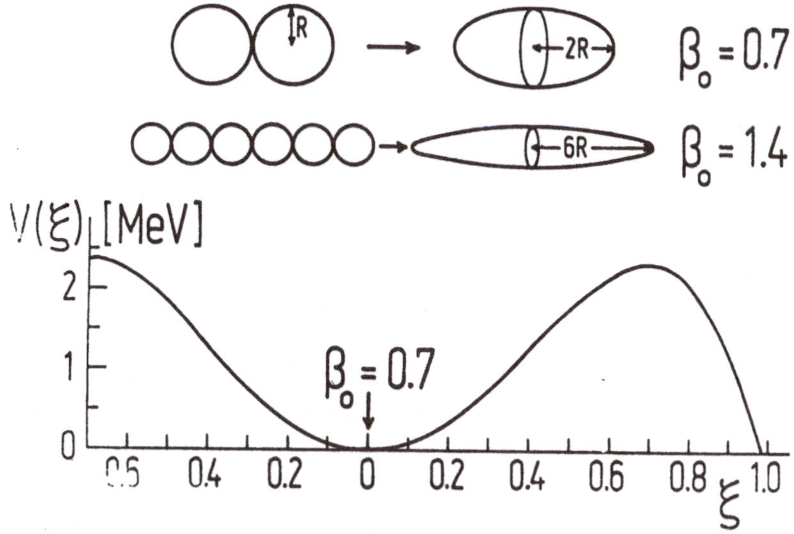

Fig. 10. Anharmonic potential around the equilibrium deformation $\beta_0 \approx 0.7$. The system breaks at $\xi \approx 0.7$ corresponding to an extremely stretched shape with equivalent quadrupole deformation $\beta_0 \approx 1.4$ (Ref. 18).

$$E(J,n_0) = \frac{3}{2} G + E_\beta (n_0 + \frac{1}{2}) + 6G(n_0 + \frac{1}{2})^2 + \frac{1}{2} \varepsilon [J(J+1)], \qquad (9)$$

which is the same as expression (7). Here E_β is, again, the β-vibration zero-point energy, $\frac{1}{2} \varepsilon = \frac{1}{2} \frac{\hbar^2}{2\mathfrak{J}}$ and G is related in a simple way to the coefficients of the potential (8).

Fig. 9a shows the fit of the same $^{12}C + ^{12}C$ data as in Fig. 9b, using $E_\beta = 1.7$ MeV and $\frac{1}{2} \varepsilon = 0.1$ MeV (both values from previous work[8]) and G = -0.019 MeV, the only fitting parameter[18]. The two fits (Figs. 9a and 9b) are about equally good and cannot decide between the two approaches (geometrical and group-theoretical ones).

The potential generating the fit in Fig. 9a is shown in Fig. 10, where ξ is the elongation from the equilibrium shape, the latter taken as a rotational ellipsoid equivalent to a $^{12}C + ^{12}C$ dumb-bell. An intriguing feature of this potential is that it bends at $\xi \approx 0.7$; this corresponds to an extremely stretched shape with an effective quadrupole deformation $\beta_0 \approx 1.4$. It is tempting to associate this shape with a chain of 6 alpha particles: in fact, the equivalent rotational ellipsoid (see figure) has $\beta_0 = 1.4$. Although such agreement is probably fortuitous, it is nevertheless indicative of the fact that the nuclear system undergoes strong deformation before breaking up or emitting alpha or ^{8}Be particles. These strongly deformed nuclear shapes can be viewed as a kind of stretched molecule that can be associated with nuclear molecular isomerism. In this sense, the rotational-vibrational level structure revealed through the classifying schemes in Fig. 9 would be the expression of the isomerism.

A step forward, undertaken recently by Satpathy and Faessler[19], is to find the internuclear potential $V(\xi)$ and then solve the corresponding Schrödinger equation. Using a Morse potential as the starting point, Satpathy and Faessler derived an effective potential for the internuclear interaction; solving the Schrödinger equation with this potential yields an expression for the energy spectrum that, in addition to the usual rotation-vibration terms present in expressions (7) and (9), contains coupling terms proportional to $(n_o + \frac{1}{2})J(J+1)$ and $J^2(J+1)^2$. Fitting the ^{12}C+^{12}C resonance data to this expression, Satpathy and Faessler determined the bonding potential. Closer scrutiny of this potential reveals that (a) it has a minimum around 7 fm, near the touching distance of two ^{12}C spheres and (b) it has a relatively long range, of the order of 20 fm. This long range can be reconciled with our present knowledge of nuclear structure only if one assumes that the configuration of the two ^{12}C nuclei is a linear chain of 6 alpha clusters!

Support for a strongly elongated configuration present in ^{12}C+^{12}C collisions also comes from TDHF calculations of the collective ion-ion potential for carbon nuclei[20]. Fig. 11 shows that the potential calculated for two ^{12}C nuclei interacting with their flat surfaces (left figure) has a much shorter range than the one calculated for ^{12}C nuclei touching with prolate ends. Inverting the argument, one concludes that the long-range potential found in Refs. 18 and 19 speaks clearly in favour of a strongly deformed, elongated configuration present in a

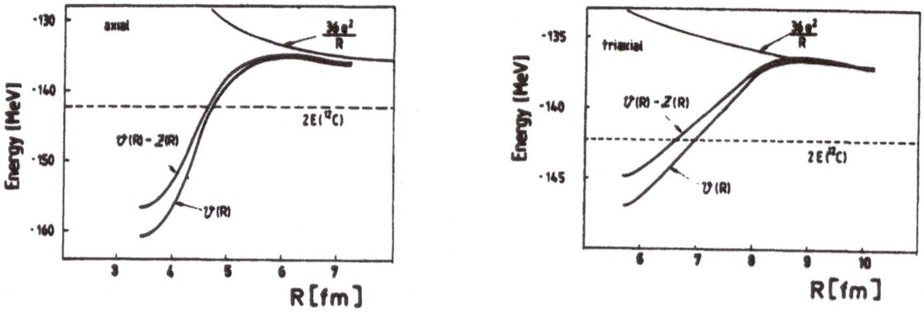

Fig. 11. Potentials for two ^{12}C nuclei touching with their flat surfaces (left) and their prolate ends (right) (Ref. 20).

resonant nuclear collision.

The above picture sheds new light on the orbiting process. Once the composite system is formed, the colliding nuclei undergo strong prolate deformation in the separation process (fission "neck" formation). The very thick Coulomb barrier prevents the decay of the system into a complex (fusion) configuration and creates the necessary condition for resonance formation. During the process the nuclei rotate and vibrate like a diatomic molecule, giving rise to molecular resonance spectra. Finally, they separate (quasi)restoring the initial state (elastic or collective inelastic scattering)[19].

4. Diffraction models

The global success of the molecular models did not prevent physicists from questioning the very nature of the resonant phenomenon in HIR, i.e. looking for non-structural causes of resonant behaviour. Such "heretic" ideas could be found even in books entitled "Nuclear Molecular Phenomena"[21]. It is, indeed, a fact that trivial causes, such as interference between smooth components of the scattering amplitude can, under certain conditions, cause resonantlike behaviour. In a more formal language, if the scattering amplitude $f(\theta,E)$ can be decomposed into several contributions $f_1, f_2, \ldots$, oscillations can appear in the cross section

$$\sigma(\theta,E) = |f(\theta,E)|^2 = |\sum a_i f_i(\theta,E)|^2 ,$$

even if all of the f_i are smooth functions of θ and E. The argument is reasonable and shows the necessity of disentangling effects of a genuine resonance behaviour from such interference effects.

The question, then, is a quantitative one: Can diffraction effects produce the correct energy dependence and scattering amplitude to account for resonantlike behaviour and do it for many exit channels? Some time ago, Phillips et al.[22] raised the point that strong energy variations in a surface-peaked reaction would naturally appear whenever the nuclear surface was transparent and the nuclear interior black. It is not unreasonable to assume that such conditions are present for $^{12}C+^{12}C$ or $^{16}O+^{16}O$ collisions. Using the Austern-Blair formalism[23], they calculated the scattering amplitude for these reactions and showed that the latter displayed prominent structure for well-matched channels. The structure strongly depends on the absorption "skin depth" Δ defined in the nuclear reflectivity

$$\eta_\ell = [1 + \exp(\frac{E-E_G}{\Delta})]^{-1} \tag{10}$$

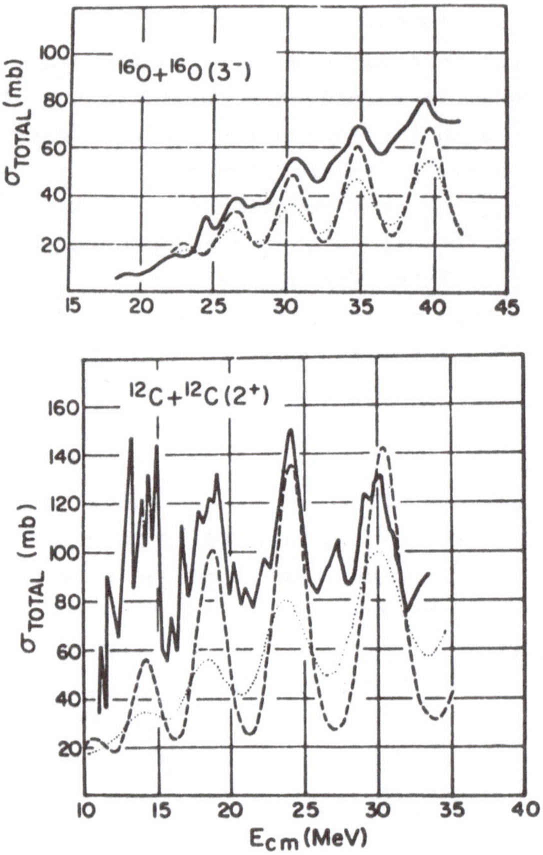

Fig. 12. Comparison of experimental (solid lines) and calculated total inelastic cross sections for $^{12}C+^{12}C$ and $^{16}O+^{16}O$, respectively. Calculations were performed using the Austern-Blair diffraction model with Δ = 1 MeV (dashed lines) and Δ = 1.2 MeV dotted lines) (Ref. 22).

with $E_G = E_o + \frac{\hbar^2}{2\mathcal{J}} \ell(\ell+1)$. A value Δ = 1 MeV yields prominent structure that matched inelastic scattering experimental data for $^{12}C+^{12}C$ and $^{16}O+^{16}O$; however, already a value Δ = 1.2 MeV smears out the oscillations (respectively, dashed and dotted lines in Fig. 12).

The sensitivity of the calculated values to Δ raises several questions as to the significance of the above results. Furthermore, the choice Δ = 1 MeV is too small to be consistent with the proximity potential and hence in apparent contradiction with the bulk of heavy-ion scattering data[24].

The DWBA method used earlier to calculate the energy dependence of

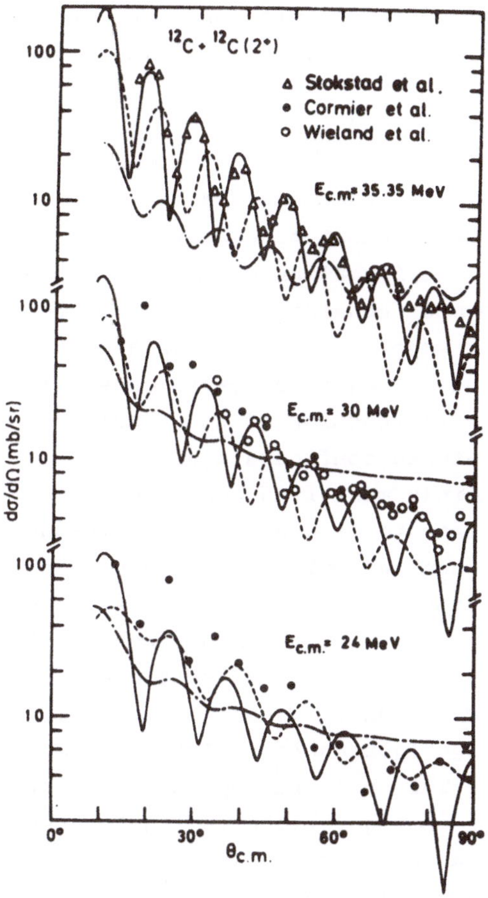

Fig. 13. Angular distributions of the $^{12}C+^{12}C$ (2^+) inelastic scattering at E_{CM} = 24, 30 and 35.35 MeV. The three sets of curves are the DWBA calculations with different potentials, the solid, dashed and dashed-dotted lines are calculations of Refs. 25, 26 and 24, respectively. (From Ref. 25).

the inelastic $^{12}C+^{12}C$ (2^+) scattering, was employed to calculate the angular distributions of this process at energies of 24, 30 and 35.35 MeV, corresponding to maxima of the angle-integrated cross sections. The results are shown in Fig. 13. While the calculation from Ref. 25 fits the data at 35.35 and 30 MeV fairly well, all other fits are rather poor and can hardly be used as a proof of a non-resonant nature of the oscillations. It is thus fair to say that the presently available DWBA analyses of resonantlike behaviour in HIR cannot be used as a decisive argument against interpreting such behaviour in terms of structural resonances. It is also fair to say that the DWBA-type analysis has so far been applied to a limited set of data and that it misses the global agreement shown by, for instance, the orbiting-cluster model.

5. Conclusions

From the discussion in sections 2-4 it turns out that an impressive set of data on the resonant behaviour of HI collisions can be interpreted in terms of the formation of long-lived configurations in the composite system. The most consistent picture of this configuration is based on the molecular model in its various aspects; one should, however, be aware that the molecular concept means only that a considerable fraction of the wave function of the resulting composite system is given by a simple, diatomic configuration.

An open and certainly most fundamental question is whether this imagery is a general feature of heavy-ion collisions or not. Do we see or feel the effect of nuclear molecules in other heavy-ion phenomena? Do we see it in collisions of very heavy nuclei at high energies? Is it there more than a figure of speech? In a recent article Thiel et al.(27) showed that the structure in the cross sections of $^{28}Si+^{28}Si$(4) can be traced back to the excitation of resonance molecular configurations that are formed at large intranuclear distances in the barrier region of the real potential. This picture represents a novel effect in comparison with molecular configurations in lighter systems, where the interacting nuclei show a large overlap of their surfaces. It is reasonable to extrapolate that molecular configurations at distances lying in the barrier range may also exist in systems heavier than $^{28}Si+^{28}Si$. Phenomena that can be explained as due to long-lived states in giant systems such as U+U or U+Cu have in fact recently been observed(28).

As to the diffractional effects in explaining the resonant behaviour of HI collisions, although they cannot be lightly dismissed, it is unlikely that they can provide a general and consistent frame for understanding the phenomenon.

This paper was partly written while one of the authors (N.C.) was at the Institut für Theoretische Physik der Universität Frankfurt/M. The financial support of the Internationales Büro der KFA Jülich and the hospitality of Prof. W. Greiner are gratefully acknowledged.

References

(1) E. Almqvist, D.A. Bromley and J.A. Kuehner, Phys. Rev. Lett. 4 (1960) 515
(2) K.A. Erb, R.R. Betts, S.K. Kortky, M.M. Hindi, P.P. Tung, M.W. Sachs, S.J. Wilett and D.A. Bromley, Phys. Rev. C 22 (1980) 507
(3) N. Cindro, Riv. N. Cimento No 6 (1981) 1
(4) R.R. Betts, B.B. Back and B.G. Glagola, Phys. Rev. Lett. 47 (1981) 23
(5) E. Vogt and H. McManus, Phys. Rev. Lett. 4 (1960) 518
(6) B. Imanishi, Nucl. Phys. A125 (1969) 37
(7) W. Greiner and W. Scheid, J. Phys. (Paris) C6 (1971) 91
W. Scheid, W. Greiner and R. Lemmer, Phys. Rev. Lett. 25 (1971) 1043
(8) N. Cindro and B. Fernandez, in Nuclear Molecular Phenomena, ed. by N. Cindro, North Holland (Amsterdam) 1978, p. 428
(9) N. Cindro and D. Počanić, J. Phys. G6 (1980) 351
(10) D.L. Hanson et al., Phys. Rev. C9 (1974) 1760
(11) J. Eisenberg, W. Greiner, Nuclear Theory, North Holland (Amsterdam) 1975, p. 147
(12) D. Počanić and N. Cindro, to be published
(13) T. Matsuse, Y. Abe and Y. Kondo, Progr. Theor. Phys. 59 (1978) 1904
Y. Kondo, Y. Abe and T. Matsuse, Phys. Rev. C19 (1979) 1356
(14) F. Haas and Y. Abe, Phys. Rev. Lett. 46 (1981) 1667
(15) F. Iachello, Phys. Rev. C 23 (1981) 2778
(16) K.A. Erb and D.A. Bromley, Phys. Rev. C 23 (1981) 2781
(17) J.L. Dunham, Phys. Rev. 41 (1932) 721
(18) N. Cindro and W. Greiner, J. Phys. G9 (1983) L175
(19) L. Satpathy and A. Faessler, Univ. Tübingen preprint 1983, unpublished
(20) K. Goeke, KFA Jülich, Priv. Comm.
(21) H. Doubre and C. Marty, in Nuclear Molecular Phenomena, ed. by N. Cindro, North Holland (Amsterdam) 1978, p. 291
(22) R.L. Phillips, K.A. Erb, D.A. Bromley and J. Weneser, Phys. Rev. Lett. 42 (1974) 566
(23) N. Austern and J.S. Blair, Ann. Phys. 33 (1965) 15
(24) S.Y. Lee, Y.H. Chu and T. Kuo, Phys. Rev. C 24 (1981) 1502
(25) O. Tanimura, R. Wolf and U. Mosel, Phys. Lett. 120 B (1983) 275
(26) L.E. Cannell, R.W. Zurmühle and D.P. Balamuth, Phys. Rev. Lett. 43 (1979) 837
(27) A. Thiel, W. Greiner and W. Scheid, to be published
(28) M. Clemente et al., Contrib. Int. Conf. on Nuclear Physics, Florence (1983), p. 693; H. Bokemayer et al., ibid. p. 694

AN ANALYTICALLY SOLVABLE MULTICHANNEL SCHRÖDINGER MODEL FOR HADRON SPECTROSCOPY

E. van Beveren, C. Dullemond and T.A. Rijken
Institute for Theoretical Physics, University of Nijmegen
NL-6525 ED Nijmegen, The Netherlands
and
G. Rupp
Zentrum für interdisziplinäre Forschung
Universität Bielefeld
D-4800 Bielefeld 1, FR Germany
(presented by G. Rupp)

I. Introduction

Over the past decade many phenomenological quark models have been presented for the description of hadrons. Their essence was the imposition of quark confinement - the notion that no quarks can exist in isolation - either by means of boundary conditions [1] or by rising potentials (see e.g. refs. [2], [3]). In most of these approaches, however, the strong interactions are only treated on the level of valence quarks. Processes as strong decay, involving the creation and annihilation of quark-antiquark pairs, are, if considered at all, usually taken into account perturbatively. The effects on the spectra of hadrons may nevertheless be too large (see e.g. ref. [4]) to rely on such a treatment, as the sizable widths of many resonances already indicate.

In this note a potential model based on the coupled-channel Schrödinger formalism will be reviewed, that, when applied to meson spectroscopy [5] [6], gives a simultaneous description of both the confinement and the decay sector. The model is formulated in such a way that the coupled-channel problem can be solved analytically, yielding an explicit expression for the S-matrix, which is either exact or a good approximation depending on the chosen interactions. In section II a two-channel toy model is studied in order to demonstrate the behaviour of resonances and bound states for suchlike systems. Section III then deals with a multichannel generalization of the model. In section IV the application to meson spectroscopy is shown and some results for two different versions of the model are presented.

II. Two-Channel Schrödinger Model [7]

Consider the two-channel radial Schrödinger equation

$$\left\{\frac{1}{2}\, M^{-1}\left[-\frac{d^2}{dr^2} + \frac{L(L+1)}{r^2}\right] + V(r)\right\} \phi(r) = E\phi(r) \ , \tag{1}$$

with the reduced mass matrix $M = M_1M_2(M_1+M_2)^{-1}$ where

$$M_1 = \begin{pmatrix} m_q & 0 \\ 0 & m_1 \end{pmatrix}, \quad M_2 = \begin{pmatrix} m_{\bar{q}} & 0 \\ 0 & m_2 \end{pmatrix} \quad \text{and} \quad M = \begin{pmatrix} \mu_c & 0 \\ 0 & \mu_f \end{pmatrix},$$

with the orbital angular momentum matrix

$$L = \begin{pmatrix} \ell_c & 0 \\ 0 & \ell_f \end{pmatrix},$$

and with the potential

$$V(r) = M_1 + M_2 + \begin{pmatrix} \frac{1}{2}\mu_c\omega r^2 & 0 \\ 0 & 0 \end{pmatrix} + \frac{g}{2\mu_c r_0}\,\delta(r-r_0)\begin{pmatrix} 0 & 1 \\ 1 & 0 \end{pmatrix}.$$

Equation (1) may describe a system of a harmonically bound quark-antiquark pair, that couples to a free channel containing two mesons. The transition potential is taken to be proportional to a spherical delta function. The S-matrix for this system can be solved in closed form (see sec. III). Bound states and resonances, showing up as poles in S, do then appear for energies E that are given by the implicit equation

$$\frac{i}{2}\, g^2 A(\ell_f, kr_0)\, \mathcal{B}(\ell_c, s, \mu_c\omega r_0^2) = 1 , \tag{2}$$

where

$$A(\ell,z) \equiv z j_\ell(z) h_\ell^{(1)}(z) ,$$

$$\mathcal{B}(\ell,s,x) \equiv x^{\ell+\frac{1}{2}} e^{-x} \frac{\Gamma(-s)}{\Gamma(\ell+\frac{3}{2})} \Phi(-s, \ell+\tfrac{3}{2} ; x)\, \Psi(-s, \ell+\tfrac{3}{2} ; x) ,$$

and

$$E = \frac{k^2}{2\mu_f} + m_1 + m_2 = \omega(2s + \ell_c + \tfrac{3}{2}) + m_q + m_{\bar{q}} .$$

Here j_ℓ is a spherical Bessel function, $h_\ell^{(1)}$ a spherical Hankel function of the first kind, and Φ, Ψ are regular respectively irregular confluent hypergeometric functions [8]. For small coupling g one can, using second-order perturbation theory, write down an approximate *explicit* expression for the positions of the S-matrix poles, reading

$$E \approx E_0 + \Delta E^{(2)} \tag{3}$$

with $\Delta E^{(2)} = -i\omega g^2 \, A(\ell_f, k_o r_o) \, \bar{B}(\ell_c, s_o, \mu_c \omega r_o^2)$,

where $E_o = \dfrac{k_o^2}{2\mu_f} + m_1 + m_2 = \omega(2s_o + \ell_c + \tfrac{3}{2}) + m_q + m_{\bar{q}}$, $s_o = 0,1,2,\ldots,$

and $$\bar{B}(\ell,s,x) \equiv x^{\ell+\frac{1}{2}} e^{-x} \frac{\Gamma(\ell+s+\frac{3}{2})}{\Gamma(s+1)} \left[\frac{\Phi(-s,\ell+\frac{3}{2};x)}{\Gamma(\ell+\frac{3}{2})} \right]^2 .$$

This offers a starting point for the application of an iterative procedure like Newton's method to fix a pole precisely, thereby explicitly making use of the analyticity of the solutions. For increasing g the pole can then be traced carefully by in turn extrapolating in g^2 and employing Newton's method. When the energy is above threshold, i.e. $E > m_1 + m_2$, the pole moves as a function of g into the lower half of the complex energy plane, representing a resonance in the coupled-channel system. This should mimic the situation where the $q\bar{q}$-system (meson) can decay into two other mesons. For $E < m_1 + m_2$ the pole moves along the real E-axis, retaining its quality of bound state. In this way the original equidistant harmonic-oscillator (H.O.) spectrum gets deformed for growing g , since different radial and angular excitations are subject to different shifts. In Fig. 1 these shifts are shown for the four lowest radial S-states, whereat the threshold is chosen such that the latter are all bound states. The most noteworthy feature we can see from this figure is that the ground state shifts most of all in spite of its largest energy gap to threshold. This property, which is qualitatively shared by the more realistic version of the model to be discussed later, may have far-reaching consequences for meson spectroscopy, as will be shown in section IV. Mainly responsible for this phenomenon is the peaked structure of the transition

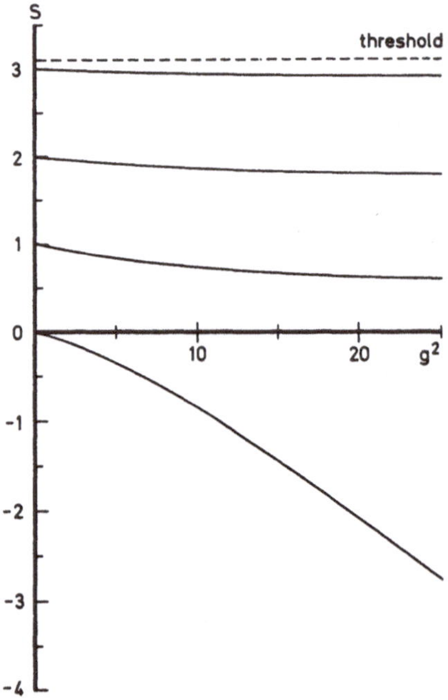

Fig. 1: Radial quantum number s as a function of g^2

potential, which tends for increasing g to deform the H.O. wave function in such a way that a node, if present, will move towards this peak. That puts a strict upper bound to the mass shift in the case of a delta function, or at least curbes the shift when the coupling potential has a more general peaked form. For the ground-state wave function, having no nodes, this effect of course does not occur. Furthermore, there is no special enhancement of the shift for states close to threshold, since

$$\lim_{k\to 0} i\, A(\ell, kr_0) = \frac{1}{2\ell+1} \tag{4}$$

which is just a constant. The above-mentioned deformation of the wave functions largely amounts to the following: they become more compact and they decrease at the origin except for the ground state. Also this is of considerable importance when the model is shaped to describe mesons (see again sec. IV). Summarizing we may already say on the basis of the results of the above toy model that a further investigation from a somewhat more realistic approach is highly desirable.

III. Generalized Multichannel Model [9]

In order to be able to handle systems with several confined and free channels and with modified potentials especially as for the coupling part, we are now going to work out a method for the obtainment of analytic though approximate solutions of a multichannel Schrödinger equation with, in a sense, arbitrary potentials. To that end we write down formally the matrix equation

$$\left[-\frac{d^2}{dr^2} + U_0(r) + U_1(r)\right] \phi(r) = 0 \,, \tag{5}$$

and let us suppose that we can solve analytically

$$\left[-\frac{d^2}{dr^2} + U_0(r)\right] \begin{cases} \phi_0(r) & = 0 \\ G(r,r') & = -\delta(r-r') \end{cases} \,. \tag{6}$$

Then (5) can be written as

$$\phi(r) = \phi_0(r) + \int_0^\infty dr'\, G(r,r')\, U_1(r')\phi(r') \,. \tag{7}$$

If we assume that the integrand falls off rapidly enough for large r, then we may approximate the infinite intregration by a finite sum (to be checked numerically afterwards), which of course equals the integral over a sum of delta functions:

$$\phi(r) \approx \phi_0(r) + \sum_{i=1}^{N} \Delta(r_i)\; G(r,r_i)\; U_1(r_i)\; \phi(r_i)$$

$$= \phi_o(r) + \int_0^\infty dr'\; G(r,r')\; \phi(r') \sum_{i=1}^{N} \Delta(r_i)\; U_1(r_i)\; \delta(r'-r_i). \qquad (8)$$

In its turn (8) is equivalent to the differential equation

$$\left[- \frac{d^2}{dr^2} + U_0(r) + \tilde{U}_1(r)\right] \phi(r) = 0 \;, \qquad (9)$$

where the approximated potential $\tilde{U}_1$ is given by

$$\tilde{U}_1(r) = \sum_{i=1}^{N} \Delta(r_i)\; U_1(r_i)\delta(r-r_i) \;. \qquad (10)$$

With this procedure we can tackle now a system of n confined channels coupled to m free channels, described by the equation

$$\left\{\frac{1}{2} M^{-1} \left[- \frac{d^2}{dr^2} + \frac{L(L+1)}{r^2}\right] + V(r)\right\} \chi(r) = E\chi(r) \;, \qquad (11)$$

where the potential $V(r) = D + V_0(r) + V_1(r)$ consists of a diagonal threshold part D, an exactly solvable diagonal potential matrix V_0, and V_1 containing all the rest of the interactions. For (11) we write as an approximation

$$\left\{- \frac{d^2}{dr^2} + \frac{L(L+1)}{r^2} + 2M^{1/2}(V_0 + \tilde{V}_1)M^{1/2} - K^2\right\} \phi(r) = 0 \;, \qquad (12)$$

with

$$\phi(r) = M^{-1/2}\chi(r) \;, \quad K^2 = 2M(E - D) \;,$$

$$\tilde{V}_1(r) = \sum_{i=1}^{N} \Delta(r_i)\; V_1(r_i)\delta(r-r_i) \;.$$

Equation (12) can be solved analytically. A detailed derivation of the S-matrix is given in ref. [9]. Here we merely present the result. For that purpose a few definitions have to be made. Let L_f and K_f be diagonal $m \times m$ submatrices of resp. L and K for the scattering channels. The two independent solutions of (12) without $\tilde{V}_1$ for every channel we range in two diagonal matrices $F(r)$ and $G(r)$: the regular ones in F and the irregular ones in G. Furthermore, these solutions are normalized such that their Wronskian equals unity, i.e.

$$W(F(r),\ G(r)) = F^T(r)\ \frac{d}{dr}\ G(r) - \left[\frac{d}{dr}\ F(r)\right]^T G(r) = 1\ . \tag{13}$$

With this we can define the following (2n+2m)×(2n+2m) matrix at the position of the ith delta function

$$X^{(i)} \equiv \begin{pmatrix} 1-2\Delta(r_i)G(r_i)M^{\frac{1}{2}}V_1(r_i)M^{\frac{1}{2}}F(r_i) & -2\Delta(r_i)G(r_i)M^{\frac{1}{2}}V_1(r_i)M^{\frac{1}{2}}G(r_i) \\ 2\Delta(r_i)F(r_i)M^{\frac{1}{2}}V_1(r_i)M^{\frac{1}{2}}F(r_i) & 1+2\Delta(r_i)F(r_i)M^{\frac{1}{2}}V_1(r_i)M^{\frac{1}{2}}G(r_i) \end{pmatrix} \tag{14}$$

and the product of all N of these

$$X \equiv X^{(N)}\ X^{(N-1)}\ \ldots\ X^{(1)}\ . \tag{15}$$

Then we write X in terms of 16 submatrices as follows:

$$X = \begin{pmatrix} X_{11} & X_{12} & X_{13} & X_{14} \\ X_{21} & X_{22} & X_{23} & X_{24} \\ X_{31} & X_{32} & X_{33} & X_{34} \\ X_{42} & X_{42} & X_{43} & X_{44} \end{pmatrix}\ . \tag{16}$$

The matrices $X_{\alpha\beta}$ are of size $p \times q$, where p resp. q equal n or m for α resp. β odd or even. The above definitions now allow us to write down the K-matrix

$$K = -K_f^{L_f+\frac{1}{2}} \left[X_{42}-X_{41}X_{11}^{-1}X_{12}\right]\left[X_{22}-X_{21}X_{11}^{-1}X_{12}\right]^{-1} K_f^{L_f+\frac{1}{2}}\ . \tag{17}$$

The S-matrix then reads

$$S = [1 + iK][1 - iK]^{-1}\ . \tag{18}$$

Bound states and resonances again manifest themselves as poles in S, so they occur when $1 - iK$ becomes singular. In principle they can be found and traced in precisely the same way like for the simple model of the previous section, although requiring considerably more computing time according as especially the number of channels increases. When the energy is real and above all thresholds, S is a m×m unitary matrix that may provide all the scattering data for the m free channels. If any of these is closed, say the ith one, i.e. $E < (m_1 + m_2)_i$, then the reduced S-matrix where the ith row and column representing this channel are skipped is again unitary. Finally, it should be investigated to what extent the result (18) supplies us with an approximate solution of the original Schrödinger equation (11). Thereto the convergence of S for an increasing number of delta functions must be studied. This has been done for the case where $V_1(r)$ is purely off-diagonal and given by a peaked, exponentially decaying smooth function, with quite satisfactory results showing that the needed computer time can be kept within the bounds of decency [9] [10].

IV. Meson Spectroscopy [5] [6] [11]

The absolutely predominant aspect of quark models in general and quarkonium models in particular has always been the choice of the confining force, either based on perturbative QCD calculations and non-peturbative QCD speculations, or from purely phenomenological contemplations. Our approach is somewhat different, since we primarily want to investigate the influence of strong decay on meson spectra, which may affect the results of any hadron model. Nevertheless our choice of a harmonic oscillator (H.O.) with constant frequency for the confining potential is neither ad hoc nor merely a technical one. On the experimental side it is supported by striking regularities in the radial spectra of several quarkonia and the nucleon (see ref. [6]), while theoretical backing may come from a geometrical quark-gluon model for confinement [12] [13]. Having committed ourselves on this point we must now specify the coupling mechanism between the quark sector and the decay sector. To that end the $^3P_0$-model [14] for hadron decay provides a natural and suitable starting point. According to this description a quark-antiquark ($q\bar{q}$) pair is created out of the vacuum with quantum numbers 0^{++}, thus in a $^3P_0$ state. In Fig. 2 this process is depicted schematically with all

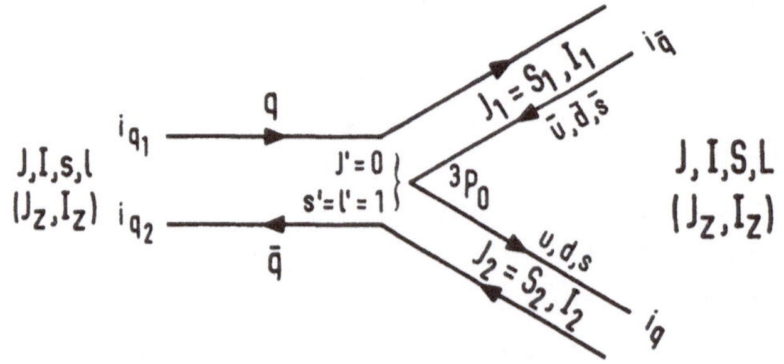

Fig. 2: Angular momenta and isospins for a $^3P_0$ *decay process*

the relevant quantum numbers. The matrix elements of the creation are proportional to 9J-symbols in terms of these quantities, i.e.

$$V_{^3P_o} \sim \begin{Bmatrix} \ell & 1 & L \\ s & 1 & S \\ J & 0 & J \end{Bmatrix} \begin{Bmatrix} \frac{1}{2} & \frac{1}{2} & S_1 \\ \frac{1}{2} & \frac{1}{2} & S_2 \\ s & 1 & S \end{Bmatrix} \begin{Bmatrix} i_{q_1} & i_{\bar{q}} & I_1 \\ i_{q_2} & i_q & I_2 \\ I & 0 & I \end{Bmatrix} \quad . \tag{19}$$

The precise radial dependence of the coupling potential, though quite essential, cannot be written down in a straightforward manner. A very simple ansatz for it of the form of a spherical delta function has been used in an application of the model to heavy quarkonia [5]. In this version two H.O. channels representing a heavy $q\bar{q}$ system in a $^3S_1$ resp. $^3D_1$ state are coupled to ten free channels containing mesons with open charm or beauty. In Fig. 3 the resulting spectrum of the $J^{PC} = 1^{--}$ $c\bar{c}$ states is shown for different values of the coupling

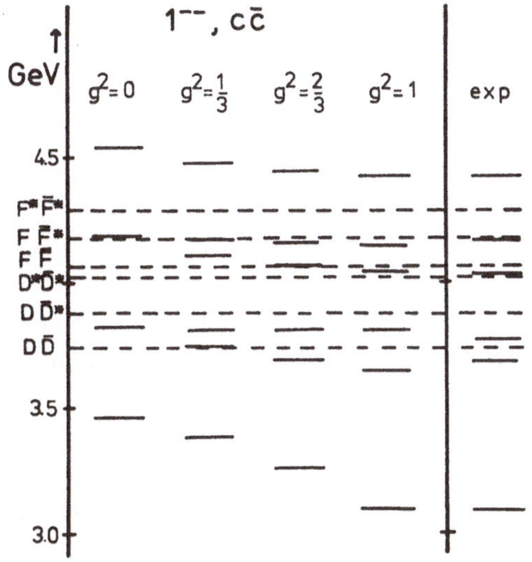

Fig. 3: Experimental and calculated (for different couplings) charmonium vector states together with the thresholds

parameter, as well as the experimental levels and thresholds. We see that the coupling to hadronic channels nicely remodels the equidistant and partly degenerate H.O. spectrum into the physical one, whereby the large shift of the ground state, already noted in section II, is very essential. The outcome is similar to what is produced by models with totally different confining potentials disregarding strong decay. Another effect of the coupled channels which was mentioned in sec. II is the deformation of the wave functions, resulting in greatly improved leptonic decay widths as compared to the ones obtained with a pure H.O. or any other convex potential [15]. On the negative side we should however point out the much too small hadronic widths of the resonant states. This deficiency turned out to be due to the singular form of the transition potential. In a more general treatment of all pseudoscalar and vector mesons [6] considerable improvement on this point has been achieved by deriving from the ^3P_o - model, in local approximation, a transition potential of the form

$$V_{^3P_o}(r) \sim \frac{r}{r_o} \exp\left(-\frac{1}{2}\frac{r^2}{r_o^2}\right) . \qquad (20)$$

In tackling the equations the formalism of sec. III is used to obtain analytic solutions. Additional ingredients of this extended model are colour splitting and relativistic kinematics for the scattering channels. The results amount to an on the whole good reproduction of the mass spectra, reasonable hadronic widths, and a fair description of meson-meson scattering data. Also when applied to scalar mesons - often considered as awkward in standard quark models - a remarkable qualitative agreement with experiment is accomplished [11]. So the model is suited to treat a wide range of mesonic data, though several modifications and refinements are still necessary.

V. Conclusions

In the foregoing a new approach to quark models within the context of the good old Schrödinger equation has been presented. Essential for the applicability to complicated hadronic systems involving many coupled channels was an approximative formulation with delta functions, which allowed to write down the S-matrix in closed form. Phenomenological realizations of the model have shown to be able to account for a multitude of experimental data in spite of the unconventional but simple form of the chosen interactions. These results also might put question-marks to the very detailed predictions of many single-channel quark models.

References

[1] A. Chodos, R.L. Jaffe, K. Johnson, C.B. Thorn and V.F. Weisskopf, Phys. Rev. D9, 3471 (1974); T.A. DeGrand, R.L. Jaffe, K. Johnson and J.J. Kiskis, ibid. 12, 2060 (1975).

[2] N. Isgur and G. Karl, Phys. Lett. 72B, 109 (1977); Phys. Rev. D18, 4187 (1978); Phys. Lett 74B, 353 (1978); Phys. Rev. D19, 2653 (1979); ibid. 20, 1191 (1979).

[3] For references concerning quarkonium potentials see e.g. A. Martin, Phys. Lett. 93B, 338 (1980).

[4] E. Eichten, K. Gottfried, K.D. Lane, T. Kinoshita and T.-M. Yan, Phys. Rev. D17, 3090 (1978); ibid. 21, 313(E) (1980); ibid. 21, 203 (1980).

[5] E. van Beveren, C. Dullemond and G. Rupp, Phys. Rev. D21, 772 (1980).

[6] E. van Beveren, G. Rupp, T.A. Rijken and C. Dullemond, Phys. Rev. D27, 1527 (1983).

[7] E. van Beveren, C. Dullemond and T.A. Rijken, Z. Phys. C19. 275 (1983).

[8] Bateman Manuscript Project, Higher Transcendental Functions, Vol. 1, eds Erdélyi et al. (McGraw-Hill, New York, 1953).

[9] C. Dullemond, G. Rupp, T.A. Rijken and E. van Beveren, Comp. Phys. Comm. 27, 377 (1982).

[10] G. Rupp, Ph.D. Thesis, Nijmegen report THEF-NYM-82.14 (1982) (unpublished).

[11] C. Dullemond, T.A. Rijken, E. van Beveren and G. Rupp, Nijmegen report THEF-NYM-83.09 (1983).

[12] E. van Beveren, C. Dullemond and T.A. Rijken, Nijmegen report THEF-NYM-84.01 (1984) (submitted to Phys. Rev. D).

[13] C. Dullemond, T.A. Rijken and E. van Beveren, Nijmegen report THEF-NYM-83.02 (1983) (to be published in Nuov. Cim. 1984).

[14] A. Le Yaouanc, L. Oliver, O. Pène and J.-C. Raynal, Phys. Rev. D8, 2223 (1973).

[15] H. Grosse and A. Martin, Phys. Rep. 60, 341 (1980).

TWO-CENTER RESONANCES AS A MEANS TO CALCULATE THE IMAGINARY PART OF THE OPTICAL POTENTIAL ASSOCIATED WITH AN OPEN BREAKUP CHANNEL

A.C. Fonseca

Centro Fisica Nuclear
Av. Gama Pinto 2, 1699 Lisboa
Portugal

It is well-known from standard quantum mechanics books [1] that in a collision process the elastic phase shifts become complex as soon as the energy rises above the first inelastic threshold. Therefore, if the aim is to develop a two-body potential model that accounts for elastic observables the natural candidate is a complex Optical Potential $V_{OP} = V(R) + i\,W(R)$ whose imganiary part is negative. For spinless particles the elastic amplitude takes the form

$$f(\theta) = \sum_{\ell}(2\ell+1)\; f_{\ell}\; P_{\ell}(\cos\theta) \tag{4}$$

where

$$f_{\ell} = \frac{1}{k}\; e^{i\delta_{\ell}}\; \sin\delta_{\ell} \tag{2}$$

satisfies the relation

$$\mathrm{Im}[f_{\ell}] \geq k|f_{\ell}|^2\;. \tag{3}$$

The Optical Potential can either be set up phenomenologically based on some chosen potential shape whose parameters are varied until agreement with data is obtained, or calculated through microscopic theories such as Folding Models, Multiple Scattering Theories, Feshbach Reaction Theory, etc. In the present contribution the work of Ref. [2] is reviewed showing how to calculate the imaginary part of an Optical Potential that may be directly associated with an existing breakup channel. This is done in the framework of the Born-Oppenheimer approximation for a process involving the collision of a heavy core "a" with another heavy particle "A" made up of a core "a" and one extra nucleon n. In order to be able to test the accuracy of the method we work with a

Fig. 1: Incoming Channel

simple model three-body problem where exact scattering results may be obtained from the solution of the appropriate Faddeev equations. Therefore, we assume that the cores "a" are inert particles of mass m_a and that the light particle, the n , is a spinless nucleon of mass m_n . The relevant mass ration of the problem is $m = m_a/m_n$. Furthermore, we assume that the light-heavy interaction V_{na} is described by a one term s-wave separable potential that supports a single n-a bound state with binding energy $\varepsilon = 4$ Mev,

$$V_{na} = \lambda |f\rangle\langle f| . \tag{4}$$

The coupling strength λ satisfies the relation

$$\lambda^{-1} = \langle f|g_o(-\varepsilon)|f\rangle \tag{5}$$

where $g_0(E)$ is the free resolvent for the light-heavy system. The form factor $|f\rangle$ is such that $\langle\vec{k}|f\rangle = (k^2+\beta^2)^{-3}$ where the range parameter $\beta = 1.85\ \mathrm{fm}^{-1}$. We also assume for simplicity that the cores do not interact ($v_{aa} = 0$). The only possible channels are therefore the elastic channel $aA \rightarrow aA$ and the breakup channel $aA \rightarrow aan$. In this way we have isolated the breakup channel as the only responsible for draining flux out of the elastic channel such that the imaginary part of the Optical Potential may be associated exclusively with the presence of such a channel. Using a separable potential for v_{na} and neglecting v_{aa} are not restrictions to the prescription we are going to develop. They are used here to make the calculations simpler and to avoid dealing with multi-channel contributions to the Optical Potential.

Our aim is to formulate a method to calculate the imaginary part of the Optical Potential associated with a breakup channel that may be used in two-body collisions whenever the ionization cross section is an important part of the total cross section. In particular, we think of nuclear heavy ion reactions such as $^{16}O + {}^{17}O$ where the breakup cross section may be important because nucleons outside closed shells are bound by only a few MeV. Therefore, as the energy rises above the threshold for the $^{16}O + {}^{17}O + n$ breakup, one expects this channel to contribute to the Optical Potential.

In the first part of this work we review some exact three-body results for the model defined above and in the second part we use the Born-Oppenheimer approximation to develop a simple but well-defined mathematical procedure to obtain a complex Optical Potential for breakup. Finally, we drew up some conclusions and give a few hints on how the method can be generalized to a realistic system such as $^{16}O + {}^{17}O$ scattering.

A) Exact Three-Body Results

The assumption of separable two-body interactions in the three-body problem reduces the Faddeev [3] equations to one variable integral equations after partial wave decomposition. In the present model this amounts to the solution of the following equation for the t-matrix

$$T_\ell(E;k',k) = B_\ell(E;k',k) + \int_0^\infty \frac{k''^2 dk''}{2\pi^2} B_\ell(E;k',k'')\,\tau(E - \frac{k''^2}{\mu_{aA}})\,T_\ell(E;k'',k) \qquad (6)$$

where $\mu_{aA} = m(m+1)/2m+1$ in units of $\hbar = 2m_n = 1$ and

$$\tau(X) = S(X)/(X+\varepsilon) \qquad (7)$$

$$[S(X)]^{-1} = -\gamma^2 \int \frac{p^2 dp}{2\pi^2} \frac{f^2(p)}{(X - \frac{p^2}{\nu})(\varepsilon + \frac{p^2}{\nu})} \quad . \qquad (8)$$

Fig. 2: Graphical representation of the three-body equation

The γ is such that $S(-\varepsilon) = 1$. The driving term B_ℓ is given by

$$B_\ell(E;k',k) = \frac{1}{2} \int_{-1}^{1} dx\, \gamma^2 \frac{f(u)\; f(v)}{E - \frac{k'^2}{\nu} - 2kk'x - \frac{k^2}{\nu}} \qquad (9)$$

$$u = k^2 + \nu^2 k'^2 + 2\nu kk'x$$
$$v = k'^2 + \nu^2 k^2 + 2\nu kk'x$$

where $\nu = m/m+1$. Equation (6) is graphically represented in Fig. 2. The driving term corresponds to the light particle exchange between the heavies which leads to a pure space exchange interaction of the Majorana type. If the heavies are bosons there is attraction in the even partial waves with the possibility of three-body bound states (see Ref. [4] for a comprehensive study of a nuclear molecular three-body-system), while the odd waves are repulsive. In the energy plane the right hand singularities of the t-matrix are a square root branch cut for $E > -\varepsilon$ that is associated with the elastic channel and, for $E > 0$, a logarithmic branch cut that accounts for the breakup channel. The elastic amplitude $f_\ell = \frac{-\mu}{4\pi} T_\ell$ relates to the elastic phase shifts through Eq. (2). Since for $E > 0$ the phase shifts become complex, we write

$$f_\ell = \frac{1}{2ik}(S_\ell - 1) \tag{10}$$

where S is the partial wave S-matrix element defined as

$$S_\ell = \eta_\ell \, e^{i2\delta_\ell} \quad . \tag{11}$$

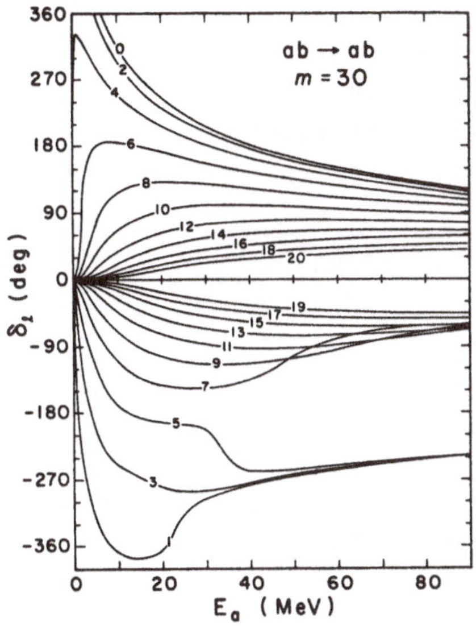

Fig. 3: Phase shift δ_ℓ for various ℓ and $m = 30$ versus lab energy E_a

The inelastic parameter η_ℓ may depart from unity for $E > 0$ and δ_ℓ is the real part of the elastic phase shift. In Fig. 3 results are shown for the elastic phase shift δ_ℓ versus the a-particle laboratory energy E_a for $m = 30$ and partial waves in the range $0 \leqq \ell \leqq 20$. We choose to normalize the phase shifts so that at threshold $\delta_\ell(0) = n_\ell \pi$, where n_ℓ is the number of three-body bound states. Accordingly, the lower even phase shifts drop from higher multiples of π whereas resonances are seen for a range of intermediate even partial waves. In contrast, the odd phase shifts are negative as is expected from the exchange interaction, but surprisingly many of them are seen to rise through multiples of $\frac{1}{2}\pi$ and thus resonate. Another feature of the results is that the inelasticity generated by the breakup channel is present only in the odd partial waves with virtually none in the even waves. We

Table I: Inelasticities for $m = 30$ and $E_a = 50$ MeV

l	η_l	l	η_l
0	0.996	10	1·000
1	0.241	11	0.607
2	0.996	12	1.000
3	0.234	13	0.783
4	0.997	14	1.000
5	0.181	15	0.889
6	0.998	16	0.999
7	0.117	17	0.946
8	0.999	18	0.999
9	0.358	19	0.974

show the inelasticity parameter η_ℓ versus E_a in Fig. 4 and it is evident that the resonances in the odd partial waves are highly inelastic. This remarkable difference between the inelasticity in the even and odd partial waves is a puzzling effect never seen before in the three-body problem. The actual numerical values of η_ℓ are given in Table I for $m = 30$ and $E_a = 50$ MeV. Argand plots of the S-matrix elements are shown in Fig. 5 for the four most inelastic partial waves. The phase shifts for $\ell = 1,3$ and 5 approach $-\pi$ as $E_a \to \infty$ while those for $\ell \geqq 7$ approach zero at high energy. Although the interaction is repulsive in ℓ odd one can find inelastic resonances in $\ell = 1,3,5$ and 7 partial waves as the phase shift rises counterclockwise through $\pi/2$ and η_ℓ approaches zero. Attempts to understand the

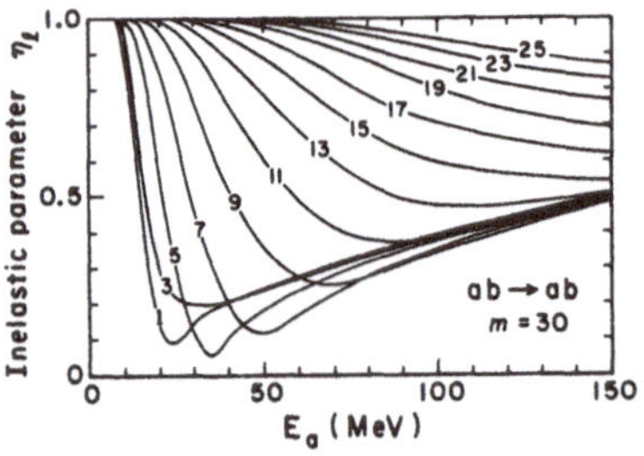

Fig. 4: Inelastic parameter η_ℓ for various ℓ and $m = 30$ versus E_a. Breakup threshold is at $E_a = 8$ MeV

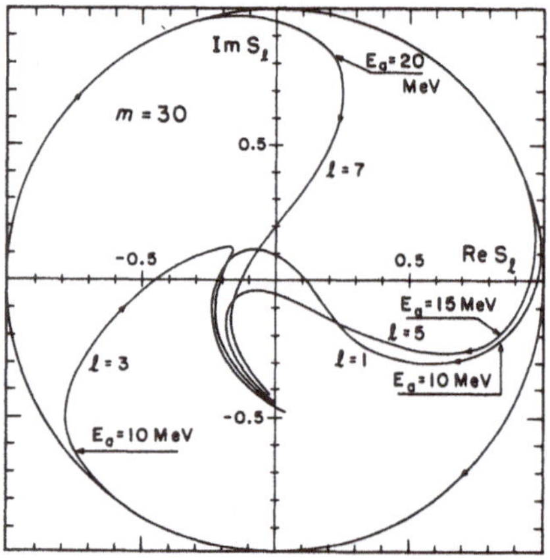

Fig. 5: Polar plot of the S-matrix versus lab energy E_a for $\ell = 1,3,5$ and 7

lack of inelasticity in even partial waves and the presence of inelastic resonances by simple analyses of the Faddeev equations have not been successful. One can only say that the Neumann series converges slowly. Thus many orders of iteration are required until the inelasticity in the odd partial waves emerges. As we show next, the interpretation of these features becomes quite transparent in the molecular approach leading to a natural prescription for a complex Optical Potential.

B) Born-Oppenheimer Solution

We can understand the nature of the three-body scattering results by solving the model problem in the Born-Oppenheimer (BO) approximation. The three-body Schrödinger equation for the problem reads

$$[(-2/m)\nabla_R^2 - (1/\mu)\nabla_r^2 + v|\vec{r}-\tfrac{1}{2}\vec{R}| + v|\vec{r}+\tfrac{1}{2}\vec{R}|]\Psi(\vec{r},\vec{R}) = E\Psi(\vec{r},\vec{R}),$$

$$\mu = 2m/(2m+1). \qquad (12)$$

The BO approach consists of approximating the wave function as a product

$$\Psi(\vec{r},\vec{R}) \approx \Psi_{BO}(\vec{r},\vec{R}) = \psi(\vec{r},\vec{R})\Phi(\vec{R}). \qquad (13)$$

where $\vec{R}$ is the a-a separation vector and $\vec{r}$ locates the light particle relative to the a-a center of mass. The wave function $\psi(\vec{r},\vec{R})$ is a negative energy solution to the two-center problem in which the heavy particles are assumed fixed at a separation $\vec{R}$. The light particle interacts with the fixed centers leading to the light particle equation

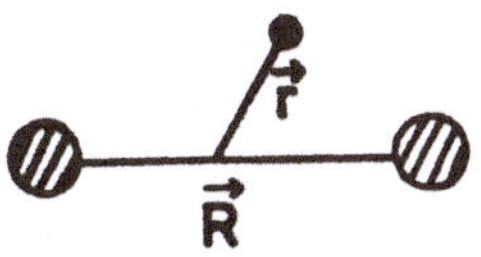

Fig. 6: Jacobian coordinates for BO

$$[-\nabla_r^2/\mu + v|\vec{r}-\tfrac{1}{2}\vec{R}| + v|\vec{r}+\tfrac{1}{2}\vec{R}|]\psi(\vec{r},\vec{R}) = \varepsilon(\vec{R})\psi(\vec{r},\vec{R}). \qquad (14)$$

Since the hamiltonian in (14) is invariant under the reflection $\vec{R}\to-\vec{R}$, the solutions may be characterized by their behavior under this symmetry operation. As before, we refer to the solution as $\varepsilon_g(R)$ if $\psi(\vec{r},-\vec{R}) = \psi(\vec{r},\vec{R})$ and as $\varepsilon_u(R)$ if $\psi(\vec{r},-\vec{R}) = -\psi(\vec{r},\vec{R})$. Given the eigenvalues $\varepsilon_g(R)$ and $\varepsilon_u(R)$ from the two-center problem, we then solve the heavy particle equation

$$[-2\nabla_R^2/m + \varepsilon(R)]\Phi(\vec{R}) = E\Phi(\vec{R}) , \qquad (15)$$

where E is the three-body energy. Under the $\vec{R}\to-\vec{R}$ reflection $\Phi(\vec{R})$ has the usual parity $(-1)^\ell$ in each partial wave. Since we are dealing with bosons we want the BO wave function to be unchanged under $\vec{R}\to-\vec{R}$ which leads to the adoption of $\varepsilon_g(R)$ in the even partial waves and $\varepsilon_u(R)$ in the odd. In the case that the a's are fermions we reverse the above prescription.

We now wish to obtain all discrete energy solutions to the two-center problem given in (14). These are easily obtained since, as shown in Ref. [2], the scattering T-matrix for the two-center problem with separable potentials may be written in a closed form as

$$T(k,k';z) = \frac{2\lambda\, f(k)f(k')}{(1-I-J)(1-I+J)}\,[(1-I)\cos\Delta_- + J\cos\Delta_+], \qquad (16)$$

where z is the energy of the light particle and

$$\Delta_{\pm} = (\vec{k} \pm \vec{k}') \cdot \tfrac{1}{2}\vec{R}, \qquad (17)$$

$$I = \lambda \int \frac{d^3q}{(2\pi)^3} \frac{f^2(q)}{z-q^2/\mu} \qquad (18)$$

$$J = \lambda \int \frac{d^3q}{(2\pi)^3} \frac{f^2(q)e^{iq\cdot\vec{R}}}{z-q^2/\mu} \quad . \qquad (19)$$

Discrete solutions of the two-center problem correspond to poles of the scattering amplitude generated by zeros in the denominator of (16). These poles are values of z for which either

$$1 - I - J = 0 , \qquad (20)$$

or

$$1 - I + J = 0 . \qquad (21)$$

As shown in Ref. [4] the solutions of (20) correspond to the attractive gerade potential $\varepsilon_g(R)$ that yields the large number of three-body bound states. Solutions to (21) yield the repulsive ungerade potential $\varepsilon_u(R)$. As $R \to \infty$ we note $J = 0$ and both curves approach the two-body binding energy as would be appropriate for separated "atoms". In the united atom limit corresponding to $R = 0$, the eigenvalue condition (20) has a solution corresponding to a separable potential with a doubled coupling strength 2λ. For intermediate R we determine numerically the values of z that solve (20) and interpret these as the attractive $\varepsilon_g(R)$ shown in Fig. 7a. The other eigenvalue condition (21) yields a real solution from $R = \infty$ up to some finite $R = R_0$ for which we have $z = \varepsilon_u(R_0) = 0$. This means that there exists no anti-symmetric bound state wave function of the two-center problem for $R < R_0$ with z real and negative. We are then left with a "hole" in the potential unless some scheme is devised for $R < R_0$.

In order to treat this region we first note that $z = 0$ corresponds to the elastic threshold for the two-center problem, so that the region $R < R_0$ must be associated with the two-center continuum and thus with breakup at the three-body level. It could then be argued that (13) should be supplemented with the continuum part of the two-center wave function which is usually neglected in the BO approximation. To do so would be quite difficult and we avoid it with the following prescription.

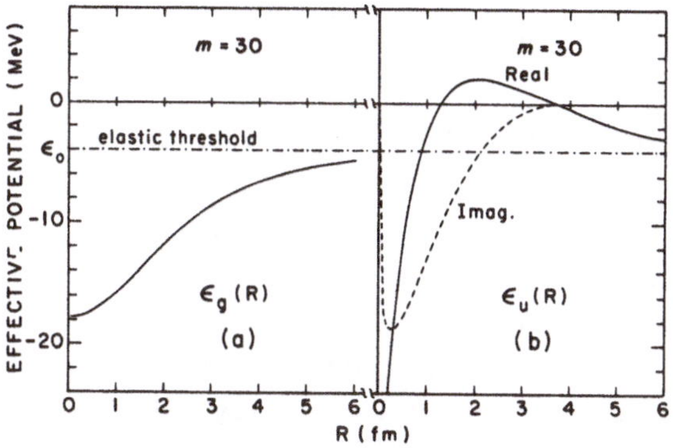

Fig. 7: *(a) Real effective potential $\varepsilon_g(R)$ versus R for m = 30.*
(b) Real and imaginary parts of the effective potential $\varepsilon_u(R)$ versus R for m = 30.

We first ask what happens to the ungerade solution for $R < R_0$? In terms of the analytic properties of the scattering amplitude, we expect that the ungerade pole moves to the elastic threshold $z = 0$, for $R = R_0$, and then onto the second Riemann sheet for $R < R_0$. From numerical study we find this to be the case, that is, if (21) is analytically continued to complex z, it is found that the ungerade pole moves through the elastic branch point onto the fourth quadrant of the second sheet for $R < R_0$. Such a solution corresponds to a two-center scattering resonance that decays into two heavy particles with a finite lifetime. The association of $R < R_0$ with three-body breakup suggest that we insert the second sheet complex solution to fill up the "hole" in $\varepsilon_u(R)$ for $R < R_0$ with the expectation that the imaginary part of the resulting "Optical Potential" will simulate three-body breakup in the BO approach. The full $\varepsilon_u(R)$ potential is shown in Fig. 7b

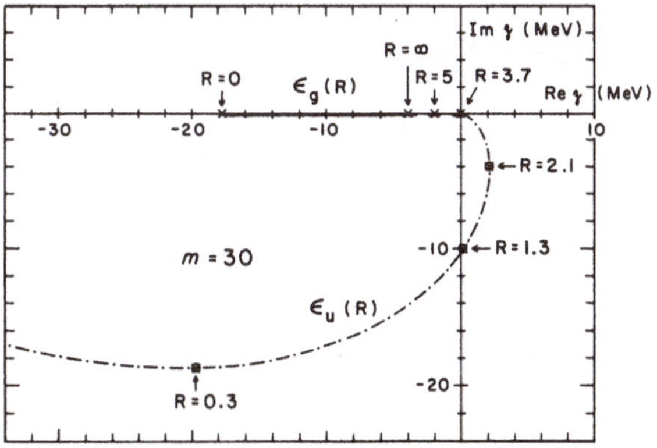

Fig. 8: *Trajectories of gerade and ungerade poles of the scattering amplitude in the complex energy plane versus R. The dashed line indicates the second Riemann sheet.*

where we note that $\mathrm{Im}\,[\varepsilon_u(R)] \leqq 0$ for $R < R_o$ and is thus necessarily absorptive.

An alternative way of showing the results is to plot the trajectories of the poles of (16) in the z-plane as is done in Fig. 8. For $R = \infty$, we have a double pole in the scattering amplitude at the two-body binding energy. For smaller R, the attractive gerade solution moves to deeper binding, whereas the ungerade solution moves to the right and through the elastic branch point onto the second sheet giving a resonance for some intermediate R. Finally, for small R, the solution ceases to be resonant and moves to the lef-half plane approaching a virtual state as $R \to 0$.

As previously discussed, the solution of the heavy particle equation proceeds with the real $\varepsilon_g(R)$ in the even partial waves and with the absorptive $\varepsilon_u(R)$ in the odd partial waves. Thus the puzzling appearance of breakup in only the odd waves emerges naturally in the BO picture. In an a-collision in the even partial waves, the valence n moves into the symmetric two-center wave function and its binding energy with respect to the two adiabatically moving heavy nuclei increases as dictated by $\varepsilon_g(R)$. For the odd partial waves the light particle moves into the anti-symmetric two-center wave function and its binding energy decreases until it becomes unbound for $R < R_o$. Therefore for $R < R_o$, the valence n has a finite probability of ionizing and in this work we give a prescription for constructing the effective ionization potential for breakup by analytic continuation of the two-center

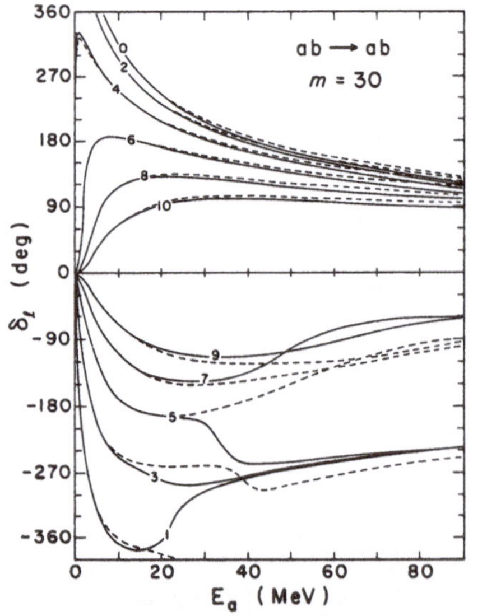

Fig. 9: Comparison of exact (-) and BO (...) phase shifts δ_ℓ for various ℓ and $m = 30$ verus lab energy E_a

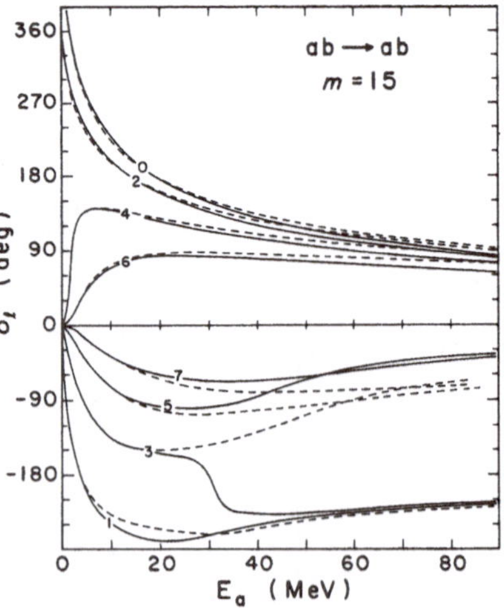

Fig. 10: Same as Fig. 9 except $m = 15$

scattering amplitude. Similar ideas find application in various aspects of atomic physics [5].

The inelastic odd parity resonances that appeared in the three-body results find a ready explanation in the BO picture. We note from Fig. 7b that $Re[\varepsilon_u(R)]$ forms a barrier for $R < R_o$ in the region where $Im[\varepsilon_g(R)] \neq 0$. The inelastic resonances may be interpreted as barrier resonances formed in the $\varepsilon_u(R)$ potential.

In Figs. 9, 10 and 11 we compare the elastic scattering phase shifts obtained in the BO approach with the exact three-body results for mass ratios of m = 30, 15 and 5. We note that for all three mass ratios the agreement is best for low energies as would be expected for the

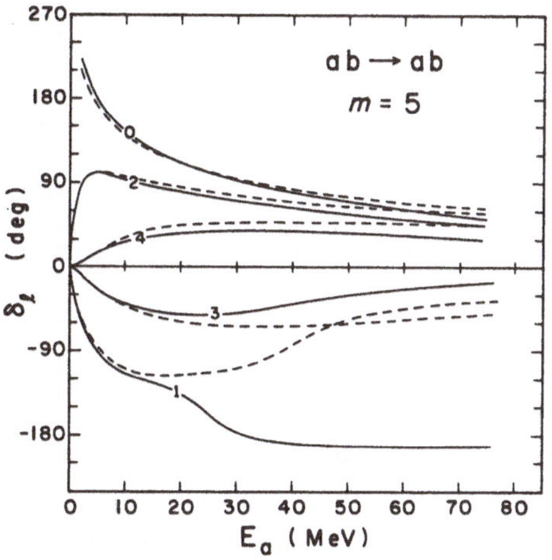

Fig. 11: Same as in Fig. 9 except m = 5

BO approximation. The largest deviations are seen in the highly inelastic odd partial waves. In Table II we also compare the inelasticity obtained for both the exact and BO calculations for m = 5, 15 and 30 at E_a = 20 MeV. Although the agreement between exact and BO results

Table II: Comparison of exact and BO inelasticities

	m = 5		m = 15		m = 30	
	exact	BO	exact	BO	exact	BO
1	0.370	0.550	0.449	0.221	0.132	0.092
3	0.880	0.975	0.481	0.681	0.295	0.276
5	0.989	0.999	0.863	0.969	0.563	0.731
7	0.999	1.000	0.976	0.998	0.859	0.964
9			0.996	0.999	0.966	0.997
11			0.999	1.000	0.933	0.999
					0.999	1.000

is best for $m = 30$ and $m = 15$ one can find that for $m = 5$ the Optical Potential method predicts the elastic phase shifts with reasonable accuracy up to 6 MeV above breakup threshold.

C) Conclusions

We find, therefore, that the BO approximation gives a natural qualitative understanding of the resonant and inelastic features of a scattering problem governed by molecular mass ratios. In addition, the BO approximation achieves surprising quantitative accuracy for even rather small mass ratio when a mechanism to account for breakup is included by analytic continuation of potential energy curves. The accuracy of BO for low mass ratios is suggested in Ref. [3] to be related to the range of the light-heavy interaction. By studying the structure of the exact wave function in a model three-body problem identical to the one presented here, it is found that the BO wave function can only remain accurate in the limit $\beta \to 0$ (range $1/\beta$) if m becomes very large. On the contrary for finite range potentials the BO wave function is accurate even for mass ratios of the order of $m = 15$. This is an interesting mathematical aspect of the BO approximation that needs further understanding.

The nuclear analogue of our model problem is the scattering of heavy ions composed of neighboring nuclei such as $^{16}O + {}^{17}O$. The molecular approach to such problems has been widely studied and the subject is reviewed by von Oertzen and Bohlen [6]. In their work the potential energy curves are obtained with the approximate LCNO solution to the two-center problem discussed above. This approximation precludes the appearance of a complex potential that simulates three-body breakup and also neglects distortion of the light-heavy bound state. To include these effects properly would require a full solution to the two-center problem for standard local interactions such as the Wood - Saxon potential. From the present work we expect that the continuation of such two-center results above the elastic scattering threshold would lead to effective Optical Potentials that simulate the three-body breakup. The location of these potential energy curves in unphysical regions would depend on the location of resonances or virtual states in the spectrum of the separated or united "atom". These Optical Potentials when combined with the complex core-core interaction would give the final results of the calculation. The importance of the breakup contribution would depend on its range and depth relative to the core-core potential. To avoid a further complication in this approach that is related to the exclusion of bound states in the nucleon-core potential ruled out by the Pauli principle one may adopt the following simplifying steps: a) consider a separable light-heavy interaction that reproduces the bound state wave function of the valence nucleon as given by the Wood - Saxon potential; b) solve the two-center potential problem in this truncated space where the full Wood - Saxon potential has been substituted by a separable potential. c) Analytically continue the potential energy curves in order to obtain a complex Optical Potential. Work on these problems is currently underway.

References

[1] C. Cohen - Tannoudji, B. Diu and F. Laluë, Mécanique Quantique, Hermann, Paris (1973)

[2] A.C. Fonseca and P.E. Shanley, Nucl. Phys. A382, 97 (1982)

[3] I.R. Afnan and A.W. Thomas, in: Modern Three-Hadron Physics, Springer-Verlag 1977 (ed. by A.W. Thomas)

[4] A.C. Fonseca and P.E. Shanley, Ann. of Phys. 117 268 (1979)

[5] A. Hertzenberg, Phys. Rev. 160, 80 (1967)
W.H. Miller, J. Chem. Phys. 52, 3563 (1970)

[6] W. von Oertzen and H.G. Bohlen, Phys. Reports 19, 1 (1975)

EXPERIMENTAL DISCOVERY OF THE LANDAU-ZENER EFFECT IN ATOMIC NUCLEI

N. Cindro
Rudjer Bošković Institute, 41001 Zagreb, Croatia, Yugoslavia
and Centre de Recherches Nucléaires, 67037 Strasbourg, France

F. Haas and R. Freeman
Centre de Recherches Nucléaires, 67037 Strasbourg, France

1. Introduction

The laws of classical mechanics and, later on, those of quantum mechanics, were discovered in trying to understand phenomena at a given scale of the physical objects involved. There is no a priori reason why such laws should hold when the scale of objects or the degree of their complexity grossly changes. Quantum mechanics, however, gives numerous examples of same or analogous effects observed at various degrees of complexity of matter and object scales. It is always exciting when such effects are found at various stages of matter organization, such as condensed matter, atoms and nuclei. The content of this paper is to report on the experimental discovery of the effects of the promotion of nucleons to higher excited states through avoided crossings of levels in a two-centre energy-level diagram and to its relation to the promotion of electrons in atoms and, to some extent, to Zener tunnelling in solids. We shall see that same or analogous quantum mechanisms, i.e. mechanisms that lift off the degeneracy of well-resolved energy levels and bands, are responsible for the phenomena on all the three niveaus.

2. Tunnelling and promotion in the presence of external fields

2.1. The Zener tunnelling in solids. A Zener diode is a well-known device in electronic technology described as a p-n junction with a sharp and well-controlled reverse bias avalanche break-down voltage (0.8 V - several hundred V)[(1)]. Thus, today it represents a generic name for devices encompassing both the avalanche break-down mechanism (higher voltages) and the Zener tunnelling proper (lower voltages). The latter is depicted in Fig. 1: At a p-n junction, electrons can tunnel through a thin barrier provided that empty states of the same energy exist on the other side of the barrier. A reverse bias (the + voltage on the n-side) raises the top of the valence band on the n-side of the junction above the bottom of the conduction band on the p-side, and electrons tunnel to it. The tunnelling probability is very large if the

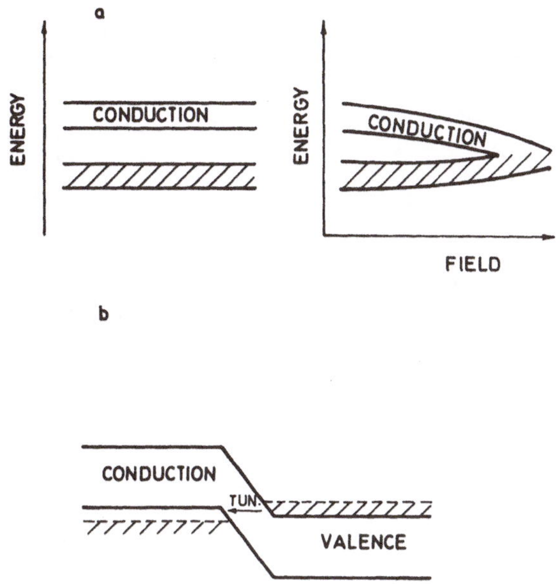

Fig. 1. (a) The energy of the electronic bands changes when an electric field is applied. (b) The Zener tunnelling through the barrier at a p-n junction.

barrier is thin, which is produced by heavy doping on each side.

The above tunnelling mechanism was predicted by Zener in 1943[2]; in 1957 Esaki made the first tunnel diode which is now commonly used to maintain a constant voltage across a load resistance.

2.2. The Landau-Zener effect in atoms and nuclei. Although proceeding via a mechanism seemingly different from the Zener effect in solids, the Landau-Zener effect in atoms can be traced back to lifting off of the energy degeneracy of electronic levels in an atom in the presence of another atom (Fig. 2, left). Under certain conditions, enhanced transitions from one such state to another can occur; promotion of electrons to higher excited states (and the subsequent emission of photons) will be particularly enhanced at points of avoided level crossings[3].

It is conceivable that a similar situation may arise in collisions between atomic nuclei; conditions for the Landau-Zener promotion should be particularly favourable for pairs of nuclei with one nucleus a closed shell and the other nucleus a closed shell + one nucleon. The

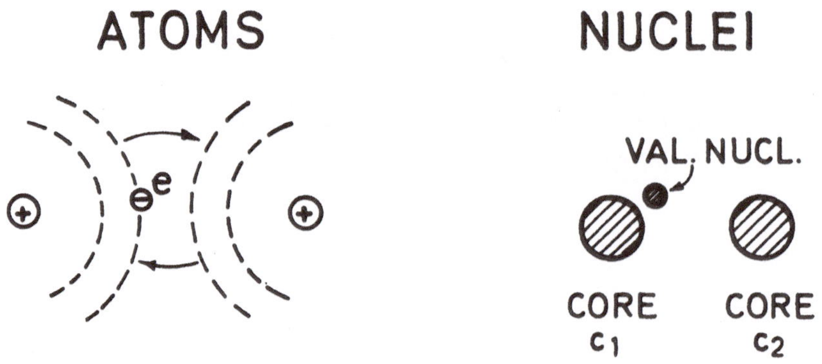

Fig. 2. Conditions for the Landau-Zener promotion in atoms (left) and electrons (right).

possibility of a nuclear Landau-Zener effect was first suggested by Park, Scheid and Greiner some years ago[4]. They also calculated the level diagrams for several nuclear systems where conditions for observing the nuclear Landau-Zener effect are met and pointed at specific transitions where the effect could be observed[5]. Recently, Abe and Park used a semiclassical approximation to calculate the effect and obtained simple expressions for the energies and the widths of the Landau-Zener "resonances"[6]. In the next section we shall discuss in more detail the mechanism of the nuclear Landau-Zener effect as well as the predictions of Refs. 4-6.

3. The mechanism of the nuclear Landau-Zener effect: avoided level crossings

As stated in the last section, the mechanism of the Landau-Zener effect in atoms and nuclei is essentially identical, the only difference being the fact that the field in atoms is the central Coulomb field, while in nuclei it is the average nuclear field. When two nu-

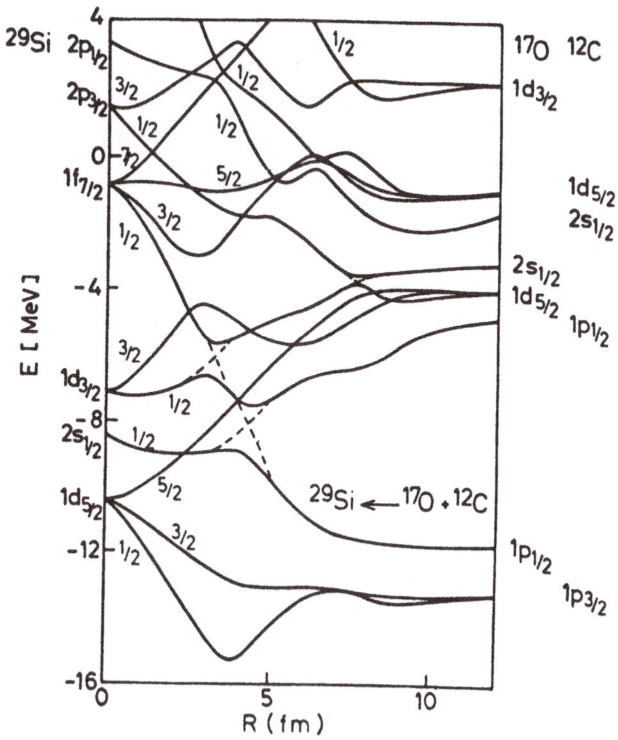

Fig. 3. The mechanism of the Landau-Zener effect illustrated for the collision of ^{12}O and ^{17}O nuclei: the TCSM diagram of the colliding ^{12}C+^{17}O system (from Ref. 5).

clei approach, the energy of their levels shifts adiabatically as a function of the distance R. At some critical distance R_C, two levels may approach so close that a nucleon is promoted to the higher level. An enhanced transition occurs, which we call a nuclear Landau-Zener transition. This mechanism of nucleon promotion into higher orbitals at avoided level crossings is shown in Fig. 3 for the $^{12}C+^{17}O$ collision.[(5)] The energy levels of ^{12}C and ^{17}O lose their Ω-degeneracy and, as their distance decreases, several avoided crossings occur. For reasons which will become clear immediately, let us concentrate on the avoided level crossing between the $\Omega = 1/2$ branch of the $1d_{5/2}$ level carrying the valence neutron in ^{17}O and the $2s_{1/2}$ level of the same nucleus. Radial coupling (i.e. a term $\partial V/\partial R$ in the Hamiltonian appearing as a consequence of the fact that we treat the system in such a way that the radial distance R is variable) will enhance the transition between these two levels when the nuclei approach a critical distance, about 8 fm in this case. Such a promotion gives rise to enhanced inelastic excitation of ^{17}O to its first excited 0.87 MeV $\frac{1}{2}^{+}$ state and will manifest itself as resonantlike, periodic peaks in the energy dependence of the total cross section.

The resonantlike behaviour of the nuclear Landau-Zener effect requires additional comment. In the atomic effect, the angle-integrated cross section is given as a function of incident energy E by[7]

$$\sigma_{21}(E) = 4\pi^2 R_c^2 \frac{\sqrt{2\mu}}{\hbar} \frac{|H'_{12}|^2}{\Delta F^{o}} \frac{(E-\bar{V})^{1/2}}{E} , \tag{1}$$

with R_c the crossing distance of the diabatic energy curves ε_2^o and ε_1^o and $|H'_{12}| = \frac{1}{2}\Delta$ the coupling matrix element between the diabatic states at R_c (see the blow-up in Fig. 4). ΔF^o is the difference of forces along the diabatic curves, i.e.

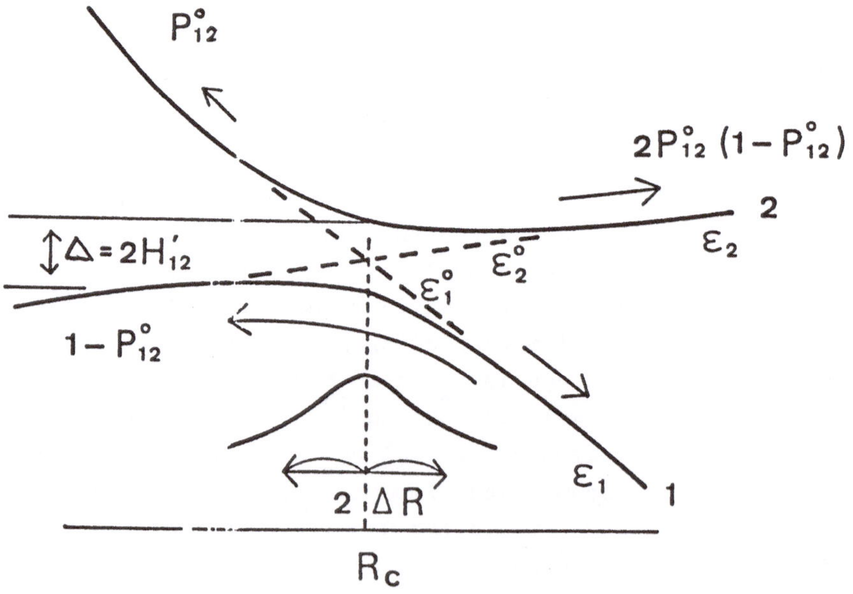

Fig. 4. The mechanism of the Landau-Zener effect: the interaction which engenders the transition (from Ref. 6).

$$\Delta F^{o} = \left| d\varepsilon_2^{o}/dR - d\varepsilon_1^{o}/dR \right|_{R=R_c}$$

and μ and $\overline{V}$ are, respectively, the reduced mass of the entrance channel and the adiabatic potential depth at R_c. Expression (1) depends smoothly on the energy with a sharp rise at the threshold $\overline{V}$. This smoothness is due to the integration over the impact parameter. However, in heavy-ion nuclear reactions, the Landau-Zener promotion mechanism is expected to act only for a very narrow range of partial waves around the grazing value. Starting from this premise, Abe and Park[6] have developed a semiclassical expression which gives the energy dependence of the cross section for the nuclear Landau-Zener effect as a series of resonantlike peaks, each one due to the advent of a new (grazing) partial wave. We present the main points of this derivation here, while for details we refer the reader to Ref. 5.

The transition probability from one adiabatic state to another by a single passage is given by[3]

$$P_{12}^{o} = e^{-2\pi G^{o}} \quad , \qquad G^{o} = \frac{1}{\hbar} \frac{|H_{12}'|^2}{v^{o} \Delta F^{o}} \quad , \tag{2}$$

with v^{o} the radial velocity at $R = R_c$. Since the system passes through the crossing point twice (once entering the interaction region and then leaving it), the transition probability in the scattering process is given by

$$P = 2P_{12}^{o}(1 - P_{12}^{o}) \quad . \tag{3}$$

In principle, all quantities used in (2) and (3) can be calculated from the TCSM diagram; the radial velocity v^{o} is the only quantity that is determined by the relative motion and is given combining the expressions

$$E = \frac{1}{2} \mu v_o^{2} + \frac{\ell(\ell + 1)\hbar^2}{2\mu R_c^2} + \overline{V}$$

and (4)

$$\overline{V} = \varepsilon^{o}(R_c) + V_c(R_c) \quad .$$

$V_c(R_c)$ is the adiabatic potential between the two centres; in the nuclear case, we assume that $\overline{V}$ is the optical potential. Expressions (4) give the ℓ-dependence of the radial velocity v^{o} and hence of the transition probability P, which, consequently, we shall denote by P_ℓ.

The angle-integrated expression for the nuclear Landau-Zener effect is obtained, as usual, by summing the contribution of all partial waves*

$$\sigma_{21} = \frac{\pi^2}{k^2} \sum_{\ell}^{\ell_{max}} (2\ell + 1) P_\ell(E) \ . \tag{5}$$

Replacing the summation over ℓ by the integration over the impact parameter, one obtains[6]

$$\sigma_{21} = 4\pi R_c^2 \left[1 - \frac{\bar{V}}{E}\right]\left[f(\gamma) - f(2\gamma)\right] \ , \tag{6}$$

with $f(\gamma) = e^{-\gamma}(1-\gamma)/2 - E_i(-\gamma)\gamma^2/2$,

$$\gamma = 2\pi g/\left[1 - \frac{\bar{V}}{E}\right]^{1/2} , \quad g = \frac{1}{\hbar}\frac{1}{v}\frac{|H_{12}'|^2}{\Delta F^o} ,$$

$$v = \left[\frac{2}{\mu} E\right]^{1/2} .$$

$E_i(-\gamma)$ is the exponential integral, $E_i(x) = -\int_{-x}^{\infty} (e^{-t}/t)dt$.

In a concrete calculation, the coupling $|H_{12}^o|$ and the crossing radius R_c are obtained from the TCSM diagram, $\bar{V}$ is the optical potential[8] and ΔF^o is replaced by the ansatz

$$\Delta F^o \approx \frac{2|H_{12}'|}{\Delta R} .$$

For each new partial wave there is a new "resonant" energy, obtained by setting $\partial P_\ell/\partial E = 0$ at $E = E^{\text{"res"}}$,

$$E^{\text{"res"}} = \frac{\pi^2}{4} \cdot \frac{1}{(\ln 2)^2} \cdot \frac{|H_{12}'|^2}{(\hbar^2/2\mu)/\Delta R^2} + \bar{V} + \frac{\hbar^2}{2\mu R_c^2}\ell(\ell + 1) \ . \tag{7}$$

In the same way, the "width" of the "resonant" peaks is given by

$$\Gamma = \frac{\pi^2}{4}\left[\ln \frac{1}{\left[\ln(1/2 + 1/2\sqrt{2})\right]^2} - \frac{1}{\left[\ln 1/2 - 1/2\sqrt{2}\right]^2}\right] \frac{|H_{12}'|^2}{(\hbar^2/2\mu)/(\Delta R)^2}$$

$$\approx 100 \frac{|H_{12}'|^2}{(\hbar^2/2\mu)/\Delta R^2} \ . \tag{8}$$

Hence, the sequence of "resonant" peaks is determined by a rotational spacing $(\hbar^2/2\mu R_c^2)(2\ell+1)$, while the width is determined by the matrix element H_{12}' and the nonadiabatic range ΔR. For values that can be read

*For the $^{12}C + ^{17}O$ collision shown in Fig. 3, a factor of 1/3 multiplies the right side of expressions (5) and (6) (only the $\Omega = 1/2$ branch of the three branches of the $1d_{5/2}$ state reaches the $2s_{1/2}$ state).

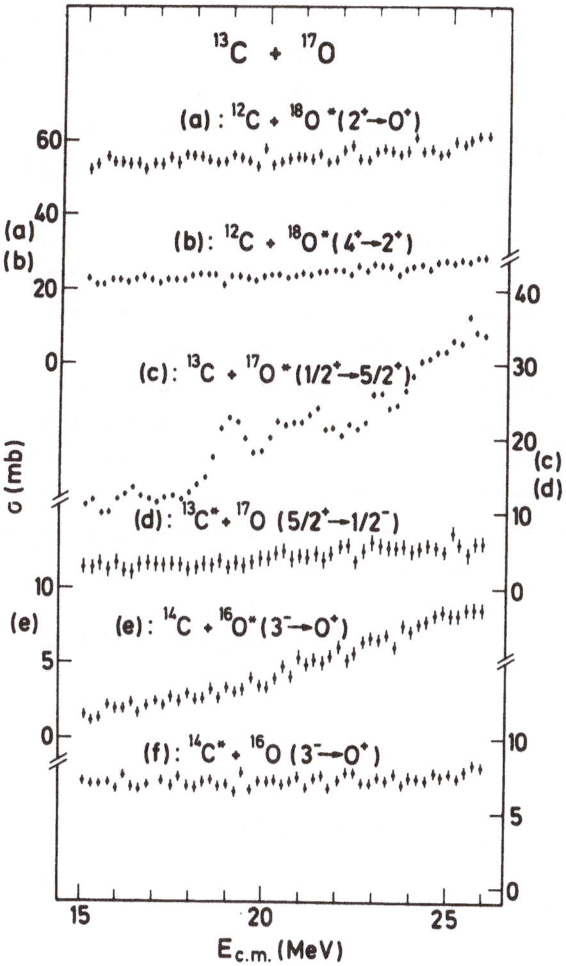

Fig. 5. Gamma-ray yield excitation functions of the $^{13}C+^{17}O$ interaction leading to various exit channels (following Ref. 9).

from Fig. 3 (E_c = 7.8 fm, H'_{12} = 0.05 MeV and ΔR = 0.1 fm), one obtains a "resonance" with $\Gamma \gtrsim 3.5$ keV, which is much smaller than the typical spacing $\Delta E^{\text{"res"}} = 0.045(2\ell+1)$ between two adjacent "resonances" with spins ℓ and $\ell+1$, respectively.*

*In atomic scattering, an order of magnitude of the width can be estimated by putting $|H'_{12}| \sim 10^{-2}\ E_{el}$ (E_{el} = typical electronic energy), $\Delta R/R_c \sim 10^{-2}$ and $(\hbar^2/2\mu)/R_c^2 \sim 10^{-3}\ E_{el}$. This gives $\Gamma \sim 10^{-3}\ E_{el}$, which would require high-resolution apparatus for experimental observation.

4. The experimental discovery of the nuclear Landau-Zener effect

The first evidence for a phenomenon that could be attributed to the nuclear Landau-Zener effect was found in the study of the $^{13}C+^{17}O$ interaction. Fig. 5 shows the excitation functions of $^{13}C+^{17}O$ leading to various exit channels measured by the γ-ray yields[9]. All studied channels show a smooth behaviour, except for $^{13}C+^{17}O \rightarrow ^{13}C+^{17}O^*$(0.81 MeV), which displays periodic oscillations. This extraordinary feature of the $^{13}C+^{17}O$ inelastic scattering led Abe and Park[6] to interpret it as due to the Landau-Zener promotion of the valence neutron from the $d_{5/2}$ (Ω = 1/2) to the $2s_{1/2}$ level. The comparison of the experimental and calculated cross sections is shown in Fig. 6. Although the agreement is far from being satisfactory, the periodicity of the experimental oscillations in the cross sections corresponds roughly to the one calculated from Eq. (6) (full line). The result of a calculation using Eq. (1) (integration over all impact parameters) is shown as a dashed line; no oscillations are present.

The interpretation of the oscillations in the γ-yield of $^{13}C+^{17}O \rightarrow ^{13}C+^{17}O^*$(0.87 $\rightarrow$ g.s.) as due to the effect of the Landau-Zener pro-

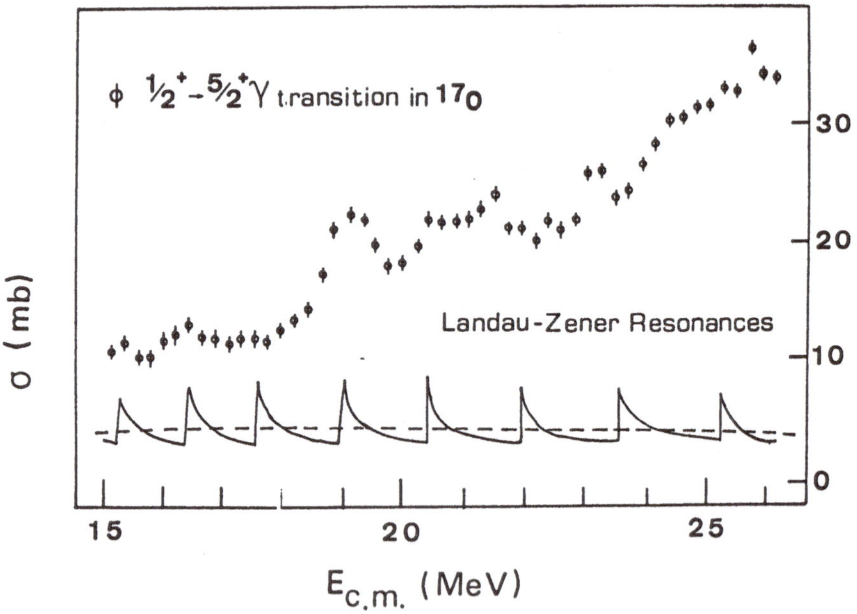

Fig. 6. Comparison of the inclusive γ-ray yield curve of the $^{13}C+^{17}O \rightarrow ^{13}C+^{17}O^*$(0.87 g.s.) transition with calculated Landau-Zener cross section. Solid line: Eq. (6); dashed line: Eq. (1). (Ref. 6).

motion mechanism is subject to some caution. The comparison of calculations and data shown in Fig. 6 is a rather unfair one, since the measured quantity is the inclusive yield of all transitions leading to the 0.87 MeV first excited level of ^{17}O. Some faith was, thus, needed to interpret the oscillations as due to the Landau-Zener promotion, the argument being rather the absence of these oscillations in other yield curves than the agreement with calculated cross sections.

In a further attempt to demonstrate experimentally the existence of the nuclear Landau-Zener effect, we turned to the study of coincident particle yield from the $^{13}C+^{17}O$ reaction(10), where the effects of the Landau-Zener promotion mechanism could be discerned more clearly. In fact, in a sharp cut-off approximation, at "resonant" energies only one partial wave will be active and at these energies the angular distributions of particles stemming from transitions due to the Landau-Zener promotion should follow a pure $P_L^2(\cos\theta)$ shape (P_L = Legendre polynomial). Hence we focussed our attention on finding angular distributions of inelastically scattered ^{17}O which follow a $P_L^2(\cos\theta)$ shape.

If the Landau-Zener promotion due to avoided crossing of levels is to be observed experimentally, the crossing should occur at a distance close to the touching distance of the two nuclei (the effects of a crossing inside the nucleus will be submerged by absorption). The avoided crossing between the $\Omega = 1/2$, $d_{5/2}$ and $s_{1/2}$ levels of ^{17}O shown in Fig. 3 occurs at 7.8 fm; the touching distance of the $^{12}C+^{17}O$ nuclei is $R_o = r_o(A_1^{1/3} + A_2^{1/3}) = 6.3$ fm ($r_o = 1.3$ fm).

The experimental search for the nuclear Landau-Zener effect via the $^{12}C(^{17}O,^{17}O^*)^{12}C$ scattering was performed using the ^{17}O beam of the CRN Strasbourg MP tandem(10). Angular distributions of all collision products were measured in energy steps of 0.5 MeV from 40-51 MeV(lab) by the kinematic-coincidence method.

Fig. 7 shows the $^{12}C(^{17}O,^{17}O^*$ 0.87 MeV$)^{12}C$ angular distributions at three incident energies. As mentioned earlier, angular distributions of the above transition should be governed by single $P_L^2(\cos\theta)$ shapes. The angular distributions at 46.5 and 49.5 MeV(lab) are well described by squared Legendre polynomials of order L=12 and 13, respectively, while the intermediary one at 48 MeV cannot be described by a single polynomial. Furthermore, as shown in Fig. 8, the angular distributions of the $^{12}C(^{17}O,^{17}O)^{12}C$ elastic scattering and the $^{12}C(^{17}O,^{16}O)^{13}C$ ground-state transfer at the "resonant" energy of 49.5 MeV show periodicities different from the periodicity of the $^{12}C(^{17}O,^{17}O^*)^{12}C$ inelastic scattering; this fact demonstrates that the cause of the periodicity in the latter is not due to the simple effects of the grazing partial wave.

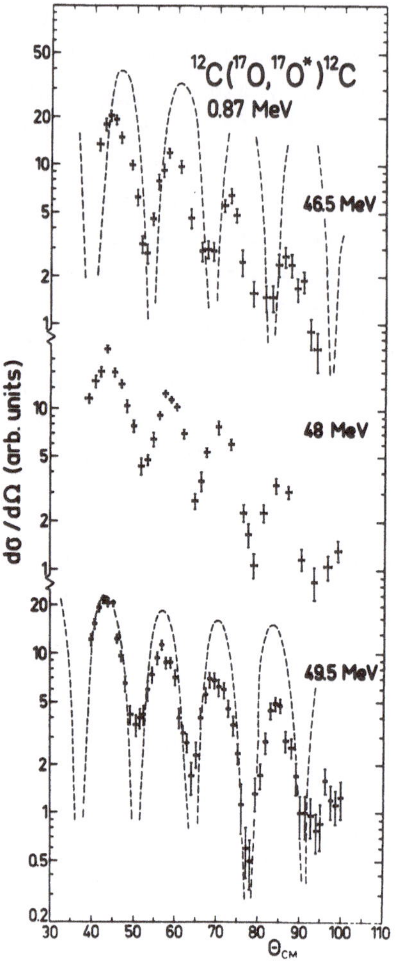

Fig. 7. Angular distributions of the $^{12}C(^{17}O,^{17}O)^{12}C$ elastic scattering at energies of 46.5, 48 and 49.5 MeV(lab). The dotted lines are $P^{2}_{12}(\cos\theta)$ (top) and $P^{2}_{13}(\cos\theta)$ (bottom) Legendre polynomial shapes.

Additional arguments in favour of interpreting the observed inelastic angular distributions as due to the Landau-Zener promotion are as follows:

(i) Using the simple sharp cut-off expression, the partial wave active in the Landau-Zener promotion in $^{12}C(^{17}O,^{17}O^{*})^{12}C$ can be calculated as

$$L_{act} = kR_C = \frac{1}{\hbar}\sqrt{2\mu(E-V_{CB})}\cdot R_C \; .$$

Using the value R_C = 7.8 fm (Fig. 3), one obtains for L_{act} the values

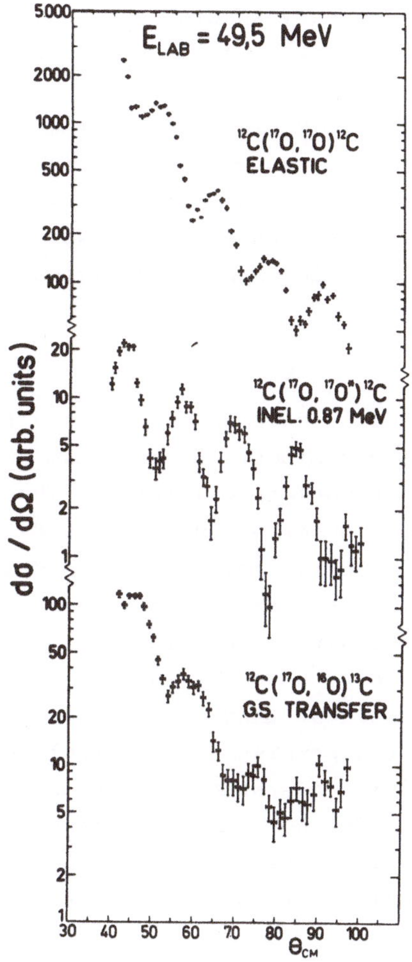

Fig. 8. Angular distributions of the $^{12}C(^{17}O,^{17}O)^{12}C$ elastic scattering (top), the $^{12}C(^{17}O,^{17}O^{*}\ 0.87)^{12}C$ inelastic scattering (middle) and the $^{12}C(^{17}O,^{16}O)^{13}C$ g.s. transfer (bottom), measured at 49.5 MeV(lab).

of 14 and 15 at incident (lab) energies of 46.5 and 49.5 MeV, respectively, quite close to the values L = 12 and 13 deduced from experimental angular distributions. However, in view of the ΔL = 2 character of the ($d_{5/2} \rightarrow s_{1/2}$) transition, one may expect $L = L_{act}-2$, giving L = 12 and 13, respectively, in perfect agreement with the observation.

(ii) The energy of the Landau-Zener "resonances" is given by eq. (7). We see that the periodicity of the "resonances" is given by the term $(\hbar^2/2\mu R_c^2)\ell(\ell+1)$. Using again the estimated critical distance R_c = 7.8 fm, one obtains a value of 0.045 MeV for the coefficient of this term. The calculated energy spacing between, say, the L = 12 and 13 "resonances", $\Delta E = 0.045\times2\times13 = 1.17$ MeV, compares well with the observed spacing of $\sim$1.2 MeV(c.m.).

The similarity of the TCSM diagrams for the $^{12}C+^{17}O$ and $^{13}C+^{17}O$ systems should show up in the Landau-Zener enhancement of the same transitions in these two systems, giving credit to Abe and Park's explanation[6] of the periodic enhancements in the excitation function of the $^{17}O(^{13}O,^{17}O^{*}\ 0.87)^{13}C$ inelastic transition discussed earlier.

5. Conclusions

The above considerations have shown evidence that under certain conditions, the Landau-Zener promotion mechanism known from atomic physics is also present in nuclear collisions. Thus the same quantum-mechanical effect has been observed in atomic and nuclear physics; furthermore, this effect is closely related to the Zener effect observed in solid semiconductors. This discovery has its intellectual attraction, as it corresponds to our desire of unification of the laws of nature. In heavy-ion reaction theory, however, this discovery has an additional fundamental aspect. The nuclear Landau-Zener effect is, in fact, a characteristic signature of the persistence of single-particle orbitals in TCSM diagrams, essential in understanding the basis of the application of the two-centre shell model to reactions between heavy ions. The observation of this effect provides direct evidence for the persistence of single-particle molecular effects in nuclear heavy-ion collisions.

References

1. McGraw-Hill Encyclopedia of Physics, McGraw-Hill (1983)
2. C. Zener, Proc. Royal Soc. A145 (1983) 523
3. L. Landau, Phys. Z. Sowietunion 2 (1932) 46;
   C. Zener, Proc. Royal Soc. A137 (1932) 696
4. G. Terlecki, W. Scheid, H.J. Fink and W. Greiner, Phys. Rev. C18 (1978) 265;
   J.Y. Park, W. Greiner and W. Scheid, Phys. Rev. C21 (1980) 958
5. J.Y. Park, W. Scheid and W. Greiner, Phys. Rev. C25 (1982) 1902.
6. Y. Abe and J.Y. Park, Phys. Rev. C28 (1983) 2316
7. For details see, e.g., E.E. Nikitin and L. Zülicke, Theory of Chemical Elementary Processes (Springer, Berlin 1978) Chap. 5
8. See, e.g., R. Siemssen, Proc. Argonne Symp. on Heavy-Ion Scattering, ANL Report ANL-7837, 1971, p. 145
9. R.M. Freeman, C. Beck, F. Haas, B. Heusch and J.J. Kolata, Phys. Rev. C28 (1983) 437
10. N. Cindro, C. Beck, R.M. Freeman and F. Haas, to be published

SEMICLASSICAL THEORY OF RESONANCES

H.J. Korsch

Fachbereich Physik, Universität Kaiserslautern
D-6750 Kaiserslautern, Federal Republik of Germany

1. Introduction

Semiclassical approximations construct the short-wavelength limit ($\hbar \to 0$) of quantum mechanics and can (in many cases quantitatively) account for quantum effects like bound state quantisation, interference oscillations, tunnelling, diffraction, etc. Besides providing numerical results these semiclassical approximations offer an indispensable way of understanding quantum processes, which is often impossible in a pure quantum treatment.

In this lecture we are concerned with semiclassical methods for characterizing and computing resonances, which will be defined as complex energy poles of the S-matrix, i.e. the corresponding solution of the Schrödinger equation is purely outgoing at infinity (a Siegert or Gamov state). It is beyond the scope of the present lecture to discuss in any detail the general features of semiclassical asymptotic approximations; many references can be found in recent reviews and textbooks /1-5/. We furthermore restrict ourselves to the first-order semiclassical treatment, which gives in many cases results sufficiently close to the exact ones. Higher-order semiclassical phase-integral methods have been developed, which can give an impressive improvement of the numerical results if neccessary /6/. There is one aspect of semiclassical techniques that should be emphasized, however, in the beginning: complex valued paths (or trajectories) appear quite naturally in semiclassical mechanics describing classically forbidden processes by continuation of classical mechanics into the complex plane. So basically there is a quite close relationship to complex-coordinate methods in exact quantum treatments of complex energy quatisation.

In Section 2 we develop the basic semiclassical ideas for the one-dimensional problem of shape resonances. Feshbach resonances in curve-crossing systems are treated in Section 3 and a more general case of semiclassical resonances in higher dimensional systems is discussed in Section 4 for the seperable case of field ionization of hydrogenic atoms. The last Section concludes with a short discussion of the general (nonseparable) case.

2. Shape Resonances

In this section, we shall derive a semiclassical approximation for one-dimensional potential scattering, where the potential has a short-range repulsive part, an attractive well and a barrier (see Fig. 1)

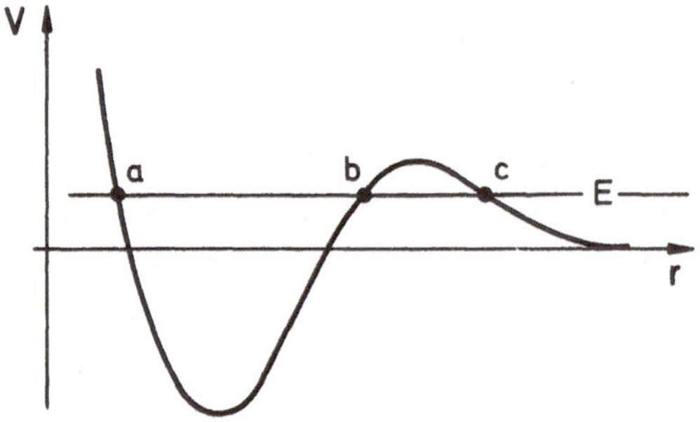

Fig. 1. Typical potential with a well and a barrier.

Potentials of this type arise often, but not exclusively, as effective potentials

$$V(r) = V_o(r) + \hbar^2 l(l+1)/2mr^2 \quad (1)$$

in atomic collisions. Here $V_o(r)$ is attractive at long-range and the barrier is caused by the centrifugal part. (Note, that in all semiclassical studies of radial problems the Langer modification $V(r) \rightarrow V(r) + \hbar^2/8mr^2$ must be used /2,7/). Such an effective potential possesses a series of resonances (in most cases an infinite number), which can be described asymptotically as purely outgoing waves at complex energies

$$E_n = \varepsilon_n - i\,\Gamma_n/2, \quad (2)$$

where the width $\Gamma(\Gamma>0)$ can be related to the 'lifetime' $\tau = \hbar/\Gamma$ of the state. Sharp resonances with small width $\Gamma_n \ll \varepsilon_n$ are important in spectroscopy (rotational predissociation) in contrast to low energy scattering, where orbiting phenomena /8/ are mainly due to relatively broad resonances.

For a real energy E below the potential barrier the classical motion possesses three turning points a, b, c where the classical momentum

$$p(r) = (2m\{E - V(r)\})^{1/2} \quad (3)$$

vanishes. More turning points may exist in the complex r-plane. (Here and in the following we assume, that the potentials can be analytically continued into the complex plane; for a discussion of some unusual semi-classical features related to non-analyticity of V(r) see Berry /9/.) For complex-valued energies (2) the turning points a, b, c are shifted into the complex plane. Typically these turning points are first-order zeros of E-V(r) = 0 and hence three Stokes lines, along which the action

$$S_{r_i r} = \frac{1}{\hbar}\int_{r}^{r_i} p(r')dr' \tag{4}$$

is purely imaginary, emanate from each turning point r_i. The active turning points are defined as an ordered sequence of turning points starting from r = o and ending at infinity (along the positive real axis) in such a way that they can be joined by a path which does not cross any Stokes line /4/. The active turning points define the active region between the successive active turning points, which is enclosed

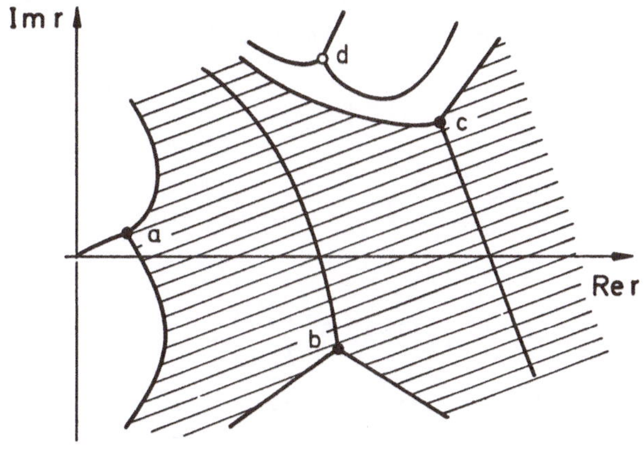

Fig. 2. Active turning points a, b, c for complex energy E. Also shown is an inactive turning point d. The active region is hatched.

by Stokes lines. The pattern of active turning points and Stokes lines shown in Fig. 2 usually is structurally stable if the energy E is slightly changed. Note, however, that the pattern can undergo abrupt changes if the energy crosses certain discontinuity lines $\mathfrak{D}_{ik}$ in the complex E-plane. $\mathfrak{D}_{ik}$ is defined by the complex energies for which the turning points r_i and r_k are linked by a common Stokes line (i.e. $\mathrm{Re}\, S_{r_i r_k} = 0$).

Using now the principle of exponential dominance ('Crossing a Stokes line, the exponentially dominant term is continuous, but the subdominant term has a discontinuity in the presence of a dominant term.

Otherwise it, too, is continuous.' /10,4/) one can continue the incoming wave by moving counter clockwise around a turning point /4/ and construct the reflected wave. Fig. 3 depicts the reflection pattern for the turning point structure of Fig. 2 as well as a two turning point

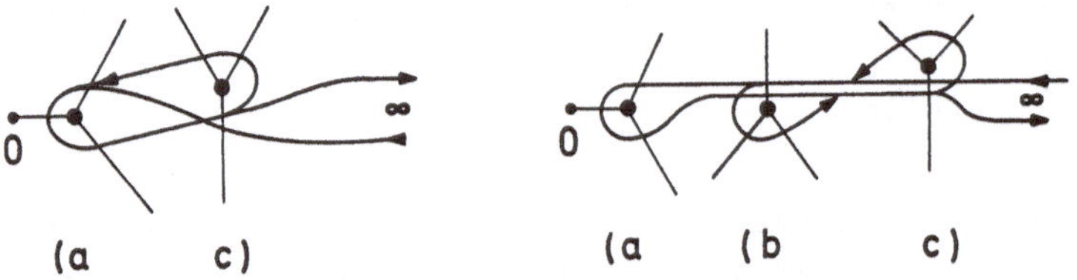

Fig. 3 Multiple reflection paths for a two turning point and a three turning point configuration.

configuration. The turning point configuration is characterized by an obvious bracket notation /4/, which we use to subscript the quantities in our equations. The multiple reflection amplitude can be summed in closed form and we obtain for the reflection amplitude

$$R_{(ac)} = -ie^{2iS_{\infty a}} \sum_{\nu=0}^{\infty} \left[-e^{2iS_{ca}}\right]^{\nu} \tag{5}$$

$$= -ie^{2iS_{\infty a}} \Big/ \left\{1+e^{2iS_{ca}}\right\}$$

and

$$R_{(a(bc)} = \left[-ie^{2iS_{\infty b}} - ie^{2iS_{\infty a}}\right] \sum_{\nu=0}^{\infty} \left[-e^{2iS_{cb}} - e^{2iS_{ca}}\right] \tag{6}$$

$$= \left[-ie^{2iS_{\infty b}} - ie^{2iS_{\infty a}}\right] \Big/ \left[1+e^{2iS_{cb}} + e^{2iS_{ca}}\right] .$$

Here $S_{\infty a}$ is a semiclassical scattering phase from a path starting at turning point a:

$$S_{\infty a} = \lim_{r\to\infty} \left\{ S_{ra} - p_{\infty} r/\hbar \right\} . \tag{7}$$

It may be worthwhile to note the relation

$$S_{ca} = S_{cb} + S_{ba} . \tag{8}$$

Poles of the reflection amplitude, i.e. purely outgoing states, are found in Eq. (5) for

$$e^{2iS_{ca}} = -1 = e^{(2n+1)\pi i} , \tag{9a}$$

which can be rewritten as a well known WKB quantisation condition

$$\mathcal{N}_{(ac)}(E) = \frac{1}{\pi} S_{ca} = n + \frac{1}{2} \quad , \; n = 0,1,\ldots \tag{9b}$$

in the two-turning point case (ac). In the three turning point configuration (a(bc) we find poles at

$$e^{2iS_{cb}} + e^{2iS_{ca}} = -1 = e^{(2n+1)\pi i} \tag{10a}$$

or

$$\mathcal{N}_{(a(bc)}(E_n) = \frac{1}{2\pi i} \ln (e^{2iS_{cb}} + e^{2iS_{ca}}) = n+\frac{1}{2} \; , \; n = 0,1,2,.. \tag{10b}$$

where the proper branch of the logarithm must be chosen by analytic continuation from the real axis. Eqs. (9b) and (10b) define implicitely the complex energy poles E_n. In particular they lie on a smooth curve (the pole string curve $\mathcal{C}$) in the complex E-plane, which is given by

$$\mathrm{Im}\, \mathcal{N}(E) = 0, \tag{11}$$

i.e. the quantum number function must be real. (Note that the quantum number function $\mathcal{N}(E)$ changes if the pole-string curve $\mathcal{C}$ reaches a discontinuity line $\mathcal{D}_{ik}$.) A higher-order semiclassical phase integral extension of the three point equation (10) has been derived by Thylwe /11/.

If two of the three turning points move far away from each other their contribution may become insignificant and Eq. (10b) reduces to the two turning point formula (9b).

The derivation of the formulae given above is based on the assumption, that the turning points are well separated. They break down when two - or more - points are close together and cannot describe the passage through the discontinuity line $\mathcal{D}_{ik}$, where two turning points are connected by a common Stokes line. For the case, that the turning points b and c (see Fig. 1 and 2) are close together and may be connected by a Stokes line, a uniform semiclassical expression has been derived by Connor /12,13/. This method uses a uniform semiclassical mapping onto a comparison equation with a closed form solution. In the present case we map onto a parabolic barrier, i.e. onto the equation for the parabolic cylinder function

$$u'' + (\nu + x^2/4)\, u = 0, \tag{12}$$

and obtain a two-turning point a parabolic connection formula, which connects the semiclassical wave between a and c. We have

$$\Psi(r) = B'\, p(r)^{-1/2}\, e^{iS_{rb}} + B''\, p(r)^{-1/2}\, e^{-iS_{rb}} \tag{13a}$$

on the left-hand side of the turning points (in the sence of the ordered sequence of active turning points) and

$$\Psi(r) = C'\, p(r)^{-1/2}\, e^{iS_{rc}} + C''\, p(r)^{-1/2}\, e^{-iS_{rc}} \tag{13b}$$

on the opposite side. The coefficients are related by a transformation matrix $\underset{\sim}{M}$ (see Figure 4), given by /12-15/

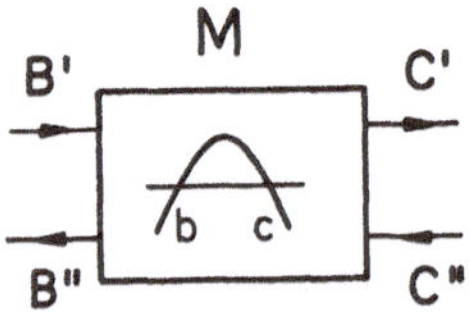

$$\begin{pmatrix} C' \\ C'' \end{pmatrix} = \underset{\sim}{M} \begin{pmatrix} B' \\ B'' \end{pmatrix}$$

Fig. 4 Two turning point connection diagram

$$\underset{\sim}{M} = \begin{pmatrix} A^- & -ie^{-\pi\varepsilon} \\ ie^{-\pi\varepsilon} & A^+ \end{pmatrix} \tag{14}$$

with

$$A^{\pm} = \frac{\sqrt{2\pi}}{\Gamma(\frac{1}{2} \mp i\varepsilon)}\, e^{-\frac{\pi}{2}\varepsilon \pm i[\varepsilon - \varepsilon \ln(-\varepsilon)]} , \tag{15}$$

where ε is given by $\varepsilon = iS_{cb}/\pi$ (i.e. ε negative real when E is real and below the barrier). The reflection at turning point a imposes the condition

$$B' = -ie^{2i\alpha}\, B'' \tag{16}$$

with $\alpha = S_{ba}$. Looking now for purely outgoing solutions (C" = 0) we find the condition

$$e^{-2i\alpha}\, e^{\pi\varepsilon}\, A^{+}(\varepsilon) = -1 \tag{17a}$$

or

$$\mathcal{N}_{(a(\overline{bc})}(E_n) = \frac{1}{\pi}\left\{\alpha - \frac{\varepsilon}{2}\left[1-\ln(-\varepsilon)\right] + \frac{i}{2}\ln\left(\frac{\sqrt{2\pi}\exp(\pi\varepsilon/2)}{\Gamma(\frac{1}{2}-\varepsilon)}\right)\right\} \qquad (17b)$$
$$= n + \frac{1}{2}, \ n = 0,1,...$$

in terms of the $(\overline{bc})$ uniformised quantum number function.

In the limit of very sharp resonances ($\Gamma \ll \mathcal{E}$) a perturbation treatment of (17b) decouples the real and imaginary parts of (17b) and we find /12/ as a quantisation condition for the resonance position

$$\alpha(\mathcal{E}_n) - \frac{1}{2}\bar{\phi}(\varepsilon(\mathcal{E}_n)) = \pi(n + \tfrac{1}{2}) \qquad (18)$$

where the small anection term $(0 \leq \bar{\phi} \leq 0.15)$ is given by

$$\bar{\phi}(\varepsilon) = \varepsilon + \arg\Gamma(\tfrac{1}{2} + i\varepsilon) - \varepsilon\ln|\varepsilon|. \qquad (19)$$

For the width we obtain

$$\Gamma_n(\mathcal{E}_n) = \frac{\hbar\,\omega(\mathcal{E}_n)}{2\pi}\ln\left[1 + \exp(2\pi\varepsilon(\mathcal{E}_n))\right], \qquad (20)$$

where $\omega(\mathcal{E})$ is the classical frequency of oscillation in the potential well:

$$\omega(\mathcal{E}) = \frac{\pi}{m} \Big/ \int_a^b \frac{dr}{p(r)} \ . \qquad (21)$$

and $\ln\left[1 + \exp(2\pi\varepsilon)\right]$ can be interpreted as the mean barrier collision number with WKB tunnelling probability $(1 + \exp(2\pi\varepsilon))^{-1}$ in a single collision. (See ref. /12,16-17/ for more details on real-energy semiclassical resonances).

In the opposite limit of large $|\varepsilon|$ (broad resonances) Stirling's approximation can be used for the Gamma function in Eq. (17b) and the resultant approximation is a two-point formula

$$\mathcal{N}_{(a(\overline{bc})}(E_n) \rightarrow \mathcal{N}_{(ac)}(E_n) = n + \tfrac{1}{2}, \ n = 0,1... \ . \qquad (22)$$

This semiclassical complex energy quantisation is closely related to the theory of complex angular momentum (Regge) poles for potentials with a centrifugal barrier (see Eq. (1)). Precisely the same semiclassical formulae can be used to define a semiclassical quantisation condition for Regge poles, i.e. $\mathcal{N}(E, l_n) = n + 1/2$ for fixed real E defines the complex angular momenta l_n. Numerical applications can be found in ref. /4,10,12,18-23/. Very recently, however, the growing interest in complex energy resonances has stimulated numerical studies of semiclassical complex energy quantisation. Computations by the author /24-26/ and later on by Connor and Smith /27-28/ for various potentials showed that semiclassical complex energy quantisation offers an efficient and reliable tool for an approximate numerical computation of resonances. Numerically the quantisation equations $\mathcal{N}(E_n) = n + \frac{1}{2}$ must be solved for the complex resonance energies E_n. This can be done efficiently by means of a complex Newton iteration; for details see ref. /24-28/.

Fig. 5 shows as a typical example /26/ the string of complex resonance poles for the potential

$$V(r) = (\tfrac{1}{2}r^2 - J)\, e^{-\lambda r^2} + J \tag{23}$$

with J = 0.8, λ = 0.1, which shows the complex threshold behaviour

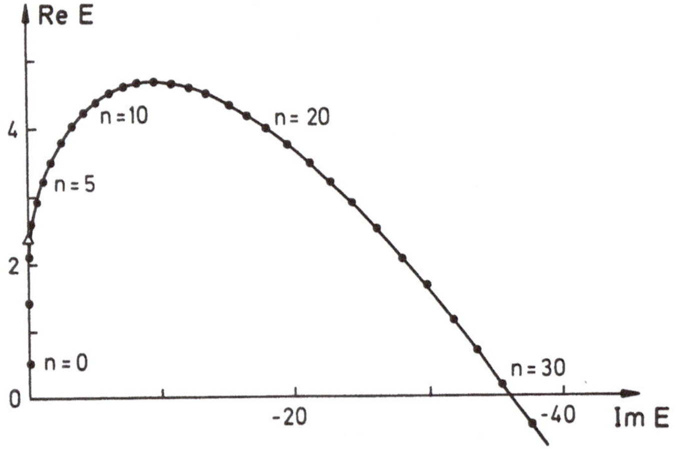

Fig. 5 Complex energy resonances for potential (23) ($\hbar$=m=1) /26/. The triangle (▲) marks the height of the potential barrier.

($\mathrm{Re}E_n$ is bounded from above). (For a discussion of the occurence of multiple spectra for this potential see ref. /26,29-31/). Note that the poles can be joined by the smooth line given by Eq. (11). Table 1 lists some of the exact quantum results obtained by a complex rotated Milne quantisation /25,26/ and the semiclassical resonances (Eq. 17b). Good agreement is observed, the difference being indistinguishable on the scale of drawing in Fig. 5. Similar agreement is found for other

n	E_n		E_n^{sc}	
3	2.585	- 0.174i	2.607	- 0.172i
7	3.824	- 2.487i	3.843	- 2.482i
11	4.523	- 6.155i	4.545	- 6.149i
15	4.644	-10.826i	4.664	-10.820i
19	4.197	-16.358i	4.211	-16.353i
23	3.185	-22.667i	3.198	-22.661i
27	1.625	-29.691i	1.637	-29.686i
31	-0.473	-37.385i	-0.461	-37.380i

Table 1. Exact and semiclassical (sc) complex energy resonance poles E_n for potential (23) ($\hbar$=m=1) for some odd values of n /26/.

potentials /24-28/, as for example the (12,6) Lennard-Jones potential

$$V(r) = D \left\{ \left(\frac{r_0}{r}\right)^{12} - 2 \left(\frac{r_0}{r}\right)^6 \right\} \tag{24}$$

with a centrifugal barrier (see Eq. (1)). Fig. 6 shows some uniform three-turning point semiclassical results taken from ref. /28/ for

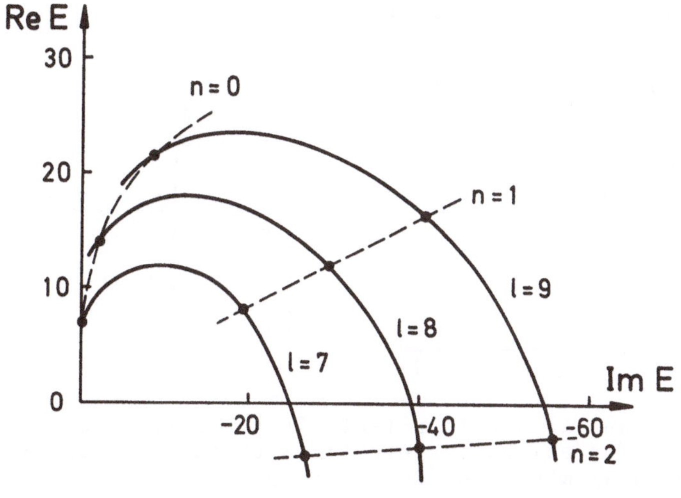

Fig. 6. Semiclassical complex energy resonances E_{nl} for potential (24) with l = 7,8,9 and n = 0,1,2.

angular momenta l = 7,8,9 and resonance numers n = 0,1,2 (for the case D=50 in the units used in /28/). The resonances $E_{n,l}$ lie on a regular mesh in the complex plane.

Finally it should be pointed out, that the semiclassical approximation has an additional advantage: The method provides automatically an unambignous assignment of a quantum number n to a given resonance state, which is not generally true in most of the quantum methods (see, however, the quantum Milne method developed in ref. /25/).

3. Curve Crossing and Feshbach Resonances

In curve crossing collisions semiclassical methods (e.g. the Landau-Zener-Stückelberg approximation) have a long history and an enormous literature exists. Here we will concentrate on some aspects of the semiclassical complex energy quantization, a field which has hardly been explored until now. We are concerned with the solutions of the two coupled equations

$$\left[\frac{\hbar^2}{2m}\frac{d^2}{dr^2} + E - V_1(r)\right]\Psi_1(r) = V_{12}(r)\,\Psi_2(r)$$

$$\left[\frac{\hbar^2}{2m}\frac{d^2}{dr^2} + E - V_2(r)\right]\Psi_2(r) = V_{21}(r)\,\Psi_1(r) \,, \qquad (25)$$

where typical diabatic potentials $V_i(r)$ and the coupling $V_{12}(r)=V_{21}(r)$

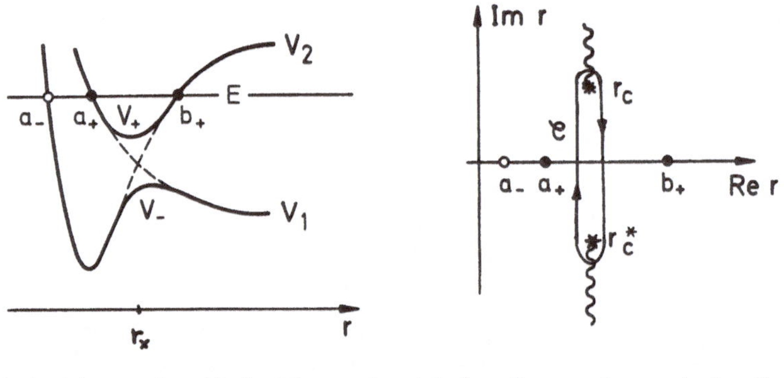

Fig. 7. Diabatic and adiabatic potentials for a two state Feshbach resonance. The adiabatic potentials cross at r_c and r_c^*, with two branch cuts emanating from these points.

are shown in Fig. 7. The diabatic potentials cross at r_x. Adiabatic potentials are defined by diagonalisation of the diabatic potential

$$V_\pm(r) = \frac{1}{2}\left\{ V_1 + V_2 \pm \left[(V_1-V_2)^2 + 4\,V_{12}^{\,2}\right]^{1/2}\right\} \;; \qquad (26)$$

the classical turning points for $V_\pm$ are denoted by $a_\pm$ and b_+.

For the case of very weak coupling V_{12} the bound levels of potential V_2 are only slightly shifted by coupling to the continuum channel V_1 and perturbation and semiclassical expressions for level shifts and decay widths are available (see for instance /32/ and references therein). As in the case of shape resonances these small width situations are mainly of interest for spectroscopic applications (e.g. predissociation line widths). More recently broader curve crossing resonance phenomena have been studied in scattering, with application of complex energy quantisation and complex rotation techniques (see e.g. ref. /30,33/). In the following we will give a semiclassical complex energy quantisation formula for predissoziating states. An important step in a semiclassical description of the curve crossing situation is to treat the transition entirely in terms of the adiabatic potentials V_+ and V_- (note that the diabatic potentials are only rarely available in actual problems and that their definition is not unique).
V_+ und V_- show an avoided crossing on the real axis and a crossing at two complex conjugate points r_c and r_c^* in the complex plane, with two branch cuts emanating from these points (see Fig.7). The adiabatic potential thus forms a Riemann surface with $V_+(V_-)$ on the upper (lower) surface and transitions can be described by trajectories passing around

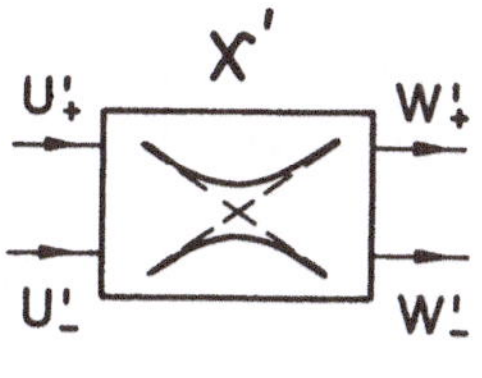

$$\begin{pmatrix} W'_+ \\ W'_- \end{pmatrix} = \underset{\sim}{X}' \begin{pmatrix} U'_+ \\ U'_- \end{pmatrix}$$

Fig. 8. Curve-crossing connection diagram

the complex intersection points and crossing at the branch cut to the other sheet of the adiabatic surface (see, e.g., ref. /34,35/). It is now possible to express the wavefunction as a multiple path series as described in Sect. 2 for the barrier transition and to write down a path-integral condition for the resonances, which is similar to Eq. (17b) /36/.

In the following we give an extension to complex energies of a diagrammatic approach following previous real energy treatments /15,34-39/. For a simple curve crossing system semiclassical mapping onto a linear crossing/constant coupling model and analysis of the connection formulae show that the amplitudes of the semiclassical wavefunction (compare Eq. 13) on the upper (+) and on the lower (-) surface left of the crossing ($U'_{\pm}$) and right of the crossing ($W'_{\pm}$) are related by the matrix

$$\underset{\sim}{X}' = \begin{pmatrix} B^- & -e^{-\pi\nu} \\ e^{-\pi\nu} & B^+ \end{pmatrix} \tag{27}$$

with

$$B^{\pm} = \frac{\sqrt{\pm 2\pi\nu i}}{\Gamma(1\pm i\nu)}\, e^{-\frac{\pi}{2}\nu \mp i[\nu - \nu \ln \nu]} \quad . \tag{28}$$

The phase-integral ν is given by

$$\nu = -\frac{1}{2\pi\hbar} \oint_{\mathcal{C}} p(r)\, dr, \tag{29}$$

where the contour $\mathcal{C}$ encircles both transition points (see Fig.7). For real energies ν is real and positive and $e^{-2\pi\nu}$ describes the nonadiabatic crossing probability in a single crossing. Eq.(27) applies to the situation, where the flux through the crossing goes toward the positive r-direction. In the opposite case the corresponding amplitudes $U''_{\pm}$ and $W''_{\pm}$ are related by

$$\begin{pmatrix} U''_+ \\ U''_- \end{pmatrix} = \underset{\sim}{X}'' \begin{pmatrix} W''_+ \\ W_- \end{pmatrix} \quad \text{with } \underset{\sim}{X}'' = \begin{pmatrix} B^- & e^{-\pi\nu} \\ -e^{-\pi\nu} & B^+ \end{pmatrix} \quad . \tag{30}$$

The Feshbach-predissociation can now be described by the diagram given in Fig. 9. The reflections at the turning points give the conditions

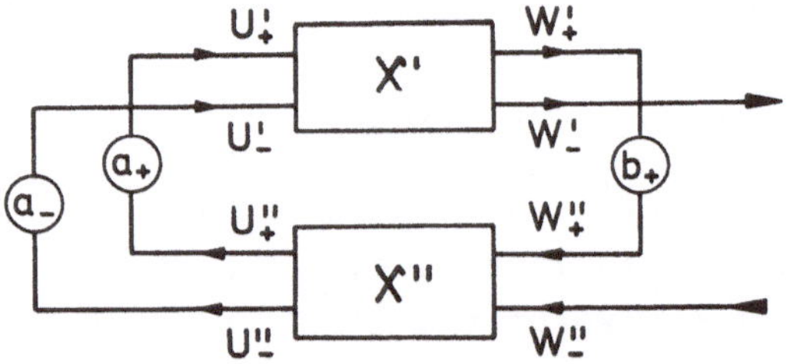

Fig. 9. Diagram for Feshbach resonances in predissoziation

$$U'_{\pm} = -ie^{2i\alpha_{\pm}}\, U''_{\pm} \quad ; \; W''_{+} = -ie^{2i\beta_{+}}\, W'_{+} \, , \tag{31}$$

where the phase integrals are given by

$$\alpha_{\pm} = \frac{1}{\hbar}\int_{a_{\pm}}^{r_x} p_{\pm}(r)dr, \quad \beta_{+} = \frac{1}{\hbar}\int_{r_x}^{b_+} p_{+}(r)dr \; . \tag{32}$$

Imposing now purely outgoing boundary conditions ($W_-'' = 0$) one easily finds from Eqs. (27), (30), (31) and the diagram in Fig. 9 the resonance condition for the curve-crossing quantum number function:

$$\mathcal{N}_x(E) = \frac{1}{\pi}\left\{\beta_+ + \frac{1}{2i}\ln\left[(B^\circ)^2\, e^{2i\alpha_+} + e^{-2\pi\nu+2i\alpha_-}\right]\right\} \tag{33}$$

$$= n + 1/2 .$$

Simplifications of Eq. (33) can be obtained in the limit of very weak or strong coupling (see for instance ref. /14,15/). Numerical applications of the complex energy resonance quantisation (33) and comparison with exact quantum results as well as a discussion of more general curve-crossing situations will be published elsewhere /36/.

4. Field Ionization

Ionization of hydrogenic atoms in static electric fields also known as 'Stark ionization' is still a problem of considerable theoretical activity, especially for highly excited (Rydberg) atoms. Semiclassical approximations for energy positions and lifetimes of decaying states in electric fields which have been extensively discussed in the literature are, however, resticted to states well below the potential barrier and only very recently a semiclassical complex energy treatment of resonant states has been successfully applied uniformly to states well below and above the barrier /40,41/.

The Schrödinger equation for hydrogen-like atoms in a homogeneous electric field F along the z-axis is given by

$$\left\{\Delta + 2\left(E + \frac{1}{r} + Fz\right)\right\}\Psi = 0 \tag{34}$$

(we use atomic units throughout this section). The potential $V = -\frac{1}{r} - Fz$ has a saddle point on the z-axis (see Fig. 10) of height $E_{sp} = -2\sqrt{F}$, i.e. classically particles with an energy above E_{sp} can escape. Quantum mechanically all states are quasibound and can ionise. Therefore all states can be described as resonances with small widths for states below the saddlepoint. It is observed, however, that long-lived states exist considerably above the saddlepoint. This is due to the existence of a conserved quantity: the z-component R_z of the generalized Runge-Lenz vector /42/ (in addition to the z-component of the angular momentum) and the conservation of R_z forbids a decay, very similar to a centrifugal barrier in rotationally symmetric systems. The right-hand

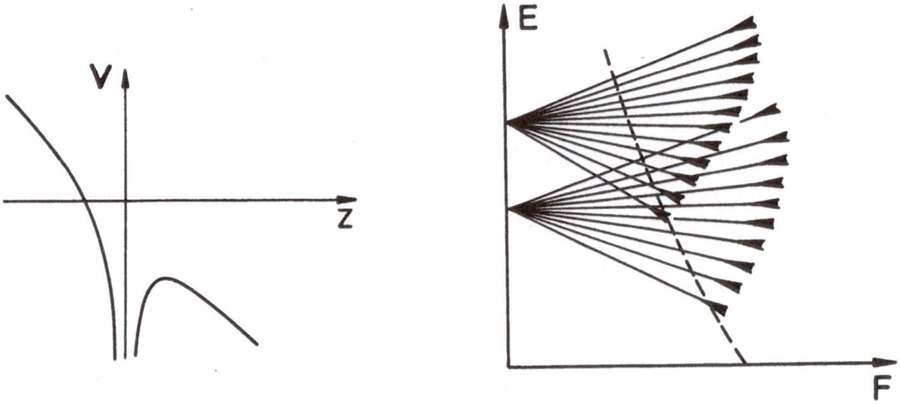

Fig. 10. Potential along the z-axis and two Stark manifolds with different principal quantum number n.

side of Fig. 10 shows schematically the field dependence of two Stark manifolds with different principal quantum number. The width Γ of the states is marked. The dashed line gives the saddle point energy.

The existence of a complete set of integrals of motion (i.e. the so-called integrability of the system) is intimately related to the separability of the Hamiltonian: In parabolic coordinates Eq.(34) reduces - after separation of the azimuthal dependence (m is the azimuthal quantum number) - to the equations

$$\left\{\frac{d^2}{d\xi^2} + \frac{1}{2}E + \frac{1-m^2}{4\xi^2} + \frac{Z_1}{\xi} - \frac{1}{4}F\xi\right\}\chi^{(1)}_{n_1,m}(\xi) = 0 \tag{35a}$$

$$\left\{\frac{d^2}{d\eta^2} + \frac{1}{2}E + \frac{1-m^2}{4\eta^2} + \frac{Z_2}{\eta} + \frac{1}{4}F\eta\right\}\chi^{(2)}_{n_2,m}(\eta) = 0 \quad , \tag{35b}$$

which are only coupled via the separation constants Z_1 and Z_2 which must satisfy

$$Z_1 + Z_2 = 1 \quad . \tag{36}$$

The solutions of (35) can be classified by two parabolic quantum numbers

n_1 and n_2 and the principal quantum number n is given by

$$n = n_1 + n_2 + |m| + 1. \tag{37}$$

Eq. (35a) describes a bound motion in ξ, the η motion is unbounded. A solution of Eq. (35) determines the resonance energy E and the separation constants Z_1 (Z_2 is then given by (36)).

The semiclassical complex energy quantisation described in Sect. 2 can be applied to the problem of field ionization by quantising the bound motion (35a) by means of the two turning point formula (9b)

$$\mathcal{N}_{(\xi_1,\xi_2)}(E,Z_1) = n_1 + \frac{1}{2} \tag{38}$$

and Eq. (35b) by one of the three turning point formulae. We have used the uniform version (17b):

$$\mathcal{N}_{(\eta_1,\overline{(\eta_2,\eta_3)})}(E,Z_1) = n_2 + \frac{1}{2} \quad . \tag{39}$$

The separation constant Z_2 has been removed from (35b) by means of (36); ξ_i and η_i in (39) denote the turning points. A solution of (38,39) defines the complex valued resonance energies $E(n_1,n_2)$. Again the quantum numbers n_1,n_2 of the resonances are unambiguously assigned by the semiclassical equations. Table 2 lists exact quantum results (obtained in

n_1	$-\mathcal{E}.10^3$	$\Gamma.10^3$	$-\mathcal{E}^{sc}.10^3$	$\Gamma^{sc}.10^3$
0	7.155	1.5-1	7.177	1.276-1
1	6.702	9.0-2	6.712	8.412-2
2	6.238	5.0-2	6.242	4.863-2
3	5.766	2.3-2	5.767	2.293-2
4	5.289	8.01-3	5.289	7.891-3
5	4.811	1.74-3	4.811	1.733-3
6	4.334	2.184-4	4.334	2.138-4
7	3.857	1.311-5	3.856	1.260-5
8	3.379	2.606-7	3.379	2.646-7

Table 2: Exact quantum /43/ and complex energy semiclassical /40/ values of resonance positions and widths for n=10 m=1 and $F=1.5\cdot10^{-5}$ a.u.

this case by real energy methods /43/) for the position $\mathcal{E}$ and the width Γ of the resonance and complex-energy semiclassical results $E=\mathcal{E}-i\Gamma/2$.

Comparisons with available exact quantum computations - also with complex-energy results /44,45/ - for other values of F, n and m showed similar agreement /40,41/. Fig. 11 shows the semiclassical complex

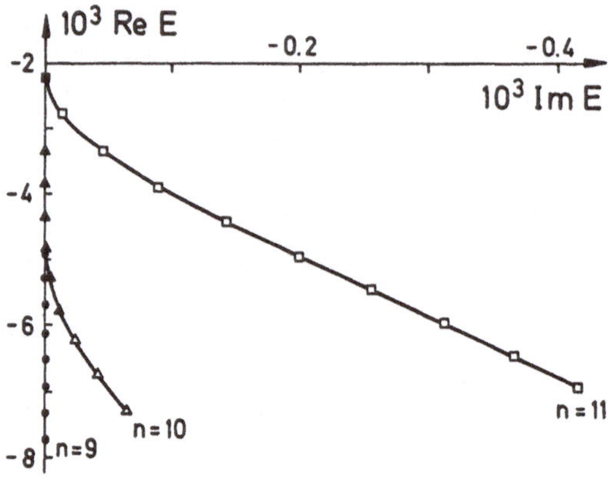

Fig. 11. Semiclassical complex energies for $F=1.5\text{-}10^{-5}$ a.u., m=1 and n=9 (o), n=10 (▲), n=11 (□).

energy resonances for $F=1.5\cdot 10^{-5}$ a.u. m=1 and n=9,10,11 in the complex plane. The states classically below the barrier are marked by open symbols. Again the resonances lie on smooth lines, which seems be characteristic for integrable systems.

5. Compound State Resonances

Let us finally briefly discuss resonances in more general systems, i.e. for nonintegrable n-dimensional Hamiltonians H, where no complete set of n compatible integrals of motion exists. Such resonances are often called compound resonances and arise quite frequently in reactive and nonreactive atom-molecule collisions. If such systems are sufficiently close to integrable ones (i.e. $H= H_0+ \varepsilon H_1$, where H_0 is integrable and ε is small), then such systems can be approximately quantised semiclassically in the same manner as integrable ones, i.e. by means of the EBK-quantisation conditions

$$\mathcal{N}_i(E) = \frac{1}{h} \oint_{\mathcal{C}_i} \sum_k p_k dq_k = n_i + \frac{1}{2} \ . \tag{40}$$

The q_K, P_K are conjugate coordinates and momenta and the $\mathcal{C}_i$ are topologically independent closed curves. The EBK-quantisation (40) has been successfully applied to many bound systems and to systems sufficiently below a potential barrier, where tunnelling was neglected, the Henon-Heiles system being a well known example. Some semiclassical studies of states above the barrier, however, have been also carried out, see, e.g. ref. /46-49/). Let us discuss as an example the Hamiltonian

$$H = p_x^2/m + p_y^2 + y^2 + De^{-\alpha(x-y)} - 2De^{-\alpha(x-y)/2} , \qquad (41)$$

which possesses no potential barrier and all quantum states with positive energy will decay. There exist, however, classically bound states for positive energies and EBK-quantisation has been used to compute real energy positions of the resonances /49/ in excellent agreement with exact ones. It seems also possible to estimate the lifetime of these states by using classically forbidden paths that tunnel through the dynamical barrier (i.e. by means of an essentially one-dimensional approximation). The development of a consistent semiclassical theory for resonance widths for such systems, especially in the region of broad resonances, as well as a complex energy semiclassical quantisation method is still a challenging task for the future.

References

/1/ N. Fröman and P.O. Fröman: "JWKB Approximation" (North-Holland, Amsterdam, 1965)

/2/ M.V. Berry and K.E. Mount, Reports Prog. Phys. 35, 315 (1972)

/3/ M.S. Child: "Molecular Collision Theory", (Academic Press, London 1974)

/4/ J. Knoll and R. Schaeffer, Ann. Phys. N.Y. 97, 307 (1976)

/5/ M.S. Child, ed.: "Semiclassical Methods in Molecular Scattering and Spectroscopy", (Reidel, Dordrecht, 1980)

/6/ N. Fröman, ref. /5/ pp. 1-44

/7/ H.J. Korsch, J. Chem. Phys. 69, 1311 (1978)

/8/ H.J. Korsch and K.E. Thylwe, J. Phys. B16, 793 (1983)

/9/ M.V. Berry, J. Phys. A15, 3693 (1982)

/10/ G.G. Stokes, Trans. Camb. Phil. Soc. 10, 106 (1857)

/11/ K.E. Thylwe, J. Phys. A16, 3325 (1983)

/12/ J.N.L. Connor, ref. /5/ pp. 45-107

/13/ J.N.L. Connor, Mol. Phys. 15, 621 (1968); 23, 717 (1973); 25, 1469 (1973)

/14/ M.S. Child, J. Mol. Spect. 53, 280 (1974)

/15/ M.S. Child, ref. /5/ pp. 127-154

/16/ R.J. LeRoy and W.-K. Liu, J. Chem. Phys. 69, 3622 (1978)

/17/ J.N.L. Connor and A.D. Smith, Mol. Phys. 43, 397 (1981); Mol. Phys. 45, 149 (1982)

/18/ J.B. Delos and C.E. Carlson, Phys. Rev. A11, 210 (1975)

/19/ J.N.L. Connor, W. Jakubetz and C.V. Sukumar, J. Phys. B9, 1783 (1976)
/20/ J.N.L. Connor, D.C. Mackay and C.V. Sukumar, J. Phys. B12, L515 (1979)
/21/ J.N.L. Connor, W. Jakubetz, D.S. Mackay and C.V. Sukumar, J. Phys. B13, 1823 (1980)
/22/ J.N.L. Connor, J.B. Delos and C.E. Carlson, Mol. Phys. 31, 1181 (1976)
/23/ K.E. Thylwe, J. Phys. B16, 1915 (1983)
/24/ H.J. Korsch, H. Laurent and R. Möhlenkamp, Mol. Phys. 43, 1441 (1981)
/25/ H.J. Korsch, H. Laurent and R. Möhlenkamp, J. Phys. B15, 1 (1982)
/26/ H.J. Korsch, H. Laurent and R. Möhlenkamp, Phys. Rev. A26, 1802 (1982)
/27/ J.N.L. Connor and A.D. Smith, Chem. Phys. Lett. 88, 559 (1982)
/28/ J.N.L. Connor and A.D. Smith, J. Chem. Phys. 78, 6161 (1983)
/29/ M. Rittby, N.Elander and E. Brändas, Phys. Rev. 26, 1804 (1982)
/30/ M. Rittby, N. Elander and E. Brändas, Int. J. Quant. Chem. 23, 865 (1983)
/31/ O. Atabek and R. Lefebvre, Nuovo Cim. 76B, 176 (1983)
/32/ M.S. Child and R. Lefebvre, Mol. Phys. 34, 979 (1977)
/33/ M. Rittby, N. Elander and E. Brändas, preprint 1984
/34/ D.S.F. Crothers, Adv. Phys. 20, 405 (1971)
/35/ A. Barany and D.S.F. Crothers, Physica Scripta 23, 1096 (1981)
/36/ H.J. Korsch, in preparation
/37/ A.D. Bandrauk and M.S. Child, Mol. Phys. 19, 95 (1970)
/38/ H. Nakamura, Phys. Rev. A26, 3125 (1982)
/39/ K.S. Lam and T.F. George, Ref./5/,pp. 179-261
/40/ H.J. Korsch and R. Möhlenkamp, Z. Phys. A314, 267 (1983)
/41/ D.F. Farrelly and W.P. Reinhardt, J. Phys. B16, 2103 (1983)
/42/ K. Helfrich, Theor. Chim. Acta 24, 271 (1972)
/43/ E. Luc-Koenig and A. Bachelier, J. Phys. B13, 1743 (1982)
/44/ M. Hehenberger, H.V. McIntosh and E. Brändas, Phys. Rev. A10, 1494 (1974)
/45/ O. Atabek and R. Lefebvre, Int. J. Quantum Chem. 19, 901 (1981)
/46/ R.A. Marcus, Faraday Discussions Chem. Soc. 55, 34 (1973)
/47/ J.R. Stine and R.A. Marcus, Chem. Phys. Lett. 29, 575 (1974)
/48/ D.K. Bondi, J.N.L. Connor, J. Manz and J. Römelt, Mol. Phys. 50, 467 (1983)
/49/ D.W. Noid and M.L. Koszykowski, Chem. Phys. Lett. 73, 114 (1980)

THE HERMITIAN REPRESENTATION OF THE COMPLEX COORDINATE METHOD: THEORY AND APPLICATION

Nimrod Moiseyev
Department of Chemistry
Technion - Israel Institute of Technology
Haifa, 3200, Israel

1. Introduction

The atomic and molecular autoionization resonances and the predissociation resonances are the two types of physical phenomena which will be discussed during this lecture.

Predissociation resonances are not rare phenomena in chemical reactions. As a matter of fact, they are obtained in any chemical reaction where an intermediate compound has a finite lifetime (known as an "activated complex") is produced. Predissociation resonances are obtained also from rotational excitations of molecules by microwaves radiation and recently were studied in van der Waals complexes auch as He-HCl, $Ar\text{-}I_2$ and $Ar\text{-}N_2$. Autoionization resonances were first observed sixty years ago by Auger[1]. However, even forty years later, the computation of autoionization resonance lifetime was considered as an art! (About "Art and Science in computation of autoionization resonances" see Ref. 2). The autoionization and the predissociation resonances are examples of physical phenomena which are more well-defined by experiment than by theory. In the scattering experiment, for example, the resonances are observed by measuring the cross-section as a function of the energy of the incoming particles, and are associated with the local maxima appearing above the threshold energy to the continuum. The resonance energies and lifetimes are defined experimentally as the positions and inverse widths of these local maxima respectively. The Lorenzian shape of the cross-section distribution is due to the fact that the S matrix (has a pole at $E_r - iE_i$) which is associated with complex eigenvalue of the Schrödinger equation,

$$\hat{H}(r)\psi(r) = W\psi(r) \qquad (1)$$

where,

$$W = E_r - iE_i \text{ and } \Gamma = 2E_i$$

The solutions of Eq. (1) cannot be obtained by conventional quantum mechanical methods. The variational methods cannot be used since they were derived for Hermitian Hamiltonian where the eigenfunctions are square integrable functions. Moreover, for the case where the asymptotic wavefunction is an outgoing plane wave, $\exp(i\sqrt{W}\ r)$, ψ diverge as $r \to \infty$ since W is a complex number (in such a case the number of the particles is not conserved in the coordinate space at any given time). However, it is easy to see that in this special case by scaling the internal coordinates of the Hamiltonian

$$H(\eta r)\psi(\eta r) = W\psi(\eta r) \tag{2}$$

and by taking η to be complex

$$\eta = \alpha \exp(i\Theta),\ \Theta > \text{Arctan}\ (E_i/E_r) \tag{3}$$

the resonance wavefunction becomes square integrable and the number of particles is conserved at any given time. This is the motivation of the complex scaling which is known as - The Complex Coordinate Method (for a recent review see Ref. 3).

2. The Complex Coordinate Method

The complex coordinate method is based on the fundamental work of Balslev, Combes[4] and Simon[5]. Following the Balslev and Combes theorem if Θ in Eq. (3) is large enough then the resonance state, like the bound states, are associated with square integrable functions and are not affected by increasing Θ . Whereas, the scattering states are affected by Θ and are not associated with eigenfunctions in L^2. The computational advantage of that theorem is obvious - it enables us to isolate the resonance state from the other states in the continuum. Yet, it is not straightforward that bound states techniques can be used to find the resonance position and width. Since $\hat{H}_\Theta \equiv H(\exp(i\Theta)r)$ is a non Hermitian Hamiltonian. However, Nuttle and Doolen[6] by using conventional variational methods successfully applied in 1973 the complex coordinate method to atomic autoionization resonances and Reinhardt[7] in 1976 applied it to atoms in external electric field (the stark effect) where the potential is not dilatation analytic.

The properties of the complex rotated hamiltonian were studied by myself, Certain and Weinhold in 1978[8]. We have shown that many of the theorems that originally have been proved in quantum mechanics for Hermitian Hamiltonians are valid also for $\hat{H}_\Theta$ if the complex product,

$$(f|g) = \int_{\text{all space}} f(r)g(r)dr \tag{4}$$

is used rather than the ordinary scaler product,

$$\langle f|g\rangle = \int_{\text{all space}} f^*(r)g(r)dr \tag{5}$$

One of the theorems that we have derived is the complex variational principle, i.e. if the optimal wavefunction is $\phi_{opt}=\psi_{exact} + \mathcal{O}(\varepsilon)$ then the complex expectation value $\tilde{W}$ given by

$$\tilde{W} = (\phi_{opt}|\hat{H}_\Theta|\phi_{opt})/(\phi_{opt}|\phi_{opt}) \tag{6}$$

deviate from the exact value by (ε^2) . That is

$$\tilde{W} = W_{exact} + \mathcal{O}(\varepsilon^2) \tag{7}$$

Since $\tilde{W}$ is complex this is a stationary condition (no upper bounds to the exact values) and the resonance is associated with the requirement that

$$d\tilde{W}/d\eta|_{\eta=\alpha\exp(i\Theta)} = 0 \tag{8}$$

This definition of a resonance can be interpretated in a more general way: <u>a resonance is a complex variational solution where η is any non-linear variational parameter</u> (for a more detailed discussion see Ref. 9). In the case of atomic autoionization resonances, on the basis of the complex-variational principle, we have derived the complex analog to the virial theorem[8,11,30]

$$\exp(-i\Theta) = -\frac{(\phi_{opt}|\hat{V}(\vec{r})|\phi_{opt})}{2(\phi_{opt}|\hat{T}(\vec{r})|\phi_{opt})} \tag{9}$$

which independently has been proved by Brändas and Froelich by using the time-dependent perturbational approach[12]. The first application of the complex virial theorem was given by us[8] for the lowest ^1S resonance state of helium. In the first step of the calculation, we gave H an arbitrary non-vanishing value and by using a finite basis set the variational value $\tilde{W}$ (defined by Eq. (6)) was obtained. The new estimate of the optimal rotation angle for which Eq. (8) is satisfied was obtained from the complex virial theorem, i.e. by calculating the ratio between the complex potential and kinetic energies. The iteration procedure was carried out to the convergence of E_r and $\Gamma=2E_i$ which were in complete agreement with previous theoretical and experimental results[8,10]. In that time it was not entirely clear if the complex coordinate can be applied to molecular autoionization resonances within the framework of the Born-Oppenheimer approximation, since the Born-Oppenheimer Hamiltonian is not dilatation analytic. However, in the spirit of the generalization of the complex coordinate method described above the resonance is associated with a complex variational solution and dilatation of the hamiltonian is equivalent to the scaling of the basis functions. That is,

$$H_{ij}= (\phi_i(r,R)|\hat{H}(r\eta,R)|\phi_j(r,R))=(\phi(r/\eta,R)|\hat{H}(r,R)|\phi(r/\eta,R)) \qquad (10)$$

Therefore the matrix elements of the scaled molecular Hamiltonian are given by,

$$H_{ij}= \eta^{-2}(i|T(r)|j)+\eta^{-1}(i|V_{11}(r)|j)+\eta^{-1}(i/V_{ne}(r,R/\eta)|j) \qquad (11)$$

The computational advantage of this procedure is that the two electronic integrals which its computation requires a long and expensive computer time should be calculated only once(!) and only the electron-nuclear attraction integrals need be recalculated for each value of η. By letting η to be complex one can find the complex stationary solution which satisfy the two following conditions (equivalent to Eq. 8):

$$\frac{dE_r}{d\Theta} = 0 \text{ and } \frac{dE_i}{d\Theta} = 0 \qquad (12)$$

The first successful application of the complex coordinate method to molecular autoionization resonances was given by Corcoran and myself[13] to $H_2{}^1\Sigma^+_g$ (σu^2) and to H_2^- in its ground state; i.e. $H_2 \rightarrow H_2^+ + e^-$, and $H_2^- \rightarrow H_2 + e^-$. The resonance position and lifetime that we have obtained for $H_2{}^1\Sigma_g^+$ autoionization resonance were in complete agreement with previous theoretical and experimental results, and the resonance position and lifetime (width) of H_2^- were recently confirmed by Cederbaum[14]. It should be stressed here that the success of our molecular calculations arose from the fact that we did not restrict ourselves to $|\eta|=1$ since we wanted to satisfy Eqs. (12). It is easy to see that if $|\eta|=1$ or near to this value, numerical difficulties can arise, since V_{ne} may become very large[13]. If for example $|i\rangle$ and $|j\rangle$ are two s-type Gaussians centered at the nucleous and β is the Gaussian exponential parameter then $V_{ne} \sim F_o(t)$ where $t=2\beta R^2(1-\eta)^2$ and F_o is the incomplete γ function. $F_o(t)$ has a prefactor exp(-t). Therefore, near to $|\eta|=1$ Real(t)<o and V_{ne} gets large values which erroneously may be interpreted as a divergence due to the non-dilation analyticity of the Born-Oppenheimer Hamiltonian. Since then the complex coordinate method has been

applied to shape[8,9,15,16] and Feshbach[17,18] predissociation resonances, to resonances where the autoionization and the predissociation channels are coupled to one another[19], to resonance states which possess several open channels[20], to resonances obtained by elastic scattering of positronium-hydrogen[21], and to rotational predissociation resonances of van der Waals molecules[22,41]. In all of these studies the complex coordinate method has been used only for calculating the resonance positions and widths. In two recent works we also analyzed the resonance wavefunctions which were obtained by the complex coordinate methods in order to get information of the time dependent process by solving the time independent Schrödinger equation.

In the first work the complex coordinate method has been used to study the possibility of having - **mode selective chemistry.** If any mode in the molecule is excited to a high vibrational energy level then always the weakest chemical bond will break-up due to a rapid energy transfer from excited mode to the other degrees of freedom. We have studied[23] the dissociation of alkyle radical reaction, $C_2H_5 \rightarrow H+C_2H_4$, by using the Hase model potential[24], $H(r,R)=H_0(r)+H_1(r,R)$ where "r" is the reaction coordinate which lead to the dissociation of the alkyl radical, and "R" is the C-C mode which is changed from a single type to a double type chemical bond during the reaction process (resulting from the coupling between the two modes) and the minimum-energy-path potential energy in H_o has a barrier which supports predissociational resonances. Following the idea proposed by Waite and Miller[25], then by the complex coordinate method one can "measure" the mixing of the excited modes: if there is no mixing between the two modes the Hamiltonian is separable and predissociation resonances are obtained by solving the one-dimensional eigenvalue problem,

$$\hat{H}_o(r)\phi_1=(E_r^{(1)}- iE_i^{(1)})\phi_1 \tag{13}$$

whereas, the energy levels of the C-C mode are obtained by solving the following one-dimensional Schrödinger equation,

$$\hat{H}_1(r_o,R)\chi_m= E_m\chi_m \tag{14}$$

where r_o is the CH bond distance at equilibrium.
Since the Hamiltonian is separable (if r is held fixed at r_0) the total energy is given by,

$$E_{total} = (E_m+E_r^{(1)}) - iE_i^{(1)}; \; 1,m = 1,2 \ldots \tag{15}$$

From Eq. (15) we see that if the two modes are not mixed then "families" of resonances will be obtained. In each "family" the resonances have the same width but different positions resulting from excitations to different vibrational levels in the non-dissociative "R" mode. The deviation of the eigenvalues from such a behaviour is an indication of the inapplicability of the normal (or local) mode approximation. The numerical results which are presented in Fig. 1 (following Simon[26] exterior scaling only the "R" coordinate was complex scaled) show very clearly that in this case there is a little mixing between the two modes.

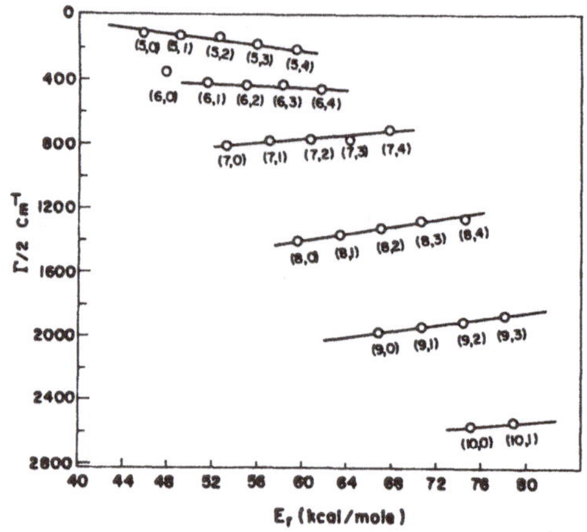

Fig. 1: The resonance width versus the resonance position for the (l,m) metastable states of the C-C-H model Hamiltonian.

Moreover, by analysing the resonance eigenfunctions we have shown[23] that in any "family" of resonances the wavefunctions have a dominant contribution from the same vibrational excited state of the dissociative "r" mode. For example, the resonance wavefunctions which yield the same width $2E_i \sim 350\ cm^{-1}$ are all dominated by the 6th vibrational level of the "r" mode (l=6 in Fig. 1). This analysis of the resonance wavefunction indicate that even in energy above the dissociation limit the mixing of the two modes is very small.
These results are in complete agreement with the classical and semiclassical quasiperiodic trajectories recently studied by Hase et al[24] and support the possible existence of narrow resonances above the potential barrier, or even in its absence, as proposed by Bloembergen et al[27], for the mechanism of the dissociation of HN_3 to NH [$^3\pi$] and N_2.
The second case where the resonance wavefunctions were studied was in the application of the complex coordinate method to **gas surface scattering resonances**, so called "selective adsorption". In such a case (neglecting the effect of surface corrugation and phonons) the resonances are obtained resulting from energy transfer from the vibration toward the surface to the rotation of the diatomic molecule. We chose to study the resonance scattering of HD from Ag(111) surface which recently were observed in scattering experiments by Yu et al[28]. By

analysing the experimental results of the angular distribution of the reflected number density for HD they determined the interacting potential of HD with the surface of Ag(111). We expanded Yu et al model potential in the Legandre polynomial and the distance of center of mass from the surface was scaled by a complex factor $\exp(i\Theta)^{29}$. The complex eigenvalue which satisfy Eq. (12) was obtained by carrying out Θ-trajectory calculations.

An illustrative representation is given in Fig. 2. The zero angle of the cusp in the Θ-trajectory calculations is an indication that we have obtained a complex variational solution which is associated with the resonance state[30]. Since we do not have boundary conditions, an error estimate could be obtained only by increasing the number of the basis functions. See for example the spiral convergent presented in Fig. 3.

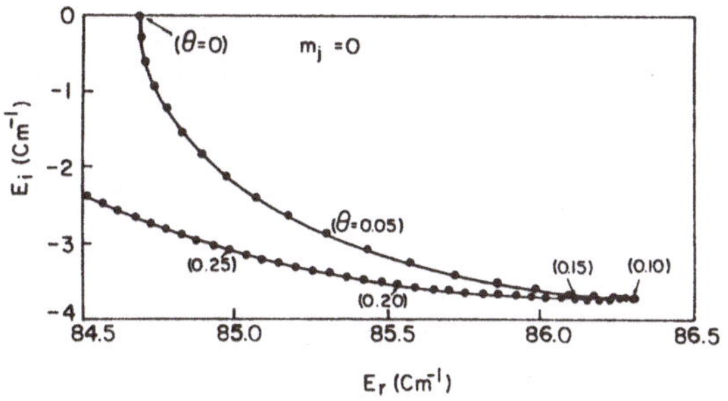

Fig. 2: Θ-trajectory calculations for a surface scattering HD-Ag(111) resonance, with a basis set constructed of 50 harmonic oscillators.

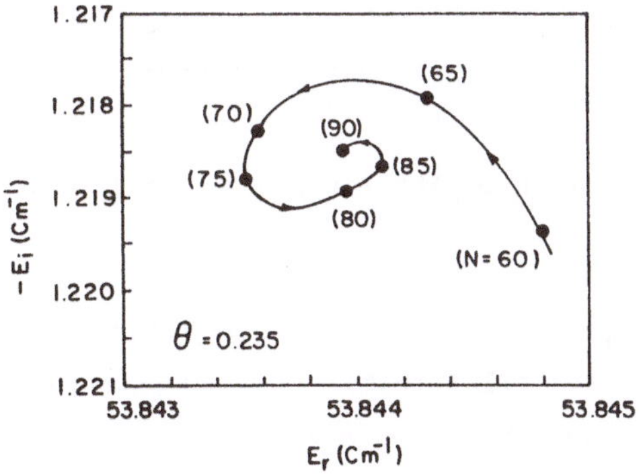

Fig. 3: Convergence of a HD-Ag(111) resonance which is dominated by a single term j=1, with the optimum rotation angle Θ.

The analysis of the resonance wavefunctions have shown[29] that they are dominated by a single value of the rotational angular momemtum j (about 90-95% of the population probabilities). Therefore, the HD molecule rotates almost freely above the silver surface and the desorbtion results from a rapid energy transfer from the rotational excited rigid rotor to the HD-Ag(111) vibrational mode (after 5-10 of rotations above the surface).

3. The Complex Coordinate Method - two computational difficulties

The complex coordinate method, in spite of its successful application to atomic and molecular resonances, has two mainly computational difficulties that arise from the fact that $\hat{H}_\Theta$ is a non Hermitian Hamiltonian.

(1) As it was indicated in the previous section the complex-variational principle, unlike the conventional one, is a stationary principle rather than a minimum principle for either the resonance position or width[8,31], and the perturbational theory can be applied only for very specific cases[32,33].

(2) The spectrum of the complex rotated Hamiltonian may be incomplete such that $(\psi|\psi)=0$.

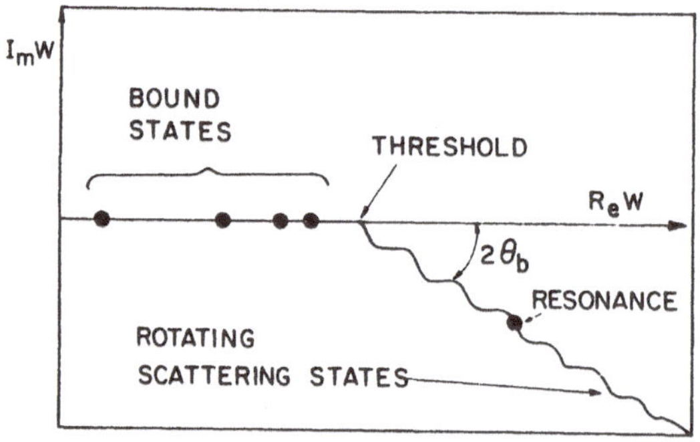

Fig. 4: Schematic representation of the eigenvalues of the complex-rotated Hamiltonian.

Within the finite matrix approach Friedland and myself[34] have pointed out that since the kinetic and the potential operators are not commute one can always find at least one value of the rotational angle for which,

$$\frac{dW}{d\eta}\Big|_{\eta=\exp(+i\Theta_b)} = \infty \qquad (16)$$

Since from the Hellmann-Feynman theorem we know that,

$$\frac{dW}{d\eta} = \frac{(\phi_{opt}|\partial H/\partial \eta|\phi_{opt})}{(\phi_{opt}|\phi_{opt})} \tag{17}$$

then we get that $dw/d\eta \to \infty$ as $(\phi_{opt}/\phi_{opt}) \to 0$. Since the resonance is associated with a stationary solution one may wonder if the problem of incomplete spectrum of $\hat{H}_\Theta$ in the matrix representation is relevant to the resonance calculations. It may happen as we proved, that a complex stationary solution will be obtained for a critical angle for which the spectrum is incomplete and the number of the independent vectors is smaller than the dimension of our matrix Hamiltonian. Yet, it seems to us that in the finite matrix approximation a branch point will be obtained as the scattering states are rotated to the complex plane by Θ_b as defined in Fig. 4.

In such a case the two necessity conditions for incomplete spectrum can be satisfied:

(a) The resonance and the scattering states are degenerated.

(b) Since the basis set is finite also the scattering states are described by wavefunctions in L^2 and coalescence of the resonance and the scattering states could occur. In order to illustrate this possibility (i.e a resonance state can be associated with a defective wavefunction which is non-complex normalizeable) we studied the lowest ^1S resonance of helium. The eigenvalues of a matrix Hamiltonian constructed of 36 Hylleraas type functions were calculated as function of a <u>real</u> scaling factor $\alpha = 1/|\eta|$. The results are presented in Fig. 5.

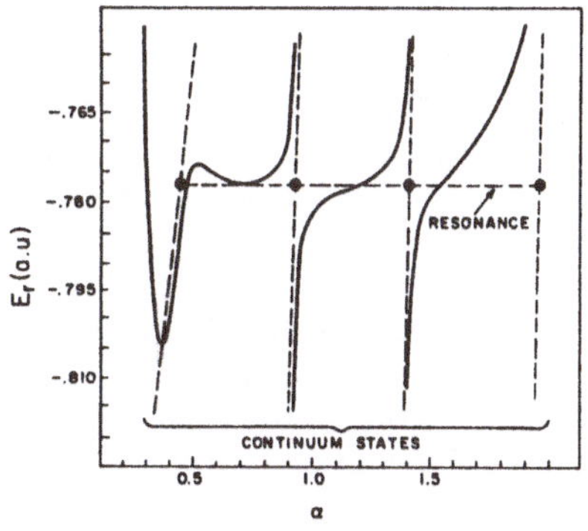

Fig. 5: The roots of 36x36 secular equations (for helium potential) as a function of the real scale factor α, showing the stabilization of eigenvalues in the resonance region.

Since the resonance wavefunction is localized in the coordinate space, the resonance eigenvalue is almost unaffected by making the basis functions more or less diffused (by varying scale factor α). On the other hand the scattering states have large amplitude even as $r \to \infty$ and therefore are strongly affected by α. Consequently the resonances are associated with the avoiding crossings presented in Fig. 5. However, if we let the scaling factor to be complex the crossings will not be avoided and branch points will be obtained (see Table II in Ref. 34). Indeed for fixed α =0.925 two eigenvalues were rotated from the real axis to the complex plane as Θ was increased and at a critical value of Θ_b a branch point was obtained and the spectrum became incomplete (denoted by a black dot in Fig. 6). As Θ was increased beyond its critical value then new branches were obtained. One of them converged to a complex stationary solution (i.e. a resonance) while the second one is strongly affected by Θ and describes the rotating continua. As the number of the basis functions were increased the resonance and the "defective" solution gets closer and closer.

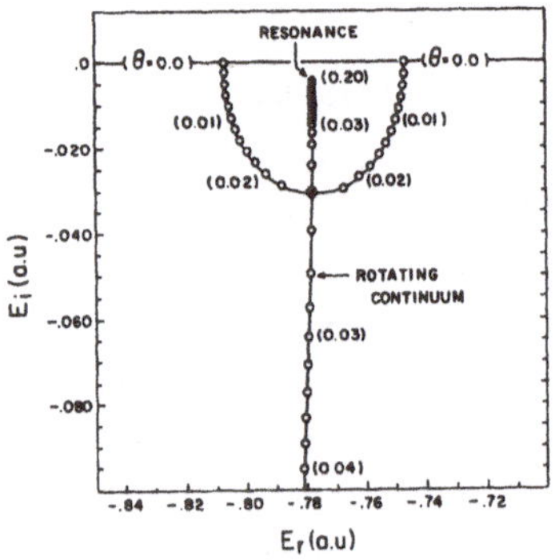

Fig. 6: Θ-trajectory for fixed α=0.925. The black dot denoted the branch point in which the spectrum is incomplete.

3. The Hermitian representation of the complex coordinate method

a. Theory[35]

The Hermitian representation of the complex coordinate method (CCM) has been proposed in order to avoid the two computational difficulties arising out of the complex arithmetic in the process of solving the eigenvalue problem. The complex Schrodinger equation obtained by rotating the coordinates to the complex plane,

$$\hat{H}_\Theta \psi = W\psi \qquad (18)$$

or

$$[(H_r + iH_i) - E_r - iE_i)](\psi_r + i\psi_i) = 0 \tag{19}$$

can be split into the following two equations of which one contains only real terms and the other only the imaginary terms:

$$\left[\begin{pmatrix} -H_i & H_r \\ H_r & H_i \end{pmatrix} - E_i \begin{pmatrix} 1 & 0 \\ 0 & -1 \end{pmatrix} + E_r \begin{pmatrix} 0 & -1 \\ -1 & 0 \end{pmatrix}\right] \begin{pmatrix} \psi_i \\ \psi_r \end{pmatrix} = 0 \tag{20}$$

Here we obtain a new Hermitian operator which is defined by,

$$\mathcal{H}(\Theta, E_r, E_i)\vec{\psi} = 0 \;;\; \vec{\psi} \equiv \begin{pmatrix} \psi_i \\ \psi_r \end{pmatrix} \tag{21}$$

where the resonance position and width are two parameters in Hamiltonian for which a zero eigenvalue of $\mathcal{H}$ is obtained. If $W \neq E_r - iE_i$ (i.e. $E_r - iE_i$ is not an eigenvalue of the complex-rotated Hamiltonian $\hat{H}_\Theta$) then the following eigenvalue problem can be considered:

$$\mathcal{H}\phi = \lambda\phi \quad ; \quad \lambda \neq 0 \tag{22}$$

Since can get a negative value also, the resonance position E_i, and width $2E_i$, for any given Θ are the <u>variational parameters</u> of H^2,

$$\mathcal{H}^2 \phi = \lambda^2 \phi \tag{23}$$

By substituting as defined in Eqs. (20-21) into Eq. (23) we get that:

$$\mathcal{H}^2 = \hat{H}_o + E_i \hat{H}_1 + E_r \hat{H}_2 + E_i^2 + E_r^2 \tag{24}$$

where,

$$\hat{H}_o = \begin{pmatrix} \hat{H}_r^2 + \hat{H}_i^2 & [\hat{H}_i , \hat{H}_r] \\ [\hat{H}_r , \hat{H}_i] & H_r^2 + H_i^2 \end{pmatrix}$$

$$\hat{H}_i = \begin{pmatrix} 2\hat{H}_i & 0 \\ 0 & 2\hat{H}_i \end{pmatrix} \tag{26}$$

$$\hat{H}_2 = \begin{pmatrix} -2\hat{H}_r & 0 \\ 0 & -2\hat{H}_r \end{pmatrix} \tag{27}$$

Since $\hat{H}_o$, $\hat{H}_1$, $\hat{H}_2$ and $\mathcal{H}^2$ are Hermitian (and real) operators by using the variational principle we can prove that the lowest eigenvalue of $\mathcal{H}^2$, i.e. λ^2, tells us where the resonance is not in the E_r-E_i plane:

$$(\Delta E_r^c)^2 + (\Delta E_i^c)^2 \geq \lambda^2 \tag{28}$$

ΔE_r and ΔE_i are the deviation of our estimate of the resonance position and width from the exact values, such that,

$$\begin{aligned} \Delta E_r^c &= E_r(\text{exact})-E_r \\ \Delta E_i^c &= E_i(\text{exact})-E_i \end{aligned} \tag{29}$$

If ΔE_r^c is known from experiment or theory and $E_i=0$ then from Eqs. (28-29) one can get that,

$$(E_i(\text{exact}))^2 \geq \lambda^2 - (\Delta E_r^c)^2 \tag{30}$$

and since $\Gamma_{exact}=2E_i(\text{exact})$ a lower bound to the resonance width is obtained,

$$\Gamma(\text{exact}) \geq 2[\lambda^2-(\Delta E_r^c)^2]^{\frac{1}{2}} \tag{31}$$

Proof of Eq. 28

Let E_r and $2E_i$ to be the approximated position and width such that,

$$\mathcal{H}^2(E_r\ ,\ E_i)\phi = \lambda^2\phi\ ,\ \lambda^2 > 0 \tag{32}$$

If ΔE_r and ΔE_i are the corrections to the resonance position and width then for each value of ΔE_r and ΔE_i the following expectation value,

$$\sigma = \langle\phi_{opt}|\mathcal{H}^2(E_r+\Delta E_r,\ E_i + \Delta E_i)|\phi_{opt}\rangle \tag{33}$$

can be optimized to yield a minimal value of .

If,

$$\begin{aligned} \Delta E_r &= \Delta E_r^c \qquad (E_r(\text{exact}) = E_r+ \Delta E_r^c) \\ \Delta E_i &= \Delta E_i^c \qquad (E_i(\text{exact}) = E_i+ \Delta E_i^c) \end{aligned} \tag{34}$$

then,

$$\sigma = 0 \tag{35}$$

and,

$$|\phi_{opt}\rangle \equiv |\psi_{exact}\rangle \tag{36}$$

Since from Eq. 24 we get that

$$\mathcal{H}^2(E_r + \Delta E_r, E_i + \Delta E_i) = \hat{H}_o + E_i\hat{H}_1 + E_r\hat{H}_2 + E_i^2 + E_r^2 + \Delta E_i\hat{H}_1 + \Delta E_r\hat{H}_2 \tag{37}$$
$$+ 2E_i\Delta E_i + 2E_r\Delta E_r + (\Delta E_i)^2 + (\Delta E_r)^2$$

then the expectation value of $\mathcal{H}^2(E_r + E_r, E_i + E_r)$ with ψ_{exact} (which is equal to zero) is given by,

$$0 = \langle\psi_{exact}|\mathcal{H}^2(E_i, E_r)|\psi_{exact}\rangle \tag{38}$$
$$+ \Delta E_i^c \langle\psi_{exact}|\hat{H}_1|\psi_{exact}\rangle + \Delta E_r^c \langle\psi_{exact}|\hat{H}_2|\psi_{exact}\rangle$$
$$+ 2E_i\Delta E_i^c + 2E_r\Delta E_r^c + (\Delta E_i^c)^2 + (\Delta E_r^c)^2$$

$\Delta E_r{}^c$ and $\Delta E_i{}^c$ can be obtained by the following variational calculations: for any values of ΔEr and ΔEi, $|\phi\rangle$ is optimized to yield a minimal value of σ. From Eq. (37) by using the Hellman-Feynman theorem[36] we get that:

$$\frac{d\sigma}{d\Delta E_r} = \langle\phi_{opt}|\frac{\partial\mathcal{H}^2}{\partial\Delta E_r}|\phi_{opt}\rangle = \langle\phi_{opt}|\hat{H}_2|\phi_{opt}\rangle + 2E_r + 2\Delta E_r \tag{39}$$

and similarly,

$$\frac{d\sigma}{d\Delta E_i} = \langle\phi_{opt}|\frac{\partial\mathcal{H}^2}{\partial\Delta E_i}|\phi_{opt}\rangle = \langle\phi_{opt}|\hat{H}_1|\phi_{opt}\rangle + 2E_i + 2\Delta E_i \tag{40}$$

At the global minimum,

$$\left.\frac{d\sigma}{d\Delta E_i}\right|_{\Delta E_i^c} = 0 \; ; \; \left.\frac{d\sigma}{d\Delta E_r}\right|_{\Delta E_i^c} = 0 \; ; \; \sigma = 0 \tag{41}$$

and the variational wavefunction is the exact one (see Eq. 36). From Eqs. (40-41) we get that,

$$E_i + \Delta E_i^c = -\tfrac{1}{2}\langle\psi_{exact}|\hat{H}_1|\psi_{exact}\rangle$$
$$E_r + \Delta E_r^c = -\tfrac{1}{2}\langle\psi_{exact}|\hat{H}_2|\psi_{exact}\rangle \tag{42}$$

and by substituting Eqs. (42) in Eq. (38) one obtains,

$$(\Delta E_i^c)^2 + (\Delta E_r^c)^2 = \langle\psi_{exact}|\mathcal{H}^2(E_r, E_i)|\psi_{exact}\rangle \tag{43}$$

Since the lowest eigenvalue of $^2(E_r,E_i)$ is λ^2 (see Eq. 32) on the basis of the variational principle it is clear that

$$<\psi_{exact}|\mathcal{H}^2(E_r, E_i)|\psi_{exact}> \geq \lambda^2 \qquad (44)$$

and the desired inequality, $(\Delta E_i{}^c)^2+(\Delta E_r{}^c)^2 \geq \lambda^2$, is obtained.

The solution of Eq. 23 by iterative procedure

The initial estimate of the resonance position, $E_r{}^{(j=0)}$, can be taken from stabilization calculations which were described in the Section 3 (see for example Fig. 5) where the initial estimate of the width can be taken as, $E_i{}^{(j=0)}=0$.

Within the framework of the finite matrix approach the basis set is truncated and from Eq. 24 we get that,

$$\underset{\approx}{\mathcal{H}}^2_{j+1} = \underset{\approx}{H}_0 + E_i{}^{(j)}\underset{\approx}{H}_1 + E_r{}^{(j)}\underset{\approx}{H}_2 + (E_r{}^{(j)})^2 + (E_i{}^{(j)})^2 \qquad (45)$$

and the lowest eigenvalue of $\underset{\approx}{\mathcal{H}}^2{}_{j+1}$ is calculated

$$\underset{\approx}{\mathcal{H}}^2_{j+1}\,\vec{c}_{j+1} = \lambda_{j+1}\,\vec{c}_{j+1} \qquad (46)$$

In this step of the computation procedure one can calculate the following expectation value,

$$\lambda_o^{2(j)} = \vec{c}_j\,\underset{\approx}{\mathcal{H}}^2{}_{j+1}\,\vec{c}_j \qquad (47)$$

The new estimate of the resonance position and width are obtained according to Eqs. (42),

$$E_r{}^{(j+1)} = -\tfrac{1}{2}\,\vec{c}_{j+1}\,\underset{\approx}{H}_2\,\vec{c}_{j+1}$$
$$E_i{}^{(j+1)} = -\tfrac{1}{2}\,\vec{c}_{j+1}\,\underset{\approx}{H}_1\,\vec{c}_{j+1} \qquad (48)$$

and are substituted in Eq. (45). The iteration is carried out upon the convergence of E_r, E_i, λ^2 and $\lambda_o{}^2$.

It has been proved by Froelich, Davidson and Brandas[37] that,

$$(\Delta E_r^c)^2 + (\Delta E_i^c)^2 \leq \lambda_o^2 \qquad (49)$$

Therefore, λ_o which is a "by-product" in the above iteration procedure together with λ (see Eq. 28) gives us the following upper and lower bounds for the resonance position.

$$E_r + \lambda_o \geq E_r(\text{exact}) \geq E_r + \lambda \tag{50}$$

or,

$$E_r - \lambda \geq E_r(\text{exact}) \geq E_r - \lambda_o$$

and width,

$$2E_i + 2\lambda_o \geq \Gamma(\text{exact}) \geq 2E_i + 2\lambda \tag{51}$$

or

$$2E_i - 2\lambda \geq \Gamma(\text{exact}) \geq 2E_i - 2\lambda_o$$

The advantages of the Hermitian representation of the CCM

1. $\mathcal{H}^2$ is a real and Hermitian operator and therefore the numerical difficulties arising from the complex arithmetic are avoided.

2. Only the ground state of the Hermitian Hamiltonian $\mathcal{H}^2$ should be calculated as function of the rotational angle.

3. The resonance width is the "natural" strength parameter in the perturbation expansion. Since the width is small with respect to the position then one may expect that the resonance width will be estimated from the second order perturbational calculations.

4. Upper and lower bounds to the resonance position and width are obtained. The exact position and width of the resonance is a point on the surface of an annular ring[38].

5. The length of the computations is truncated. If the resonance position is known from experiment or theory, the width can be estimated from λ^2 or E_i which are obtained in the first step of the iteration procedure (Eqs. 46,48).

b. Applications

b1. One dimensional model Hamiltonian

To illustrate the variational calculations by the complex-coordinate method with Hermitian Hamiltonian we studied[38] the one-dimensional model Hamiltonian

$$H = -\tfrac{1}{2}\frac{d^2}{dx^2} + (\frac{1}{2}X^2 - 0.8)\exp(-0.1X^2) + 0.8 \tag{52}$$

whose potential presented in Fig. 7 exhibits the shape predissociation type resonances, and which was used previously to illustrate the variational calculations by the complex coordinate method(8).

The computations are carried out by using a basis set constructed of 100 even-tempered Gaussians, $\{\phi_i = \exp(-0.75^i X^2)\}$. This large basis set can be referred to as a "complete" one since the resonance position and width obtained by the complex coordinate method (the internal coordinate X was complex scaled by the factor $\eta = 1.8 \exp(-i\ 0.36)$), are in agreement to six digits with the accurate results previously obtained by numerical integration of the complex-rotated Riccati equation(39). Within the framework of the finite matrix approach Eqs. 45-48 were solved. The optimal rotation angle, $\Theta_{opt} = 0.38$,Θ was associated with the cusp which was obtained by carrying out Θ-trajectory calculations. The error estimates of the resonance position and width are obtained from $\lambda_o^{(j)}$ and $\lambda_o^{(j)}$ (as defined in Eqs. (46,47)). The center of the annular ring (on its surface the exact solution is located) obtained in the jth iteration is $(E_r^{(j)}, E_i^{(j)})$ and its inner and outer radii are λ^{j+1} and λ_o^{j+1} respectively. Fig. 8 show how the error estimates are consequently reduced in the process of iteration.

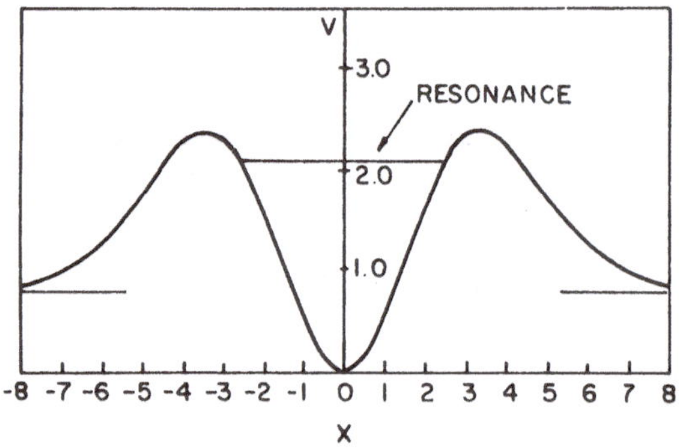

Fig. 7: The model potential given in Eq. 52.

After completion of each iteration the exact result is contained in the crescent area. Comparing two consecutive iterations confines the exact result to the area of an annular ring. The relatively large area of the first annular ring is heavily reduced to a small crescent area obtains after the second iteration, the third and etc. iterations, and illustrate clearly how the error estimates shrink to zero as the variational resonance position and width approach the exact values.

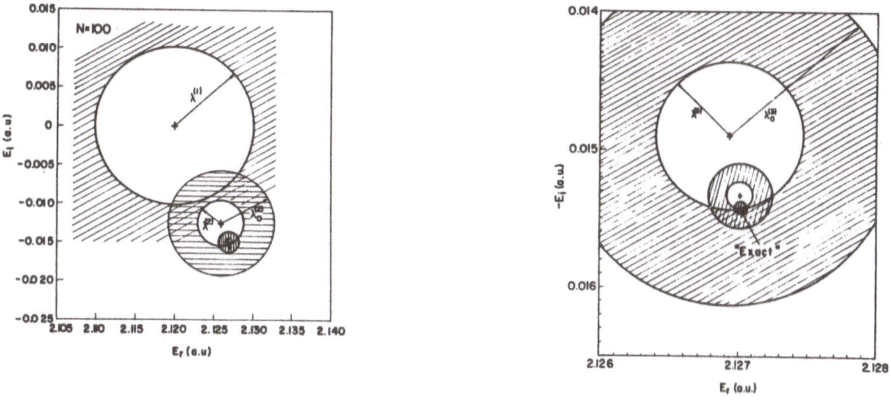

(a) (b)

Fig. 8: Upper and lower bounds of the resonance positional width representation of the complex coordination method. The "+" signs indicate the estimate of the resonance position and width in each iteration. The resonance is located within the dashed area. The intersection of the dashed areas gives the optimal estimate of the resonance location.

(a) Results obtained through the first three steps of the iteration procedure.

(b) Results obtained throughout iterations 3 to 6. The largest annular ring on this figure is repeated from Fig. (a) where it appears as the smallest annular ring.

b2. Atomic autoionization resonances

The ^1S autoionization resonance of helium, $He \rightarrow He^{+}+e$, was studied by using a basis set constructed of 36 Hylleraas type functions,

$$\phi_{l,m,n} = r_1^{\,l} r_2^{\,m} r_2^{\,n} \exp(-0.75\ r_1 - 1.25\ r_2) \qquad (53)$$

where l,m,n=0,1,2,3. The resonance position and width were obtained by solving iteratively Eqs. (45-48)[40]. The results are presented in Table I.

Iteration	$-E_r$	$E_i \equiv \Gamma/2$
0	0.77000	0
1	0.77766	0.00120
2	0.78145	0.00186
3	0.78323	0.00222
4	0.78413	0.00242
5	0.78453	0.00253
6	0.78471	0.00259

Table I: The ^1S helium resonance position and width obtained by the iterative procedure derived from the Hermitian representation of the complex coordinate.

In our calculations[40] the matrix elements of $\hat{H}_o^{\,2}$ were accurately obtained and therefore the resonance position and width differ from the values obtained by solving the complex eigenvalue problem $\hat{H}_\Theta \vec{D}=(E_r-iE_i)\vec{D}$. Note, that under the approximation $\underset{\approx}{H}^2_\Theta \simeq \underset{\approx}{H}_\Theta \cdot \underset{\approx}{H}_\Theta$ the iterative solution will converge to $\lambda=0$ and to the values of the resonance position and width obtained by solving the complex-eigenvalue problem.

b3. Rotational predissociation resonances of van der Waals complex

During the last years extensive experimental and theoretical studies of the van der Waals complexes have been carried out. Most studies were done on the collision at low energy of "inert" gas atom with a diatomic molecules. In such a case resonances are obtained because of the coupling between the rotation of the diatomic molecule (which is treated as a rigid rotor) and the vibration of the van der Waals complex (i.e. diatomic molecule - 'inert' gas atom). The complex coordinate method has been applied by Chu[22] and by Bacic and Simons[41] to atom-diatomic model Hamiltonian. It turned out[22] that by complex scaling the atom-diatomic distance, R, by exp(iΘ) where the Hamiltonian is written in the space-fixed (SF) axes) then both the resonance position and width are very sensitive to the value of Θ . To increase the numerical stability of the resonance position and width it was suggested by Lefebvre[42] to carry out numerical integration calculations, and by Chu[43] to use the body-fixed (BF) formalism. A third alternative procedure is to use the SF formalism but to scale the Hamiltonian matrix elements as suggested by McCurdy and Rescigno[44]. McCurdy and Rescigno pointed out that by scaling the internal coordinates of the system the basis functions are shifted from its "physical" centers and large basis set is required. They suggested also to scale the centers of the basis functions. In the molecular autoionization calculations the last procedure has the disadvantage that the two-electron integrals should be recalculated. In the predissociation calculations, in most cases, the integrals are simple and recalculating the integrals for different values of the rotation angle Θ are not expensive. Moreover, it has been shown by Certain and myself(45) that if the center of the basis function is also scaled then the resonance position and width become very stable and not much affected by varying Θ. To illustrate the application of the Hermitian representation of the complex coordinate method we studied the model potential,

$$V(R, \gamma) = V_o(R) + V_2(R)\, P_2(\cos\gamma) \tag{54}$$

where γ is the angle between the atom-diatomic distance, R, and the internuclear axis. The $V_o(R)$ and $V_2(R)$ we used are as of Chu(22),

$$V_o(R) = 4\varepsilon[(\sigma/R)^{12} - (\sigma/R)^6] \tag{55}$$

and

$$V_2(R) = 0.6\ \varepsilon\ (\sigma/R)^{12} \tag{56}$$

where σ=3.0 Å, ε =384.092 cm^{-1}, Brot=60.962 cm^{-1} and μ =1.34015 amu. The close coupled equations which were derived for the potential surface $V(R,\gamma)$ were solved within the framework of the finite matrix approach. The basis set was constructed of 20 orthonormal harmonic oscillator functions, $\chi(R-R_o)$, which are centered at the equilibrium distance of $V_o(R)$, that is $R_o=2^{1/6}\sigma$. The complex matrix Hamiltonian was obtained by using the Gaussian quadratune procedure(46) to calculate the complex expectation values, $(Xn[\eta(R-R_o)]|R^{-m}|Xn(\eta(R-R_o))$, where m=2,6,12 and $\eta=\exp(i\Theta)$. From the complex non-Hermitian, matrix Hamiltonian one can get(35) that,

$$(\underset{\approx}{H}{}_o^1 + E_i\, \underset{\approx}{H}_1 + E_i^2\, \underset{\approx}{H}_2)\, \vec{c} = \lambda^2 \vec{c} \tag{57}$$

where

$$\underset{\approx}{H}{}_o^1 = H_o - E_r\, \underset{\approx}{I} \tag{58}$$

and H_o, H_1, H_2 are defined in Eqs. (25-27). E_r the resonance position was estimated to be 114.456 cm^{-1} from stabilization calculations(45). Since λ^2=0 if E_i is equal to the half of the <u>exact</u> resonance width we shall estimate the width by calculating the value of E_i for which λ^2 in Eq. (57) gets a minimal value. In estimated to be 114.456 cm^{-1} from stabilization calculations[45]. Since ^{2}=0 if E_i is equal to the half of the exact resonance width we shall estimate the width by calculating the value of E_i for which λ^2 in Eq. (57) gets a minimal value. In the first step of the computations $E_i=0$ is substituted in Eq. (57) and E_i is estimated by $-1/2\ \vec{C}^T H_1 \vec{C}$ (denotes by $E_i^{(1)}$ in Fig. 9) for which $d\lambda^2/dE_i=0$. The minimal value of λ^2 can be obtained by solving iteratively Eq. (57) where $E_r^{(j+1)} = -1/2\ \vec{C}^{(j)} H1\ \vec{C}^{(j)}$ and $\vec{C}^{(j)}$ is the eigenvector which is obtained in the jth step of the iteration procedure.

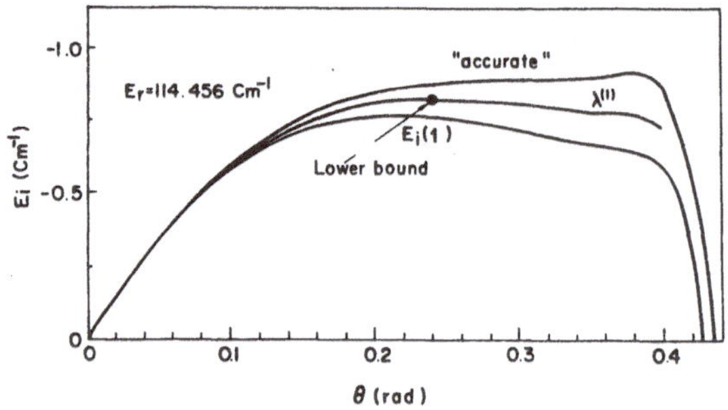

Fig. 9: The resonance width, $\Gamma=2E_i$, obtained by the Hermitian representation of the complex coordinate method with 20 harmonic oscillators. $\lambda^{(1)}$ and $E_i^{(1)}$ are defined in the text and are obtained in the first iteration of the calculations. The solid "accurate" line is obtained after convergence of the iteration computations.

As one can see from Fig. 9 the maximal value of $E_i^{(1)}$ and $\lambda^{(1)}$ provides a good estimate of the resonance width and there is no need to carry out the computation to convergence. The black dots stress the maximal value of $\lambda^{(1)}$ which yields a lower bound to the resonance width.

References

1. P. Auger, Compt. Rend. 180, 1939 (1925); J. Phys. Rad. 6, 205 (1925).
2. A. Tempkin, Autoionization (Mono Book Corp.), 1966.
3. W.P. Reinhardt, Ann. Rev. Phys. Chem. 33, 223 91982); B.R. Junker, Advan. At. Mol. Phys. 18, 207 (1982).
4. E. Balslev and J.M. Combes, Commun. math. Phys. 22, 280 (1971).

5. B. Simon, Commun. Math. Phys. 27, 1 (1972).

6. J. Nuttle, Bull. Am. Phys. Soc. 17, 598 (1972); G.D. Doolen, M. Hidalgo and R. Stagart, Atomic Physics (Plenum), 1973), see also Int. J. Quantum Chem. 14, (1978).

7. W.P. Reinhardt, Int. J. Quantum Chem. S10, 359 (1976).

8. N. Moiseyev, P.R. Certain and F. Weinhold, Mol. Phys. 36, 1613 (1978).

9. N. Moiseyev, Mol. Phys., 47, 585 (1982).

10. N. Moiseyev, P.R. Certain and F. Weinhold, Int. J. Quantum Chem. 14, 727 (1978).

11. N. Moiseyev, P. Rev. A 24, 2824 (1981).

12. E. Brändas and P. Froelich, Phys. Rev. A 16, 2207 (1977).

13. N. Moiseyev and C.T. Corcoran, Phys. Rev. A, 20, 814 (1979).

14. L.S. Cederbaum, private communication.

15. D. Doolen, Int. J. Quantum Chem. 14, 523 (1980).

16. R. Yaris, J. Bendler, R.A. Lovett, C.M. Bender and P.A. Fedders, Phys. Rev. A 18, 1816 (1978).

17. R. Lefebvre, Chem. Phys. Lett. 70, 430 (1980).

18. N. Moiseyev, Int. J. Quantum Chem. 29, 835 (1981).

19. N. Moiseyev, Mol. Phys. 42, 129 (1981).

20. Z. Bacic and J. Simons, Intern. J. Quantum Chem. 21, 727 (1982).

21. Y.K. Ho, Phys. Rev. A 17, 1675 (1978).

22. S.I. Chu , J. Chem. Phys. 72, 4772 (1980).

23. N. Moiseyev and R. Bar-Adon, J. Chem. Phys. 80, 1917 (1984).

24. W.L. Hase, J. Phys. Chem. 86, 2873 (1982); R.J. Wolf and W.L. Hase, J. Chem. Phys. 73, 3779 (1980), ibid 72, 316 (1979).

25. B.A. Waite and W.H. Miller, J. Chem. Phys. 74, 3910 (1981).

26. B. Simon, Phys. Lett. 71A, 211 (1979).

27. T.B. Simpson, E. Mazor, K.K. Lehmann, I. Burak and N. Bloembergen, J. Chem. Phys. (in press).

28. C.F. Yu, C.S. Hogg, J.P. Cowin, K.B. Whaley, J.C. Light and S.J. Sibener, Isr. J. Chem. 22, 305 (1982).

29. R. Elber, R.B. Gerber, T. Maniv and N. Moiseyev, to be published.

30. N. Moiseyev, S. Friedland and P.R. Certain, J. Chem. Phys. 74, 4739 (1981).

31. N. Moiseyev and F. Weinhold, Int. J. Quantum Chem. 17, 1201 (1980).

32. N. Moiseyev and P.R. Certain, Mol. Phys. 37, 1621 (1979).

33. N. Moiseyev and F. Weinhold, Phys. Rev. A 20, 27 (1979).

34. N. Moiseyev and S. friedland, Phys. Rev. A 22, 618 (1980).

35. N. Moiseyev, Chem. Phys. Lett. 99, 364 (1983).

36. H. Hellman, Einfuhrung in die Quantenchemie (Dentiche, Leipzing 1937); A.P. Feynman, Phys. Rev. 56, 340 (1939).

37. P. Froelich, E. Davidson and E. Brandas, Phys. Rev A 28, 2641 (1983).

38. N. Moiseyev, P. Froelich and E. Watkins, J. Chem. Phys. (in press).

39. M. Rittby, N. Elander and E. Brandas, Phys. Rev. A 24, 1636 (1981).

40. P. Froelich and N. Moiseyev, J. Chem. Phys. (in press).

41. Z. Bacic and J. Simons, Int. J. Quantum Chem. 14, 467 (1980).

42. R. Lefebvre, The Fifteenth Jerusalem Symposium, D. Reidel Publishing Co., (1982).

43. S.T. Chu, Chem. Phys. Lett. 88, 213 (1982).

44. C.W. McCurdy and T.N. Rescigno, Phys. Rev. Lett. 41, 1364 (1978).

45. N. Moiseyev and P.R. Certain, to be published.

46. A.S. Dickinson and P.R. Certain, J. Chem. Phys. 49, 4209 (1968); D.O. Harris, G.G. Engerholm and W.D. Gwinn, J. Chem. Phys. 43, 1515 (1965).

ONE-AND TWO-PHOTON FREE-FREE TRANSITIONS IN A COULOMB FIELD

Alfred MAQUET
Laboratoire de Chimie Physique(*)
Université Pierre et Marie Curie
11, Rue Pierre et Marie Curie
F - 75231 PARIS Cedex 05 - France

I - INTRODUCTION

The first studies on the so-called free-free radiative transitions were dedicated to the bremsstrahlung and, thanks to Sommerfeld,[1] at the end of the thirties everything was known on the one-photon transitions between states belonging to the Coulomb continuous spectrum. The interest for two-photon and multiphoton free-free transitions was stimulated recently by the advent of powerful laser sources which have made possible the observation of such processes. Schematically two types of experiments have been performed : either a controlled three-beam scattering process in the course of which incoming electrons collide with an atomic beam in the presence of a strong laser [2],[3], or the so -called above-threshold multiphoton ionization experiments in which the target atoms absorb more photons than the minimum number required by energy conservation for being ionized [4],[5],[6]. In both cases one observes in the final state electrons which have experienced real (or virtual) radiative transitions in the continuous spectrum of an atom or an ion. Note that, owing to the presence of the laser field and in contrast with the earlier studies on the bremsstrahlung, in which one considered the spontaneous emission of one photon, one is led here to consider stimulated processes [7]. More precisely one can equivalently discuss multiphoton absorption (inverse bremsstrahlung) or stimulated emission (stimulated bremsstrahlung).

Even in the presence of a strong laser field the perturbative approach provides an interesting basis for discussion and its results can be helpful as reference marks when discussing the case of very strong fields in which non-perturbative effects are expected to arise. On the other hand the Coulomb field (and the H-atom) represent an excellent test-case potential and, though somewhat naive, it provides a good account of most of the features of multiphoton processes in atomic systems.[8] In addition it has the recognized advantage of lending itself

(*) Laboratoire "Matière et Rayonnement", Associé au CNRS.

to "exact" calculations and there is no doubt that a precise evaluation of stimulated radiative cross-sections in a Coulomb field represents a first step towards a better understanding of the dynamics of the electron-ion laser-assisted collisions.

The organization of the paper will be as follows : In Sec II we shall briefly review the main properties of the Coulomb Green's Function (CGF) and introduce the compact representation derived by Schwinger[9]. We shall present also the so-called sturmian expansion of the CGF which has proven to be very useful in multiphoton calculations[8]. This latter expansion has been used to compute two- and three-photon above threshold ionization cross-sections in H-atom[10]. Our results presented in Sec III show unambiguously that such transitions are to be considered as one-step and not as multi-step processes. Recent related works currently in progress will be also briefly presented. The main object of the Sec.IV is to present a progress report on an "exact" calculation of two-photon transition amplitudes in the Coulomb continuous spectrum. To this end we shall first recall the main features of the one-photon case, then we shall consider the second-order case and make contact with other approximations. A brief conclusion will be presented in the Sec.V.

II - THE COULOMB GREEN'S FUNCTION (CGF)

The CGF, denoted G(z) and associated to the Coulomb Hamiltonian ($2m = e = \hbar = 1$) :

$$H_c = -\Delta - 1/r \qquad (1)$$

is solution of the inhomogeneous equation :

$$(z - H_c)G(z) = 1 \quad ; \; z < 0 \qquad (2)$$

It can be conveniently expanded onto the hydrogenic basis in the position space $\{\Phi_{nlm}(\vec{r}) = Y_{lm}(\hat{r})\, R_{nl}(r)\}$ and one has :

$$G(\vec{r},\vec{r}\,';z) = \sum_{L,M} Y^*_{LM}(\hat{r})\, Y_{LM}(\hat{r}\,')\, G_L(r,r';z) \qquad (3)$$

Where the radial component G_L for angular momentum L reads explicitly :

$$G_L(r,r';z) = \sum_n R^*_{n,L}(r)R_{n,L}(r')(z - E_n)^{-1} + \int_0^{+\infty} dE\, R^*_{E,L}(r)\, R_{E,L}(r')(z - E)^{-1}. \qquad (4)$$

Although the expansions (3) and (4) of the CGF can be easily used as they stand, it appears that the presence of the integral over the continuous spectrum makes them not so useful in practical computations.

Schwinger's representation we shall present now does not suffer from such a drawback.

A Schwinger's Representation [9]

On using the O(4) symmetry properties of the Kepler (and Coulomb) problem, Fock has shown that the integral Schrödinger equation in momentum space was more easily solved in the four-dimensional space :[11]

Schrödinger equation:

$$\left\{\begin{array}{c}\text{position space}\\ \text{differential Eq.}\\ Y_{lm}(\hat{r})R_{nl}(r)\end{array}\right\}\xrightarrow{\text{Fourier}}\left\{\begin{array}{c}\text{momentum space}\\ \text{Integral Eq.}\\ Y_{m}(\hat{p})\Phi_{nl}(p)\end{array}\right\}\xrightarrow{\text{Fock}}\left\{\begin{array}{c}\xi\in\mathbf{R}^4\\ \text{Integral Eq.}\\ Y_{nlm}(\xi)\end{array}\right\}$$

where $\xi=(\xi_0,\vec{\xi})$; $\xi_0=\dfrac{p^2-p_0^2}{p^2+p_0^2}$; $\vec{\xi}=-\dfrac{2p_0\vec{p}}{p^2+p_0^2}$; $p_0=1/n$,

and $Y_{nlm}(\xi)$ are four-dimensional hyperspherical harmonics.

Using the same set of transformations, Schwinger has been able to show that the inhomogeneous Eq. (2) verified by the CGF was also more easily solved in $\mathbf{R}^4$:

$$G(\vec{r},\vec{r}\,';z)\xrightarrow{\text{Fourier}}G(\vec{p},\vec{p}\,';z)\xrightarrow{\text{Fock}}\Gamma(\xi,\xi')$$

where $\Gamma(\xi,\xi')$ can be expanded onto the hyperspherical basis $(z<0)$:

$$\Gamma(\xi,\xi')=\sum_{nlm}(1-(np_0)^{-1})^{-1}\,Y_{nlm}(\xi)\,Y^{*}_{nlm}(\xi'),\qquad(5)$$

where $p_0=(-z)^{1/2}$.

The most salient feature of this expansion is that the infinite sum runs on the set of the discrete values of the quantum numbers nlm . In fact, this expansion represents the four-dimensional version of the so-called sturmian expansion we shall introduce below. Coming back to the momentum space, via the inverse Fock transform, Schwinger has obtained the following integral representation:

$$G(\vec{p},\vec{p}\,';z)\sim\int_1^{(0+)}d\rho\;\rho^{-i/p_0}\frac{d}{d\rho}\left[\frac{\rho^{-1}(1-\rho^2)}{[p_0^2(\vec{p}-\vec{p}\,')^2+(p^2+p_0^2)][(p'^2+p_0^2)(1-\rho^2)/4\rho]^2}\right].\qquad(6)$$

This representation has proven to be very useful for studying the analytical properties of second-order perturbative amplitudes in a Coulomb field and of the Coulomb T-matrix.[12]

B The Coulomb Sturmian Basis

The revival of the Coulomb sturmian basis is due to Rotenberg who named it so after the french mathematician Sturm (†1855) [13]. In fact it had been already used in the twenties by Epstein, Podolski and others [14]. For the sake of simplicity it may be more convenient to introduce it by reference to the usual Schrödinger problem:

Schrödinger Equation	Sturmian Equation
$(-\Delta - E - \frac{Z}{r})\Psi = 0$	$(-\Delta - E - \frac{\alpha}{r})\Psi = 0$
Z = charge of the nucleus (fixed parameter)	α = coupling constant $E<0$ is considered as a fixed parameter
Eigenvalue problem for the energy E (Z fixed).	Eigenvalue problem for the coupling constant (E fixed)
Solutions : discrete + continuous } spectrum	Solutions $\alpha = np_0$; $p_0 = (-E)^{1/2}$ discrete spectrum
a) $E<0$; $E_n = -Z^2/n^2$. $\Psi_{nlm}(\vec{r}) = Y_{lm}(\hat{r})\, R_{nl}(\frac{Z}{n} r)$.	$\Psi_{nlm}(\vec{r}) = Y_{lm}(\hat{r}) R_{nl}(p_0 r)$
b) $E>0$; $n \longrightarrow -i/k$	

The main advantage of the sturmian basis, when compared to the hydrogenic one, is that it is discrete and that it leads naturally to the so-called sturmian expansion for the CGF : when specializing to the radial part, one gets :

$$G_L(r,r';z) = \sum_{n=L+1}^{\infty} R^*_{nL}(p_0 r)\, R_{nL}(p_0 r')\, (1-np_0)^{-1} \qquad (8)$$

where $p_0 = \sqrt{-z}$. This expansion has been successfully used by several authors in multiphoton perturbative calculations[15]. We present in the next section one of these recent applications.

III - THREE-PHOTON ABOVE THRESHOLD IONIZATION OF ATOMIC HYDROGEN[10].

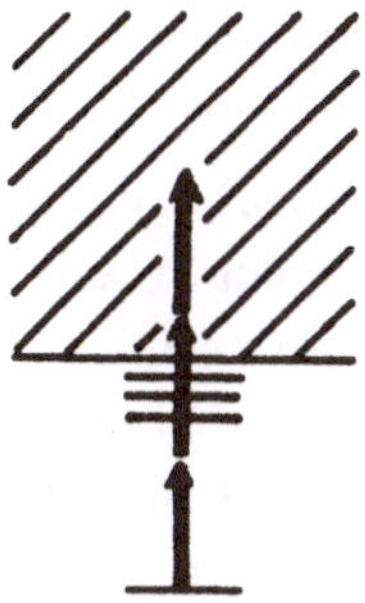

We consider here the cross-section for three-photon ionization of H-atom in the ground state when two-photon ionization is energetically possible. One can then expect to detect photoelectrons with kinetic energies : $E_n = E_{1s} + n\omega$, where $E_{1s} = -1$ Ry. is the ground state energy, ω is the frequency of the laser and $n \geqslant 2$ indicates the number of photons absorbed by the atomic system. In situations similar to the case considered here $(n = 3)$ the question has arisen if whether or not the process could be described as a two-photon ionization followed by a one-photon free-free transition, leading to the same final state of energy $E = E_{1s} + 3\omega$, as the direct three-photon process. For the sake of simplicity we shall discuss the case of a circularly polarized, monomode, laser field and assume the validity of the dipole approximation. The third-order transition amplitude then reduces, to within constant angular factors, to only one term, with radial part :

$$T^{(3)} \sim \lim_{\epsilon \to 0} \langle E_3, 3 \mid r \; G_2(E_{1s}+2\omega + i\epsilon) \; r \; G_1(E_{1s}+\omega + i\epsilon) r \mid 1,0 \rangle \; . \tag{9}$$

Note that if $E_{1s}+\omega < E_{1s}+2\omega < 0$ this radial amplitude is real. In the case considered here however, one has $E_{1s}+2\omega > 0 > E_{1s}+\omega$ and consequently the amplitude becomes complex. This can be verified by making explicit the expression of G, when expanded onto the physical hydrogenic basis :

$$\lim_{\epsilon \to 0} G_2(r,r'; E_{1s}+2\omega + i\epsilon) = \sum_n \frac{R^*_{n2}(r) R_{n2}(r')}{E_{1s}+2\omega - E_n} + P\!\int dE \, \frac{R^*_{E,2}(r) R_{E,2}(r')}{E_{1s}+2\omega - E}$$

$$- i\pi \, R^*_{E,2}(r) \, R_{E,2}(r') \Big|_{E = E_{1s}+2\omega} \; . \tag{10}$$

The imaginary part of $T^{(3)}$ then reads :

$$Im(T^{(3)}) = -\pi \underbrace{\langle E_3,3| r |E_{1s} + 2\omega ,2 \rangle}_{\text{one-photon free-free transition amplitude}} \underbrace{\langle E_{1s}+2\omega ,2| rG_1(E_0+\omega)r | 1,0\rangle}_{\text{two-photon ionization transition amplitude}} . \quad (11)$$

This result shows that the imaginary part factors into two terms corresponding respectively to a two-photon ionization amplitude and a one-photon free-free transition amplitude. It was thus interesting to determine the relative magnitudes of $Re(T^{(3)})$ and $Im(T^{(3)})$: if the latter was dominant ($Im(T) \gg Re(T)$) this would lend support to the hypothesis of a two step process (i.e. two-photon ionization + one-photon free-free transition) for interpreting the corresponding experiments. As a matter of fact, within the framework of this hypothesis, the total cross section could be expressed as the product of two cross-sections associated with each of the lower order processes.

We have performed the computation of the amplitude Eq.(9) by replacing the CGF by their sturmian expansions Eq. (8). The amplitude T (3) is then expressed as a double sum running on the sturmian spectrum. However owing to the fact that the argument $E_{1s}+2\omega$ of G_2 becomes positive the series strongly diverges but can be nevertheless resummed by using Padé-related techniques. The power of such techniques is illustrated in the following table, where we compare the sequence of partial sums S_n to the corresponding diagonal Padé Approximants (PA):

Partial sums versus diagonal PA to the radial transition amplitude T_{123} at λ=1500 Å

n	S_n	(n/n)
0	.164(5) +i.269(3)	.164(5) +i.269(3)
1	.118(6) +i.195(6)	-.112(5) +i.352(4)
2	-.939(6) +i.168(7)	.292(4) -i.458(4)
⋮	⋮	⋮
11	.368(13) +i.716(13)	.576(3) +i.376(3)
12	-.223(14) +i.215(14)	.574(3) +i.377(3)
13	-.105(15) -i.592(14)	.574(3) +i.377(3)
⋮	⋮	
25	.572(21) -i.421(21)	

$$T^{(3)} = \lim_{n\to\infty} S_n ;$$

$$S_n = \sum_{n_1=0}^{n} \sum_{n_2=0}^{\infty} u_{n_1 n_2} ; \qquad (12)$$

The PA sequence converges towards a stable limit, providing the numerical values of the real and imaginary part of the amplitude. One can easily check that they are of the same order of magnitude, which demonstrates that the two-step hypothesis is invalid at least in the domain of applicability of the perturbative approach. We have checked this result in different conditions of laser frequencies and polarization and verified that neither the real nor the imaginary part become dominant with respect to the other : both terms have to be coherently added and do contribute to the overall amplitude.

IV - TWO-PHOTON FREE-FREE TRANSITIONS.

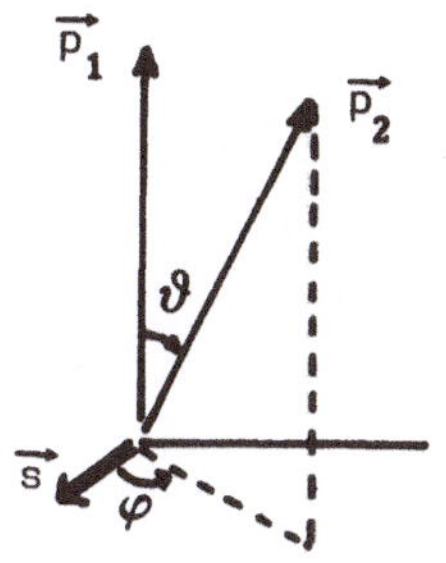

The calculation presented in this section correspond to the experimental situation schematized in the figure : while scattered in the Coulomb field of a nucleus, incoming electrons absorb (or emit) one or two photons of a monomode laser field. For the sake of comparizon with other results we shall adopt the following definition of the differential cross-section [16]:

$$\frac{d\sigma_\nu}{d\Omega(\hat{p}_2)} = \frac{\text{Probability of scattering in } d\Omega\,(\hat{p}_2) \text{ with absorption of } \nu \text{ photons.}}{\text{density of incoming electronic current}}$$

$$= \frac{p_2}{p_1}\,|f_\nu|^2 \qquad (13)$$

where $\vec{p}_1$ (resp. $\vec{p}_2$) is the momentum of the incoming (resp. outgoing) electron ; $p_2^2/2m = p_1^2/2m + \nu\hbar\omega$; $|\nu| = 1, 2, \ldots$ and f_ν is the scattering amplitude. Before going further, let us recall the main features and present the basic formalism used in the simpler one-photon calculation.

A - One-photon transitions

If $\nu = 1$, i.e. if one photon of the field is absorbed, one has explicitely, in the dipole approximation :

$$f_1 = \frac{m}{2\pi\hbar^2}\left(\frac{e\,a}{2\,m\,c}\right)\langle\vec{p}_2^{\,(-)}|\vec{s}\cdot\vec{p}\,|\vec{p}_1^{\,(+)}\rangle \qquad (14)$$

where a is the amplitude of the vector potential associated to the laser field, with polarization $\hat{s}$; $|\vec{p}_i^{(\pm)}\rangle$ are momentum space Coulomb wavefunctions with asymptotic momenta $\vec{p}_i$ (i= 1 , 2) :

$$\langle \vec{p}\,'|\vec{p}^{(+)}\rangle \sim \lim_{\eta\to 0} \oint \left(\frac{\zeta}{\zeta-1}\right)^{-i/p_i} \frac{d\zeta}{\left\{(\vec{p}\,' - \vec{p}_i\zeta)^2 + [\eta - ip_i(1-\zeta)]^2\right\}^2} \quad (15)$$

and the contour encircles the points$\{0 , 1\}$. One has also : $\langle \vec{p}\,'|\vec{p}_i^{(+)}\rangle = \langle \vec{p}_i^{(-)}|\vec{p}\,'\rangle$. The amplitude f_1 is then expressed in terms of a triple integral of the general form :

$$f_1 \sim \oint d\zeta_1 \ldots \oint d\zeta_2 \ldots \int d\vec{p}\,' \ldots \quad (16)$$

The $\vec{p}\,'$ integration reduces to a Dalitz integral [17], the ζ_2 integral yields an algebraic expression in terms of ζ_1 and the remaining integration results in a Gauss Hypergeometric function. One recovers in that way a result similar to Sommerfeld's [1] :

$$\frac{d\sigma_1}{d\Omega} \sim \left(\frac{I}{I_0}\right)\frac{1}{K^4} \left|{}_2F_1(1 - \frac{i}{p_1} , 2 - \frac{i}{p_2} ; 2 ; z)\right|^2 (\vec{s} \cdot \vec{p}_2)^2 \quad (17)$$

where I is the time averaged field strength intensity of the laser , $I_0 \sim 3.5\ 10^{16}$ W/cm^2 is an averaged electric field strength intensity, characteristic of the atomic field, $K = \hbar\omega/1Ry.$, $z = - 4p_1p_2 \sin^2(\theta/2)(p_1 - p_2)^{-2}$ and we have specialized the result to the case of the experimental geometry above displayed ($\vec{p}_1 \perp \vec{s}$).

This perturbative result,which is valid in the limit $I/I_0 \ll 1$, is general, provided the conditions for both the non-relativistic ($p_i \ll mc$) and the dipole ($K \lesssim 1$) approximations are fulfilled . It is nevertheless interesting to study the Born approximation ($p_1 , p_2 \gg 1$). In this approximation the ${}_2F_1$ function entering the expression (17) of the cross section simplifies to :

$${}_2F_1(\ldots) \longrightarrow {}_2F_1(1, 2; 2; z) = (1 - z)^{-1} = (p_1 - p_2)^2/(\vec{p}_1 - \vec{p}_2)^2$$

and the cross section itself becomes accordingly :

$$\frac{d\sigma_1}{d\Omega} = 4\left(\frac{I}{I_0}\right)(\vec{s} \cdot \vec{p}_2)^2 \frac{1}{K^4} \frac{p_2}{p_1} \frac{4}{(\vec{p}_1 - \vec{p}_2)^4} a_0^2 \ . \quad (18)$$

This result displays two interesting features connected to the behaviour of the cross section : i) for large values of the incoming p_1 and outgoing p_2 electron momenta the cross section is proportional to the Rutherford scattering cross section for momentum transfer

$\vec{Q}=\vec{p}_1-\vec{p}_2$ ii) in the limit of small photon energy, $K\rightarrow 0$, the cross section diverges. One recovers in that way the so-called Low theorem connected to the divergence of the bremsstrahlung cross-section in the soft photon limit [18] :

$$\frac{d\sigma_1}{d\Omega} = 4\ \frac{I}{I_0}\ (\vec{s}\cdot\vec{p}_2)\ \frac{1}{K^4}\ \left(\frac{d\sigma}{d\Omega}\right)_{el} \tag{19}$$

where $(d\sigma / d\Omega)_{el}$ is the Rutherford elastic scattering cross-section :

$$\left(\frac{d\sigma}{d\Omega}\right)_{el} = \frac{4\ a_0^2}{(\vec{p}_1-\vec{p}_2)^4}\ . \tag{20}$$

These results will permit us to display the connection existing between our perturbative results and the low intensity, low frequency, limit of the non-perturbative approach (see below).

B <u>Two - photon transitions</u>

The second order amplitude f_2, corresponding to the absorption of two photons of the laser field, reads, in the dipole approximation :

$$f_2 = \frac{m}{2\pi\hbar^2}\left(\frac{e\ a}{2\ m\ c}\right)^2 \langle\vec{p}_2(-)|(\vec{s}\cdot\vec{p})\,G(\vec{p},\vec{p}\,';E_1+\omega)(\vec{s}\cdot\vec{p})|\vec{p}_1{}^{(+)}\rangle, \tag{21}$$

where G is the momentum space representation of the CGF. If one chooses Schwinger's representation, Eq.(6), the second order amplitude f_2 can be expressed as a multiple integral of the following form :

$$f_2 \sim \int_1^{(0+)} d\rho\ \rho^{-i/p_0} \oint d\zeta_1 \ldots \oint d\zeta_2 \ldots \int d\vec{p}_1{}' \int d\vec{p}_2{}' \ldots \tag{22}$$

where $p_0 = (p_1{}^2 + K)^{1/2}$; $p_2{}^2 = p_1{}^2 + 2K$.

The double integral on $\vec{p}_1{}'$, $\vec{p}_2{}'$ has been evaluated in a similar context by Gavrila et al., Ref.(19) and turns out to be an algebraic function of the variables ζ_1, ζ_2 and ρ. In the case of the special geometry $(\vec{s}\perp\vec{p}_1)$ chosen here, f_2 becomes :

$$f_2 = A + (\vec{s}\cdot\vec{p}_2)^2\ B$$

where A and B are triple integrals of the general form :

$$\begin{pmatrix}A\\B\end{pmatrix} \sim \int_1^{(0+)} d\rho\ \rho^{-i/p_0} \oint d\zeta_1 \oint d\zeta_2 \begin{pmatrix}a(\zeta_1,\zeta_2,\rho)\\ b(\zeta_1,\zeta_2,\rho)\end{pmatrix} \tag{24}$$

After some algebra the two contour integrals can be expressed in terms of Gauss ${}_2F_1$ functions and the reduced amplitudes A and B contain integrals of the general form :

$$J = \int_0^1 d\rho\, \rho^{q-i/p_0} (1 - x_0\rho)^{\alpha} (1 - x_1\rho)^{\beta} (1 - x_2\rho)^{\gamma}\, {}_2F_1(a,b;c;z(\rho)) \tag{25}$$

which can be conveniently computed numerically.

The differential cross section becomes :

$$\frac{d\sigma_2}{d\Omega} = \left(\frac{I}{I_0}\right)^2 \frac{1}{K^8} \left[a' + b' (\vec{s} \cdot \vec{p}_2)^2 + c' (\vec{s} \cdot \vec{p}_2)^4 \right] a_0^2 , \tag{26}$$

where the coefficients a' , b' , c' are expressed in terms of integrals similar to J [20].

Again it is interesting to study the limits corresponding to the Born and to the soft-photon approximations. In the limit $(p_1 , p_2) \to \infty$ and $K / p_1^2 \ll 1$ one shows that the cross section reduces to :

$$\frac{d\sigma_2}{d\Omega} \to 16 \frac{p_2^5}{p_1} \frac{1}{K^8} \left(\frac{I}{I_0}\right)^2 \frac{(\vec{s} \cdot \vec{p}_2)^4}{(\vec{p}_1 - \vec{p}_2)^4} a_0^2 . \tag{27}$$

Again one observes that the Rutherford scattering cross-section for momentum transfer $\vec{Q} = \vec{p}_1 - \vec{p}_2$ can be factorized and that in the low frequency limit ($K \to 0$), $d\sigma_2 / d\Omega$ diverges. This result represents a generalization of the Low theorem [18] to the case of multiphoton bremsstrahlung. In this limit p_1 p_2 and the cross section becomes :

$$\frac{d\sigma_2}{d\Omega} \sim \frac{16}{K^8} \left(\frac{I}{I_0}\right)^2 \frac{(\hat{s}\cdot\hat{p}_2)^4}{(\vec{p}_1 - \vec{p}_2)^4} a_0^2 , \tag{28}$$

result which is independent of the magnitude of p_1. These expressions are useful for establishing the correspondance between the perturbative and the non-perturbative approaches.

C - Non-perturbative approximation

In the limit of very strong fields one can expect that the perturbative approach is no longer valid and, indeed, it is then necessary to take into account the "dressing" of the scattered electron by the laser field. The cross-sections are then modified and, in the soft-photon approximation one gets [16]:

$$\frac{d\sigma_\nu}{d\Omega} = \frac{p_2}{p_1} \left| J_\nu\, (\vec{\alpha}.\vec{Q}) \right|^2 \left(\frac{d\sigma}{d\Omega}\right)_{el} \qquad (29)$$

where J_ν is a Bessel function ; $\vec{\alpha} = \frac{4}{K^2} \left(\frac{I}{I_o}\right)^{1/2} \vec{s}$ is a parameter characteristic of the field strength; $\vec{Q} = \vec{p}_1 - \vec{p}_2$ is the momentum transfer and $(d\sigma / d\Omega)_{el}$ is the elastic scattering cross section without field, for momentum transfer $\vec{Q}$.

In the case of low or moderate laser intensities, i.e. for relatively small magnitudes of the coupling between the electron and the field, the argument of the Bessel function becomes small ($\vec{\alpha}.\vec{Q} \ll 1$) and one can retain only the first term of its power series expansion. One recovers in that way the perturbative results Eq. (18), for $\nu = 1$ and Eq.(27) for $\nu = 2$. This provides an independent check of our general expressions Eqs.(17) and (26) and on comparing the numerical results obtained by the two methods one can assess their respective limits of validity. Detailed results obtained along these lines will be published elsewhere [20].

V - CONCLUSION

We have presented recent applications of two different representations of the Coulomb Green's function in the field of atomic physics. The so-called sturmian expansion of the CGF has been used for computing two-and three-photon "above threshold" ionization cross sections, processes in the course of which the photoelectrons experience (virtual) free-free transitions. Our computation has permitted to discard a recent hypothesis which had been formulated in order to interpret recent experimental results. We have presented also a progress report on the calculation of two-photon transitions amplitudes in the continuous spectrum of a Coulomb field. Our calculation, performed within the framework of a perturbative theory, relies on the use of the Schwinger representation of the CGF. Our "exact" results will permit to assess the respective ranges of validity of both the perturbative and non-perturbative approaches. These two examples clearly illustrate the usefulness of the compact representation of the CGF presented here, for performing perturbative computations in a Coulomb field.

Acknowledgments This work is the result of several collaborations with S.Klarsfeld (Orsay), M.Gavrila (FOM , Amsterdam) and V.Véniard (Université Pierre et Marie Curie, Paris). Partial support from the FOM Instituut (Amsterdam) is gratefully acknowledged.

REFERENCES

(1) A.Sommerfeld, " Atombau und Spektrallinien" (Vieweg, Braunschweig, 1939) Vol.II, p.495
(2) A.Weingartshofer , J.K. Holmes, J.Sabbagh and S.L.Chin, J.Phys. B, 16, 1805 (1983) and references therein.
(3) D.Andrick and L.Langhans, J.Phys. B 11, 2355 (1978).
(4) F.Fabre, G.Petite, P.Agostini and M.Clement, J.Phys. B 15, 1353 (1982) and references therein.
(5) P.Kruit, H.G.Muller, J.Kimman and M.J.Van der Wiel, J.Phys. B 16, 937 (1983).
(6) R.Hippler, H.J.Humpert, H.Schwier, S.Jetzke and H.O.Lutz,in press.
(7) Though the cross section for two-photon spontaneous emission is expected to be extremely small there is still some hope to observe such a process in the field of a heavy nucleus (H.Kleinpoppen private communication).
(8) A.Maquet, Phys.Rev. A 15, 1088 (1977).
(9) J.Schwinger, J.Math.Phys. 5, 1606 (1964).
(10) S.Klarsfeld and A.Maquet, Phys.Lett. 78A, 40 (1980); see also A.Tip (this volume) who presents a complementary aspect of this type of processes.
(11) V.Fock, Z.Phys. 98, 145 (1935).
(12) J.C.Y.Chen and A.C.Chen in "Advances in Atomic and Molecular Physics" edited by D.R. Bates and I.Estermann (Academic, New York, 1972) Vol. 8.
(13) M.Rotenberg, in "Advances in Atomic and Molecular Physics" edited by D.R.Bates and I.Estermann (Academic, New York, 1970) Vol.6.
(14) P.S.Epstein, Proc. Natl. Acad. Sci. (USA) 12, 637 (1926); B.Podolsky, Proc. Natl. Acad. Sci. (USA) 14, 253 (1928).
(15) Recent references include : J.T.Broad, Phys. Rev. A26, 3078,(1982); in "Electron-Atom and Electron-Molecule Collisions", J.Hinze ed. (Plenum, London, 1983); S.Jetzke et al. 1984 (unpublished).
(16) N.M.Kroll and K.M.Watson, Phys.Rev. A8, 804 (1973).
(17) R.H.Dalitz, Proc.Roy. Soc. (London) A206, 509 (1951); R.R.Lewis, Phys.Rev. 102, 537 (1956).
(18) F.E. Low, Phys. Rev. 110, 974 (1958).
(19) M.Gavrila and A.Costescu, Phys. Rev. A2, 1752 (1970).
(20) M.Gavrila and A.Maquet, to be published.

RESONANCES IN ATOMIC PHOTO-IONIZATION

A. Tip
FOM-Institute for Atomic and Molecular Physics, Kruislaan 407, 1098 SJ Amsterdam, The Netherlands

ABSTRACT

We discuss the theoretical interpretation of recent experiments dealing with many-photon ionization of atoms. Some essential features are explained by means of a simple model of which the mathematical and physical properties are studied. In particular it is shown that resonances are an essential feature of the formalism.

I. INTRODUCTION AND EXPERIMENTAL RESULTS

It is sometimes possible to create resonances in a physical system by means of some external device. Here we shall discuss the resonance structure associated with the photoionization of atoms, the external device being a laser that produces a high intensity, nearly monochromatic electro-magnetic field. Of course photo-ionization of atoms is an old subject but only with the advent of powerful lasers new processes could be studied, in particular the multi-photon ionization of atoms, where multi does not stand for two or three but rather ten or more. If we denote the bound state energies of an atom by ε_n, ε_o being the ground state energy, and the ionization potential by ε_∞ we expect in a photo-ionization experiment to observe ionized electrons with energies

$$E_N = N\hbar\omega - \varepsilon_\infty + \varepsilon_o \tag{1}$$

if the atom is originally in its ground state. Here is N the number of photons absorbed by the atom. Obviously N must be so large that $E_N > 0$. Of course different processes are possible as well; the remaining ion can be left in an excited state or more than one electron can be ionized. Experiments by Boreham and Hughes[1]) on helium with field intensities of 10^{14} to $6 \cdot 10^{16}$ W/cm² did however not show such a peak structure. They also found a shift in the electron energy towards higher energies which they attributed to the so-called ponderomotive force. The latter is connected to the following. Consider an electron in a time-periodic ($T = 2\pi/\omega$)) electro-magnetic field ($\underline{x}$ and $\underline{p}$ are the electronic canonical variables, $\underline{a}(\underline{x},t)$ the vector potential, $\underline{e}(\underline{x},t) = \partial_t \underline{a}(\underline{x},t)$ the electric field vector). Then, in atomic units and in the Coulomb gauge, its Hamiltonian is

$$H(t) = \tfrac{1}{2} [\underline{p} - \underline{a}(\underline{x},t)]^2 \tag{2}$$

The ponderomotive potential is then $V_p(\underline{x}) = \tfrac{1}{2} < \underline{a}(\underline{x},t)^2 >$, where the average is over one period of the field. Since V_p is positive it pushes the ionized electron out of the field. At the field intensities employed V_p ($\sim$ field intensity I) is appreciable and could account for the observed shift in electron energy. Also, since I is not constant during an experiment different electrons will experience different ponderomotive energy gains and the peak structure (1) is washed out. Later experiments by Kruit, Kimman and van der Wiel[2]) at more moderate field intensities ($I \approx 10^{13}$ W/cm², this corresponds to about 1/10 of the nuclear field strength experienced by an electron in the first Bohr orbit in hydrogen) show a different pattern. These authors studied the multi-photon-ionization, $N > 11$, of xenon in its ground state and in figure 1 their results, showing the number of observed electrons as a function of final electron energy, are plotted. It is found that within the experimental inaccuracy no ponderomotive shift is observed (it would amount to about 1 eV at 10^{13} W/cm2). Also no smearing out of the peaks due to field inhomogeneities is found. A second striking observation is that the first ($N = 11$) peak disappears with increasing field intensity (10^{13} W/cm² in the bottom picture, $3 \cdot 10^{13}$ W/cm² in the top one). Thus we see that these experimental results lead to the following two questions: 1. Why is there no ponderomotive shift of the peaks?; 2. Why does the lowest peak disappear with increasing field intensity? As will be discussed in the following sections the ponderomotive shift <u>does</u> play a role but its action is more subtle and in fact counteracted by a second phe-

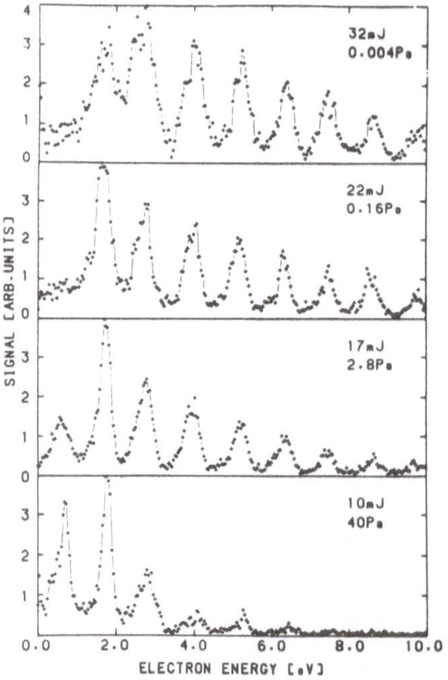

Figure 1

nomenon; a shift in the ionization potential.

II. A MODEL FOR CIRCULARLY POLARIZED FIELDS

The simplest model for atomic photo-ionization is one where the field is supposed to be spatially homogeneous and monochromatic. Since in an actual experiment the detector is always outside the field this presupposes that we can factorize the process in an actual ionization process in such a field followed by an escape of the ionized electron from the field region. We adopt this point of view in the following. So far it has turned out to be consistent with the available experimental information. There are two ways to consider homogeneous monochromatic fields. The first is to study an atom coupled to a single quantized field mode. This model has been treated by Grossmann and Tip [3]) who, among other matters, consider the dilatation-analytic properties of the associated Hamiltonian for hydrogen. The extension of the model to other atoms is straightforward as long as only Coulomb interactions are taken into account. Since actual experiments are performed in high intensity fields it is also possible to use a semi-classical model where the field is taken to be an external, spatially homogeneous one. A general account of this case is given in the talk of Dr. Graffi at this conference. Here we shall consider a special case; an electron in a spherically symmetric potential and a circularly polarized field. The Hamiltonian of such a system is given by

$$H(t) = \tfrac{1}{2}[\underline{p} - \underline{a}(t)]^2 + V(r), \quad \underline{a}(t) = \{a \cos \omega t,\ a \sin \omega t,\ 0\} \qquad (3)$$

It is well known [4,5]) that the time-dependence of $H(t)$ can be removed by means of a simple unitary transformation. In fact the time-evolution operator $U(t)$ $(U(0) = 1)$ associated with (3) can be represented as

$$U(t) = \exp[-i\omega l_3 t]\, \exp[-i Ht] \qquad (4)$$

where l_3 is the third component of the angular momentum operator $\underline{x} \times \underline{p}$ and H, the Floquet Hamiltonian, is given by $(\underline{a} = \{a,o,o\})$

$$H = \tfrac{1}{2}(\underline{p} - \underline{a})^2 - \omega l_3 + V(r) = H^{at} + \tfrac{1}{2}a^2 - \omega l_3 - \underline{p} \cdot \underline{a} = H_o + V(r) \qquad (5)$$

Equation (4) has the structure predicted by Floquet theory, $\exp[-i\omega l_3 t]$ being periodic in time with period $T = 2\pi/\omega$. The advantage of this model is that the operators in (4) are explicitly known. Although it is in principle possible to write U(t) in the Floquet form in the general case of periodic time dependence [6]) the operators replacing l_3 and H are not known even in the linearly polarized case; $\underline{a}(t) = \{a \cos \omega t, o, o\}$. This problem can be circumvented at the expense of working in a larger Hilbert space (see Graffi's talk) but then (as well as in the second quantized version) calculations can become more complicated since basis sets of functions depending on at least one additional variable are needed. We now turn back to (5). Disregarding the $\underline{p}.\underline{a}$-term for the moment we note that a shift over $\frac{1}{2}a^2$ (the ponderomotive shift, which is constant for this model) takes place in addition to a Zeeman splitting due to $-\omega l_3$ of the bound states supported by V(r). But also the continuum threshold (zero for $a = o$) changes into a set of thresholds, located at $\frac{1}{2}a^2 - m\omega$, $m \in \mathbb{Z}$, i.e. the bound states become continuum-embedded Since the $\underline{p}.\underline{a}$-term couples the various m-subspaces (eigenspaces of l_3) we expect the bound states to turn into resonances. In fact Enss and Veselič [7]) conjecture that for $a \neq o$, $\omega \neq o$ all bound states become resonances. We also expect that, although the continuum threshold (modulo $-m\omega$) is shifted over $\frac{1}{2}a^2$ the real part of the resonances (modulo $-m\omega$) will shift much less. This can be seen by considering the Floquet Hamiltonian in the gauge containing the electric field vector $\underline{e} = \{o, \omega a, o\}$; $\tilde{H} = H^{at} - \underline{e}.\underline{x} - \omega l_3$. Since the bound states of H^{at} are strongly localized the perturbation $-\underline{e}.\underline{x}$ will only lead to minor shifts as compared to $\frac{1}{2}a^2$. Thus the distance between the resonance $\varepsilon_o(\underline{a},\omega)$ associated with the atomic ground state and the ionization threshold will increase along with a and this suggests that the disappearance of the first peak in the experiment discussed in section 1 is associated with the fact that the continuum threshold can no longer be reached through the absorption of eleven photons. We recently studied some mathematical properties of H [8]). It turns out that the domain $\mathcal{D}(H_o)$ of $H_o = H_o(\underline{a},\omega)$ is strictly larger than $\mathcal{D}(\underline{p}^2) \cap \mathcal{D}(l_3)$, the intersection of the domains of $\underline{p}^2$ and l_3 and also depends on ω. If V(r) is relatively compact with respect to $\frac{1}{2}\underline{p}^2$, it is also $H_o(\underline{a},\omega)$-compact. The complex dilated form of H_o is

$$H_o(\underline{a},\omega,\zeta) = \tfrac{1}{2}[\underline{p}.\exp[-\zeta] - \underline{a}]^2 - \omega l_3 \tag{6}$$

and for $\mathrm{Im}\zeta \neq o$ its domain is $\mathcal{D}(\underline{p}^2) \cap \mathcal{D}(l_3)$.
Its spectrum consists of a set of half-lines starting at $\frac{1}{2}a^2 - m\omega$, $m \in \mathbb{Z}$ and going off under an angle $-2\,\mathrm{Im}\zeta$ with the positive real axis. If V(r) is dilatation-analytic with respect to $\frac{1}{2}\underline{p}^2$ then it is dilatation-analytic with respect to $H_o(\underline{a},\omega,\zeta)$. In view of the domain properties mentioned above, we cannot expect to find a type A family in

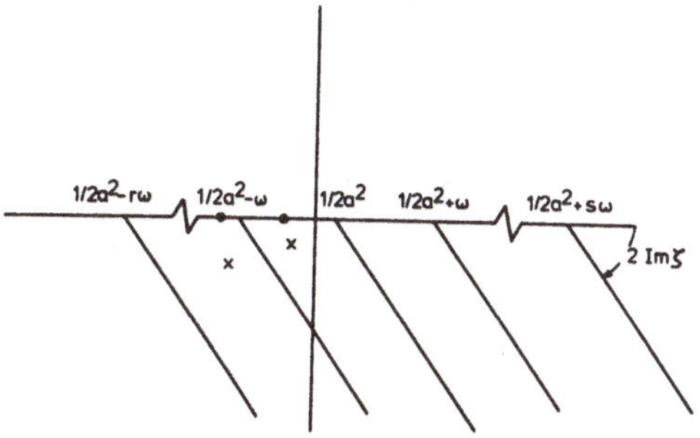

Figure 2

Kato's sense [9]). Nevertheless the complex dilatated Hamiltonian converges to the real dilated one in the strong resolvent sense as $\text{Im}\zeta \downarrow o$ (or $\uparrow o$). The same situation has been found in the dc-Stark case [10]). This is sufficient for physical applications since the relevant physical objects can usually be expressed in terms of inner products containing the resolvent of H. The spectrum of $H(\zeta) = H_o(\zeta) + V(r\exp[\zeta])$ is sketched in figure 2. Half-lines are essential spectrum and crosses resonances originating from the atomic eigenvalues indicated by dots.

III. AMPLITUDES FOR MULTI-PHOTON IONIZATION

We consider an atom in its ground state ϕ_o (eigenvalue ε_o) at time t=o and are interested to find the probability $w(\Delta)$ to find the electron ionized with kinetic energy in the interval Δ as $t \to \infty$. The first step is then to establish the existence of the wave-operator

$$\Omega = \lim_{t\to\infty} \exp[iH\,t]\,\exp[-i\,H_o\,t] \tag{7}$$

In principle we should start with U(t), given by (4) and the corresponding evolution operator without the potential. This, however, is equivalent to (7). For short range V(r) the existence of (7) follows by Cook's method, whereas for potentials with a Coulomb tail Dollard's modiciation of the free evolution is required. Details are given in [11]). Then ($\chi_\Delta(y) = 1$ for y in Δ and zero otherwise and $\underline{h} = \{o, a/\omega, o\}$)

$$w(\Delta) = \langle\phi_o|\Omega\,\exp[-i\underline{h}.\underline{p}]\,\chi_\Delta(\tfrac{1}{2}\,\underline{p}^2)\,\exp[i\underline{h}.\underline{p}]\,\Omega^*|\phi_o\rangle =$$

$$= \int_\Delta dk \sum_{lm} |F_{lm}(k)|^2 \tag{8}$$

where

$$F_{lm}(k) = \langle klm\,|\,\exp[i\underline{h}.\underline{p}]\,\Omega^*|\phi_o\rangle \tag{9}$$

Here the $|klm>$ are spherical free-wave states and the $\exp[\pm i\underline{h}.\underline{p}]$, which commute with $\chi_\Delta(\frac{1}{2}p^2)$, are inserted to obtain continuum eigenstates of H_o, i.e.

$$H_o \exp[-i\underline{h}.\underline{p}] \,|klm> = E_{km} \exp[-i\underline{h}.\underline{p}] \,|klm> , \quad E_{km} = \tfrac{1}{2}(k^2 + a^2) - m\omega .$$

Using an integral representation for the wave-operators [11]) we then have ($\varepsilon \downarrow o$ being understood);

$$F_{lm}(k) = F^B_{lm}(k) + \Phi_{lm}(k) = <klm \,|\exp[i\underline{h}.\underline{p}]|\phi_o>$$

$$+ <klm\,|\exp[i\underline{h}.\underline{p}] \; V[E_{km} + i\varepsilon - H]^{-1} \,|\phi_o> . \qquad (10)$$

The "Born" part F^B gives a back-ground contribution but Φ shows a resonance structure, i.e. it peaks when E_{km} passes through the real part of one of the resonances associated with H. Indeed if $-\underline{p}.\underline{a}$ is removed from H there is a pole at $E_{km} = \varepsilon_o$ (ϕ_o is an eigenfunction in that case) i.e. at $\frac{1}{2}k^2 = \varepsilon_o + m\omega - \frac{1}{2}a^2$. Here we note that (10) refers to an electron in the field; on its way to the detector it will acquire a ponderomotive energy gain of $\frac{1}{2}a^2$. Restoring the $\underline{a}.\underline{p}$-term the poles will move from the real axis and turn into resonances. In order to get access to these poles we need a method to continue $\Phi_{lm}(k) = \Phi$ analytically and this can most conveniently be done by means of the dilatation-analytic method. Here we encounter a complication since $\exp[-i\underline{h}.\underline{p}] \; V|klm>$ is in general not a dilatation-analytic vector. One way out of this problem is to apply the Feshbach projection formula with $P = |\phi_o> <\phi_o|$. Then ($Q = 1 - P$)

$$\Phi = <klm\,|\exp[i\underline{h}.\underline{p}] \; V\{[E_{km} + i\varepsilon - H_Q]^{-1} \, H_{QP} + 1\} \,|\phi_o> \,.$$

$$. <\phi_o| \, [E_{km} + i\varepsilon - H]^{-1} \,|\phi_o> \qquad (11)$$

Since Q removes the ground state from H^{at} the corresponding resonance will be removed from H_Q and we expect the first factor in (11) to vary smoothly if E_{km} passes through the real part of $\varepsilon_o(\underline{a},\omega)$, the resonance associated with ε_o. The second factor will peak at this value but since it can be dilated (ϕ_o is dilatation-analytic) we can disentangle resonance and background contributions in this quantity. The case that the field is resonant between ε_o and another atomic eigenvalue (energy difference a multiple of ω) is more complicated, since ε_o and one of the Zeeman-split components of the second eigenvalue become degenerate. The remedy is to increase P so that both states are removed from H^{at} in H_Q.

We can now obtain an approximate expression for the area under a resonance peak by taking k in the first factor of Φ such that $E_{km} = \mathrm{Re}\ \varepsilon_o(\underline{a},\omega)$ and making a pole approximation for the second factor of (11). We then find that this integrated peak intensity is proportional to $R_m(\underline{a},\omega) = \mathrm{Re}\ [\varepsilon_o(\underline{a},\omega) + m\omega - \frac{1}{2} a^2]^{\frac{1}{2}}$, for details see ref. 11.
Here we have found the mechanism that suppresses the first peak for large field intensity since R_m decreases rapidly once $\mathrm{Re}\ \varepsilon_o(\underline{a},\omega) + m\omega - \frac{1}{2} a^2$ approaches zero and becomes negative. This happens if $\varepsilon_o(\underline{a},\omega) + m\omega$ is originally (for a close to zero) positive and $\varepsilon_o(\underline{a},\omega)$ increases less with increasing a than $\frac{1}{2}a^2$. Obviously other peaks also can shrink if the field intensity becomes larger and larger. Indeed it is seen in figure 1 that at the highest field intensity employed the second peak also starts disappearing.
Separable potentials and in particular the zero-range potential (point interaction potential) give rise to significant simplifications. In fact the Feshbach trick is not needed in this case. We made a numerical evaluation for the case of the zero-range potential (which supports one bound state). Thus we calculated the corresponding resonance for a number of values of a and ω. In figure 3 the orbit of the resonance pole in the complex plane is plotted for the case N = 5 (ε_o = -1, ω = 0.24). In fact we plotted $\varepsilon_o(\underline{a},\omega) - \frac{1}{2} a^2$. It takes off from the real axis with a slope $\sim a^{10} \sim I^5$ in accordance with perturbation theory. At the maximal value of a used in the computation ($a \approx 1$) $\Gamma = \mathrm{Im}\ \varepsilon_o(\underline{a},\omega) \approx 10^{-3}$. We also determined some orbits of other poles in different Riemman sheets. The sharp bend in the curve in figure 3 turns out to be related to an avoided crossing with such an orbit.

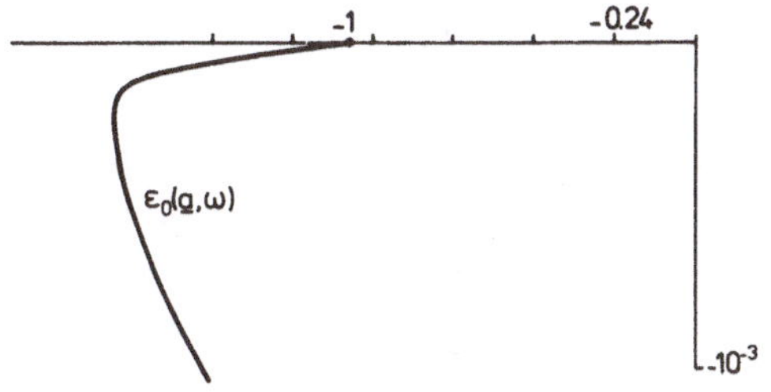

Figure 3

IV. CONCLUDING REMARKS

We finally come back to the questions raised in section 1. We have seen that the increase in the ionization potential with $\frac{1}{2} a^2$ leads to a mechanism that suppresses the lowest peaks if the field intensity is increased. The corresponding energy loss, however, is compensated by the ponderomotive force mechanism when the ionized electron is on its way to the detector. The net effect is a suppression of the lowest peaks but no shift of order $\frac{1}{2}a^2$ but instead the shift associated with the a.c. Stark shift of the ground state which is much smaller. A mathematically more precise treatment, including a spatially inhomogeneous but asymptotically vanishing field, would show an essential spectrum of the corresponding Hamiltonian without the $\frac{1}{2}a^2$-shift (Weyl's theorem on the location of the essential spectrum). In figure 4 we present a suggestive picture of the actual situation. The ponderomotive potential gives rise to a barrier of macroscopic dimensions (the laser focus) that essentially prevents the electron to escape from the nucleus if it has not absorbed a sufficient amount of photons. Thus either no ionization takes place during the time-interval the laser is acting or the atom picks up an additional photon and then passes over the barrier. We are at present investigating whether these notions can be made mathematically precise as asymptotic results for small ε for the case of fields $\underline{a}(\varepsilon\underline{x},t)$ with smooth spatial dependence. In principle the formalism presented here can also be set up for fields with polarization different from the circular one. Then the use of a larger Hilbert-space seems unavoidable. In such a Hilbert-space separable potentials in the original space are no longer separable, however.

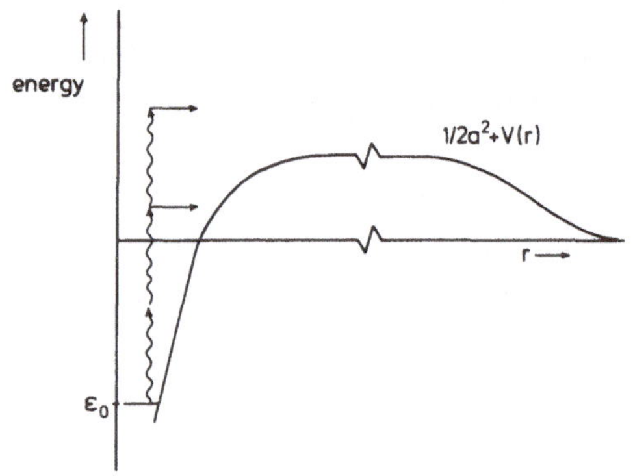

Figure 4

ACKNOWLEDGEMENTS

The work reported here was done in collaboration with H.G. Muller. It benefitted from many discussions with J. Kimman and M. van der Wiel. This work is part of the research program of the Stichting voor Fundamenteel Onderzoek der Materie (Foundation for Fundamental Research on Matter) and was made possible by financial support from the Nederlandse Organisatie voor Zuiver-Wetenschappelijk Onderzoek (Netherlands Organization for the Advancement of Pure Research).

REFERENCES

[1]) B.W. Boreham and J.L. Hughes: Sov.Phys.JETP 53, 252 (1981).

[2]) P. Kruit, J. Kimman, H.G. Muller and M.J. van der Wiel: Phys.Rev. A 28, 248 (1983).

[3]) A. Grossmann and A. Tip: J.Phys.A Math.Gen. 13, 3381 (1980).

[4]) F.V. Bunkin and A.M. Prokhorov: Sov.Phys.JETP 19, 739 (1964).

[5]) W.R. Salzman: Chem.Phys.Lett. 25, 302 (1974).

[6]) E. Prugovečki and A. Tip: J.Phys.A Math.Gen. 7, 572 (1974).

[7]) V. Enss and K. Veselić: Ann.Inst.H. Poincaré A 39, 159 (1983).

[8]) A. Tip: J.Phys.A Math.Gen. 16, 3237 (1983).

[9]) T. Kato: Perturbation theory of linear operators (Springer, Berlin 1966).

[10]) I.W. Herbst: Comm.Math.Phys. 64, 279 (1979).

[11]) H.G. Muller and A. Tip: Multi-photon ionization in strong fields (preprint).

RESONANCES AND PERTURBATION THEORY FOR N-BODY ATOMIC SYSTEMS IN EXTERNAL AC-ELECTRIC FIELDS

S. Graffi

Dipartimento di Matematica, Università di Bologna
40127 Bologna, Italy

1. Introduction. The purpose of this talk is to describe some recent results on the basic mathematical aspects of phenomena, such as AC-Lo Surdo-Stark effect, photoionization, stimulated emission-absorption, occurring in N-body non-relativistic, spinless atomic systems under the action of an external, spatially homogeneous, AC-electric field. These problems and related ones are reviewed in Reinhardt[1], and further discussions can be found in the talks of Maquet and Tip in these Proceedings. Postponing the exact formalization to the subsequent section, let

$$T = -\frac{1}{2}\sum_{k=1}^{N}\Delta_k + \sum_{k=1}^{N} V_k(\vec{r}_k) + \sum_{i<k=1}^{N} V_{ik}(\vec{r}_i-\vec{r}_k) \equiv -\frac{1}{2}\sum_{k=1}^{N}\Delta_k + W(\vec{r}_1,\dots,\vec{r}_N)$$

$\vec{r}_k \in \mathbb{R}^3$, $k = 1,\dots,N$, be the N-body Schrödinger operator describing (in atomic units) the quantum motions of N points in the field generated by the sum of the two-body dilation analytic potentials $V_k(\cdot)$, $V_{ik}(\cdot)$ (in particular, the Coulomb ones). Let in addition the system be under the action of an external electric field $t \to F\vec{E}(t)$, $F > 0$, $\|\vec{E}(t)\| \leq 1$, homogeneous in space and time-periodic of period $2\pi/\omega$, $\omega > 0$, for example $\vec{E}(t) = (\cos\omega t, 0, 0)$. Writing the e.m. potentials in the Coulomb gauge $(\vec{0},\Phi)$, $(\vec{E}(t) = \vec{\nabla}\Phi$, $\Phi = \sum_{k=1}^{N} \langle \vec{r}_k, \vec{E}(t) \rangle$, the time-dependent Schrödinger equation is:

$$H_C\psi_C \equiv (T + F\Phi)\Psi_C = i\partial\psi_C/\partial t \tag{1.1}$$

while in the radiation gauge $(\vec{A}(t),0)$, $\vec{E}(t) = -d\vec{A}(t)/dt$ the Schrödinger equation takes the form:

$$H_R\psi_R \equiv (\sum_{k=1}^{N}(i\vec{\nabla}_k - F\vec{A}(t))^2 + W)\psi_R = i\partial\psi_R/\partial t \tag{1.2}$$

Equations (1.1) and (1.2) go of course into each other through the unitary transformation $U_{\psi_R} \equiv e^{i\chi}\psi_R = \psi_C$, $U^{-1}\psi_C \equiv e^{-i\chi}\psi_C = \psi_R$, where $\chi = F \sum_{k=1}^{N} \langle \vec{A}(t), \vec{r}_k \rangle$ is the generating function of the gauge transormation $\Phi \to \Phi - \partial\chi/\partial t \equiv 0$, $\vec{O} \to \vec{O} + \vec{\nabla}\chi \equiv \vec{A}(t)$.

Assuming, as we shall do from now on, $\vec{A}(t) \in C^{\infty}(T_\omega; \mathbb{R}^3)$, $T_\omega \equiv \mathbb{R}\setminus(2\pi/\omega)$, with mean 0, $\int_0^{2\pi/\omega} \vec{A}(t)dt = 0$, we see that at the limit $\omega \to 0$ (1.1) goes into the Schrödinger equation for the N-body DC-Lo Surdo-Stark effect analyzed in Refs.(2-5):

$$(T + F \sum_{k=1}^{N} \langle \vec{A}, \vec{r}_i \rangle)\, \psi_C = i\partial\psi_C/\partial t \tag{1.3}$$

$\vec{A} = \sum_{n=-\infty}^{+\infty} \vec{a}_n$, $\{\vec{a}_n : n \in \mathbb{Z},\ \vec{a}_o = 0\}$ the Fourier coefficients of $\vec{A}(t)$, while (1.2) goes into

$$\left(\sum_{k=1}^{N} (-i\vec{\nabla}_k - F\vec{A}t)^2 + W\right)\psi_R = i\partial\psi_R/\partial t \tag{1.4}$$

which is equivalent to (1.3) in the moving frame $\vec{r}_k \to \vec{r}_k - \frac{1}{2} F\vec{A}t^2$, $k = 1,\dots,N$.

The non-autonomous, time-periodic Schrödinger equations (1.1),(1.2) can be reduced to a stationary problem by the same argument of Floquet theory for ODE with periodic coefficients, i.e. one looks for quasi-periodic solutions of the form:

$$\begin{aligned} \psi_R(\vec{r}_1,\dots,\vec{r}_N;t) &= e^{-i\lambda t}\phi_R(\vec{r}_1,\dots,\vec{r}_N;t) \\ \psi_C(\vec{r}_1,\dots,\vec{r}_N;t) &= e^{-i\lambda t}\phi_C(\vec{r}_1,\dots,\vec{r}_N;t) \end{aligned} \tag{1.5}$$

where $\phi_R(\cdot,t), \phi_C(\cdot,t)$ are $2\pi/\omega$-periodic $L^2(\mathbb{R}^{3N})$-valued functions of time. Formally, it is easy to check that solutions of this type exist if and only if $\lambda = \lambda(F,\omega)$ is eigenvalue of the unitarily equivalent operators

$$K_C(F) = T + F\Phi(\vec{r}_1,\dots,\vec{r}_N;t) - i\partial/\partial t \equiv H_C - i\partial/\partial t \tag{1.6}$$

$$K_R(F) = \sum_{k=1}^{N} (-i\vec{\nabla}_k - F\vec{A}(t))^2 + W(\vec{r}_1,\dots,\vec{r}_N) - i\partial/\partial t \equiv H_R - i\partial/\partial t \tag{1.7}$$

with eigenvectors $\phi_C(\cdot,t), \phi_R(\cdot,t)$, respectively.(1.6) and (1.7) act on the Hilbert space $K = L^2(\mathbb{R}^{3N}) \otimes L^2(T_\omega)$, and the dependence on ω has been omitted because this quantity will be always kept fixed at some positive value. We thus see that the Floquet Hamiltonian, equivalently represented by (1.6) and (1.7) in the Coulomb and radiation gauge, respectively, plays the role of the time-independent Schrödinger operator in the autonomous case because its spectral properties determine the time evolution.

When $F = 0$ we have of course that, for each eigenvalue λ of $T, \lambda + n\omega$, $n = 0, \pm 1, \ldots$, is an embedded eigenvalue of

$$K = T - i\partial/\partial t \tag{1.8}$$

because $\sigma(K) = \sigma(T) + n\omega$, $n = 0, \pm 1, \ldots$ .However, as in the DC-field[2-5] case , we do not expect that the eigenvalues $\lambda + n\omega$ of K keep stable as eigenvalues of $K_C(F)$, $K_R(F)$ as $F \neq 0$: rather, we expect them to turn into resonances $\lambda(F)$, whose imaginary part is to be proportional to the ionization rate.

Since the potentials are dilation analytic, to describe resonances complex scaling has to be introduced in the Floquet formalism, first implemented in Schrödinger operator theory by Yajima[6] in the framework of two-body scattering with short-range, time-periodic potentials. We will see that the choice of gauge is essential to this effect.

More specifically, consider the formal complex-scaled version of (1.1), (1.2)

$$H_C(F,\theta)\psi_C(\theta) = i\partial\psi_C(\theta)/\partial t \tag{1.9}$$

$$H_R(F,\theta)\psi_R(\theta) = i\partial\psi_R(\theta)/\partial t \tag{1.10}$$

and the formal complex-scaled Floquet Hamiltonians $K_C(F,\theta)$, $K_R(F,\theta)$, obtained by (1.1), (1.2); (1.6), (1.7), respectively, replacing $\vec{r}_k$ by $e^{\theta}\vec{r}_k$, $k = 1,\ldots,N$. Then the basic mathematical problems consist in proving, for $|\mathrm{Im}\theta| < \alpha$, $\alpha > 0$, existence and uniqueness of solutions for at least one of the temporally inhomogeneous equations (1.9), (1.10), in realizing the corresponding complex-scaled Floquet Hamiltonian, and in performing its spectral analysis to the effect of showing existence of resonances so that the formal connection (1.5) can be implemented in the sense of resonances for θ complex.

Even more than in the DC-field case, the main difficulties in implemen ting this program lie in the high singularities introduced by the time-dependent perturbation, expecially strong in the form (1.1),(1.9) to which not even the standard existence theorems for temporally inhomogeneous Schrödinger equations can be applied.

However, even though in the two-body case and for a restricted class of very smooth potential, a complete mathematical justification of the above picture has been obtained by Yajima[7]: he first remarked that the unitary transformation[8]

$$Q_D\psi_C(\vec{r},t) \equiv \psi_D(\vec{r},t) = \tag{1.11}$$
$$= \exp(-iF<\vec{A}(t),\vec{r}> + F^2 \int_0^t \|A(\tau)\|^2 d\theta)\psi_C(\vec{r} - F\int_0^t \vec{A}(\tau)d\tau)$$

transforms (1.9) $N=1$ into:

$$H_D(F,\theta)\psi_D \equiv (-\frac{1}{2}e^{-2\theta}\Delta + V(e^{\theta}\vec{r} + F\int_0^t \vec{A}(\tau)d\tau))\psi_D = i\partial\psi_D/\partial t \tag{1.12}$$

which eliminates the highly singular term $F<\vec{E}(t), e^{\theta}\vec{r}>$. The results thus obtained by Yajima[7] assuming V translation analytic, bounded and short range can be summed up as follows:

(1) For $0 \le \text{Im}\theta < \alpha < \frac{\pi}{4}$ ($0 \ge \text{Im}\theta \ge -\alpha \ge -\frac{\pi}{4}$) equation (1.12) generates a unique propagator $\{U_D(t,s;F,\theta): \pm t \ge \pm s\}$ in $L^2(\mathbb{R}^3)$. U_D is unitary for $\theta \in \mathbb{R}$, strongly continuous in $(t,s;F,\theta)$ for $\pm t \ge \pm s$, $0 \le F < \bar{F}$, $\text{Im}\theta \ge 0$ ($\text{Im}\theta \le 0$); holomorphic in θ for $\text{Im}\theta > 0$ ($\text{Im}\theta < 0$), $U_D(s,s;F,\theta) = I$, $U_D(t,r;F,\theta)\, U_D(r,s;F,\theta) = U_D(t,s;F,\theta)$ for $\pm t \ge \pm r \ge \pm s$, $U_D(t+2\pi/\omega, s+2\pi/\omega;F,\theta) = U_D(t,s;F,\theta)$.

(2) The complex-scaled Floquet Hamiltonian

$$K_D(F,\theta) = -\frac{1}{2}e^{-2\theta}\Delta + V(e^{\theta}\vec{r} + F\int_0^t \vec{A}(\tau)d\tau - i\partial/\partial t \tag{1.13}$$

can be realized as an operator family in K such that

(i) For $(F,\theta) \in \mathbb{R}$ $K_D(F,\theta)$ is self-adjoint and $\sigma(K_D(\cdot)) = \mathbb{R}$.

(ii) $K_D(F,\theta)$ is a holomorphic family in (θ,F) for $0 < \text{Im}\theta < \frac{\pi}{4}$ ($0 > \text{Im}\theta > -\frac{\pi}{4}$), $F \in \mathbb{C}$ and for any such (θ,F) $\sigma_{ess}(K_D(\cdot)) = \bigcup_{n=-\infty}^{+\infty}(n\omega + e^{-2\theta}\mathbb{R}_+)$.

(iii) The discrete eigenvalues of $K_D(F,\theta)$, $\text{Im}\theta > 0$, have non-positive imaginary parts and are resonances of $K_D(F) \equiv K_D(F,0)$ in the standard sense of dilation analyticity. If $\lambda(F)$ is any such resonance, with eigenvector $\phi(F,\theta,\vec{r},t) \in K$, then $f(\cdot,t) = e^{-i\lambda(F)t}\phi(\cdot,t)$ solves the Schrödinger equation (1.12), and thus (1.5) is realized in the sense of resonances through (1.11). Furthermore $e^{-(2\pi i/\omega)\lambda(F)}$ is an eigenvalue of the Floquet operator $U_D(s+2\pi/\omega, s;F,\theta)$.

(iv) If λ is an isolated eigenvalue of T, and $\text{Im}\theta > 0$, $\lambda + n\omega$ is an isolated eigenvalue of $K_D(\theta) \equiv K_D(0,\theta) = -\frac{1}{2}e^{-2\theta}\Delta + V(e^{\theta}\vec{r}) - i\partial/\partial t$, and for $|F|$ small $K_D(F,\theta)$ admits a number of eigenvalues near λ equal to the total multiplicity of all eigenvalues λ_j of $K_D(\theta)$ such that $\lambda_j = \lambda + n_j\omega$, $j = 1,\ldots,\ell$. All such resonances $\lambda(F)$ are (branches of) analytic functions near $F = 0$. If the multiplicity of λ is 1, and $n_j = 0$ for all j, $\lambda(F)$ admits a convergent perturbation expansion in powers of F^2 (generated by expanding V in powers of F).

(v) If $\lambda < 0$, and $\vec{A}(t) = \sum_{k=-p}^{p} \vec{a}_k e^{ik\omega t}$, the first non-vanishing order in perturbation theory for $\text{Im}\,\lambda(F)$ is determined by the condition $\lambda + np\omega > 0$, and $\text{Im}\,\lambda(F)$ is given by the Fermi Golden Rule at that order.

(vi) Let $\lambda_1 < 0, \lambda_2 < 0$ be simple eigenvalues of T, with eigenvectors ϕ_1, ϕ_2, respectively, and let $\lambda_1 - \lambda_2 = n\omega$. Let $\lambda_1(F), \lambda_2(F)$ the nearby resonances of K(F) for F small. Then, as $t \to \infty$:

$$\langle U_D(t,s;F)\phi_1,\phi_2\rangle = \frac{1}{2}(e^{-i\lambda_1(F)(t-s)} - e^{-i\lambda_2(F)(t-s)})e^{-in\omega t} + O(F) \quad (1.14)$$

uniformly in $\pm t \geq \pm s$, where $U_D(t,s;F) \equiv U_D(t,s;F,0)$ is the propagator of $H_D\psi_D = i\partial\psi_D/\partial t$.

We remark that statement 2.(v) above justifies the well known fact that no ionization takes place unless the external field supplies the minimum number of photons needed to remove any given bound state into the continuum, and that the forced oscillation (1.14) between two unperturbed bound states yields a mathematical description of the stimulated emission-absorption phenomenon (see e.g.Merzbacher[(9)]).

Extending these results to more realistic potentials such as the Coulomb one, not to mention the N-body case, is a non trivial problem because $V(e^{\theta}\vec{r} + F\int_0^t \vec{A}(\tau)d\tau)$ is not analytic in θ nor in F when $V = |\vec{r}|^{-1}$, for example. Existence of resonances and validity of the Fermi Golden Mule for a class of potentials including the Coulomb one were obtained in Ref.(10) by implementing exterior complex scaling[11] into Yajima's formalism. It appears however, even though the form (1.1), (1.6) is more convenient for computational purposes[1], that the natural gauge for exploiting dilation analyticity in this kind of problems is the radiation one[12-15]. In this case indeed the Kitada-Yajima transformation (1.11) is unnecessary and equations (1.10) can be directly discussed in the N-body case. This time the singularity introduced by the time-dependent perturbation, i.e. essentially the term $Fe^{\theta}\sum_{k=1}^{N} < \vec{A}(t),\vec{\nabla}_k >$, is less violent so that, as we shall see later, the standard existence conditions for the propagator generated by (1.10) can be verified and hence statement (1) above holds. To go further, the Floquet Hamiltonian $H_R(F,\theta)$ has to be realized as a holomorphic operator family acting in K for F and θ in suitable domains. Here, as in the DC-field[2-5] case, the main difficulty comes from the lacking of analyticity at $\theta=0$; furthermore the perturbation $Fe^{\theta}\sum_{k=1}^{N} < \vec{A}(t),\vec{\nabla}_k>$ is not even $K_R(0,\theta)$-bounded for θ real. However, a fundamental difference between the DC-case and the AC-one is that in the former field strength analyticity and dilation analyticity are the same thing[2-5] because by scaling there is an overall dependence on $e^{3\theta}F$, while in the latter this scaling property does not hold. Thus it turns out that in the AC-case the Floquet Hamiltonian $K_R(F,\theta)$ is actually a holomorphic operator family near $F=0$ for θ non real (and the continuity as $\mathrm{Im}\theta \to 0$ holds in the sense of strong resolvent convergence as in the DC-case). Hence not only statements 2(i-v) can be established in the general N-body case with two-body dilation analytic potentials, but this is possible without an a priori determination of $\sigma_{ess}(K(F,\theta))$, which has been obtained by Howland[15] in the two-body case and represents a highly non trivial point in the N-body one. The most striking difference between AC- and DC-cases is of course the convergence of the perturbation expansion in the former versus its divergence in the latter[2-5]: however, we expect divergence also in the AC-case if $\int_0^{2\pi/\omega} \vec{A}(t)dt = \vec{A}_0 \neq 0$, because this would correspond to a superposition between a static and a time-variable field. In the next Section we describe in some detail, essentially following

Ref.(16), the proof of statements (1) and (2)-(i-v) above in the general N-body case with dilation analytic potentials. We also mention the extension of statement 2-(vi) to a class of potentials more general than those considered by Yajima. For the proof of this last statement, for further details and any further notation not explicitly defined here the reader is referred to Graffi-Grecchi-Silverstone[16].

2. Resonances and Perturbation Theory. Consider eq.(1.10), rewritten as:

$$\sum_{k=1}^{N} (-ie^{-\theta}\vec{\nabla}_k - F\vec{A}(t))^2\psi + W(e^{\theta}\vec{r}_1,\dots,e^{\theta}\vec{r}_N)\psi = i\,\partial\psi/\partial t \tag{2.1}$$

for $\theta \in \mathbb{C}_a = \{\theta \in \mathbb{C}: |\mathrm{Im}\,\theta| < a;\ a < \frac{\pi}{4}\}$. The subscript R has been omitted because, as already mentioned, we consider only the radiation gauge. Denote now by $H(F,\theta,t)$ the operator family in $\mathcal{H} \equiv L^2(\mathbb{R}^{3N})$ defined by the ℓ.h.s. of (2.1) on the domain $H^2(\mathbb{R}^{3N}) \equiv H^2$. The solution of (2.1), rewritten as $H(F,\theta,t) = i\partial\psi/\partial t$, is specified as follows (compare with statement (1) in Sect.1):

2.1. Proposition. Let $\theta \in \overline{C}_a^{\pm} \equiv \{\theta \in \mathbb{C}: 0 \le \mathrm{Im}\,\theta \le a;\ 0 \ge \mathrm{Im}\,\theta \ge -a\}$, $F \in \mathbb{R}$. Then the equation $H(F,\theta,t)\psi = i\partial\psi/\partial t$ defines a unique propagator $U(t,s;F,\theta)$ such that:

(1) $U(s,s;\cdot) = I$, $U(t,r;\cdot)\,U(r,s;\cdot) = U(t,s;\cdot)$, $\pm t \ge \pm r \ge \pm s$

(2) For $\theta \in \mathbb{R}$ $\{U(t,s;F,\theta): (t,s) \in \mathbb{R}\}$ is a unitary propagator and, if $\phi \in \mathbb{R}$:

$$U(t,s;F,\theta+\phi) = S(\phi)\, U(t,s;F,\theta) S(\phi)^{-1} \tag{2.2}$$

where $S(\phi)$ is the unitary dilation in $\mathcal{H}$: $(S(\phi)f)(\vec{r}_1,\dots,\vec{r}_N) =$ $= e^{3N\phi/2} f(e^{\phi}\vec{r}_1,\dots,e^{\phi}\vec{r}_N)$.

(3) The propagator $U(\cdot)$ is time-periodic

$$U(t+2\pi/\omega,\ s+2\pi/\omega;\cdot) = U(t,s,\cdot) \tag{2.3}$$

(4) $$\|U(t,s;F,\theta)\| \le e^{M|t-s|} \quad \text{for some } M>0 \tag{2.4}$$

(5) $U(t,s;\cdot)H^2 \subset H^2$; $U(t,s;\cdot)f$ is differentiable in (t,s) for any $f \in H^2$,

and:

$$i \frac{\partial}{\partial t} U(t,s;\cdot)f = H(\cdot,t)\, U(t,s;\cdot)f \tag{2.5}$$

$$-i \frac{\partial}{\partial t} U(t,s;\cdot)f = U(t,s;\cdot)\, H(\cdot,s)f \tag{2.6}$$

(6) $U(t,s;\, F,\theta)$ is strongly continuous in $(t,s;\, F,\theta)$ for $\pm t \geq \pm s$, $\theta \in \overline{\mathbb{C}}_a^{\pm}$, $F \in \mathbb{R}$, and is analytic in $\theta \in \mathbb{C}_a^{\pm}$ for any fixed $\pm t \geq \pm s$, F.

Proof. Write, for $u \in H^2$, $\theta \in \overline{\mathbb{C}}_a$, $F \in \mathbb{C}$:

$$H(F,\theta,t) = T(\theta)u + ie^{-\theta} F < \vec{A}(t) \cdot \sum_{k=1}^{N} \vec{\nabla}_k u > + \frac{1}{2} N \|\vec{A}(t)\|^2 \tag{2.7}$$

where $T(\theta)$, the action of $-\frac{1}{2} e^{-2\theta} \sum_{k=1}^{N} \Delta_k + W(e^{\theta}\vec{r}_1, \ldots, e^{\theta}\vec{r}_N)$ on H^2, is a type A-holomorphic operator family in H for $\theta \in \mathbb{C}_a$ (see e.g. Reed-Simon [17] XII.2, XIII.10). Now $\|A(t)\|$ is bounded, and since $\sum_{k=1}^{N} \vec{\nabla}_k$ is continuous from H^2 to H^1 for any $b > 0$ there is $a > 0$ independent of $(\theta,t) \in \overline{\mathbb{C}}_a \times \mathbb{T}_\omega$ such that $\|- i \sum_{k=1}^{N} \vec{\nabla}_k u\| \leq b\, \|T(\theta)u\| + a\, \|u\|$ for any $u \in H^2$. Hence by well known arguments $H(F,\theta,t)$ is for any fixed $t \in \mathbb{T}_\omega$ a self-adjoint holomorphic family of type A in $(F,\theta) \in \mathbb{C} \times \mathbb{C}_a$, with $H(F,\theta+\phi,t) = S(\phi)\, H(F,\theta,t)\, S(\phi)^{-1}$, $\phi \in \mathbb{R}$. Furthermore the above bound also implies (see Ref.(16) for more detail) that for some $M > 0$ $\pm i\, H(F,\theta,t) + M$ ($\pm$ according $\mathrm{Im}\theta > 0$, $\mathrm{Im}\theta < 0$) generates a C_0-semigroup $\exp(\mp i\sigma\, H(F,\theta,t))$, $\sigma \geq 0$, such that $\exp(\mp i\sigma\, H(F,\theta+\phi,t)) = S(\phi)\exp(\mp i\sigma\, H(F,\theta,t))\, S(\phi)^{-1}$, $\phi \in \mathbb{R}$, and that the function $(F,\phi,t) \rightarrow (\pm i\, H(F,\theta,t) - z)^{-1}$ is norm differentiable in $(F,\theta,t) \in \mathbb{C} \times \overline{\mathbb{C}}_a \times \mathbb{T}_\omega$. This is enough to verify the condition of a well known existence and uniqueness result for solutions of temporally-inhomogeneous Schrödinger equations (see e.g. Reed-Simon [17], Thm. X.70). Thus (1)-(5) above follow; in particular, (3) is due to the time-periodicity of $H(F,\theta,t)$. Finally (6) is a consequence of the fact that the above bound yields the strong continuity of $\exp(\mp i\sigma H(F,\theta,t))$ in $(\sigma,F,\theta,t) \in \mathbb{R}_+ \cup \{0\} \times \mathbb{R} \times \overline{\mathbb{C}}_a^{\pm} \times \mathbb{T}_\omega$ and its analyticity in $\theta \in \mathbb{C}_a^{\pm}$ at (F,t) fixed. For more detail see Refs.(7,16). ∎

The next result deals with the realization of the Floquet Hamiltonian $K(F,\theta)$ in $\mathcal{K} = \mathcal{H} \otimes L^2(\mathbb{T}_\omega)$ and with its relation to the propagator $U(t,s;F,\theta)$, i.e. to the solution of the Schrödinger equation. By Prop. 2.1, the one-parameter operator family in $\mathcal{K}$ $\{\mathcal{U}(\pm\sigma:F,\theta): \pm\sigma \geq 0\}$ defined for $\theta \in \overline{C}_a^{\pm}$, $F \in \mathbb{R}$ by

$$\mathcal{U}(\sigma;F,\theta)f(\cdot,t) = U(t,t-\sigma;F,\theta)\ f(\cdot,t-\sigma),\quad f = f(\cdot,t) \in \mathcal{K} \tag{2.8}$$

is a C_0-semigroup, extending to a unitary group $\{\mathcal{U}(\sigma;F,\theta): \sigma \in \mathbb{R}\}$ for $\theta \in \mathbb{R}$, and by (2.6):

$$i\frac{d}{d\sigma}\,\mathcal{U}(\sigma,F,\theta)f\Big|_{\sigma=0} = H(F,\theta,t)f - i\partial f/\partial t \tag{2.9}$$

if $f \in \mathcal{D} \equiv C^1(\mathbb{T}_\omega;\mathcal{H}) \cap C(T_\omega;H^2)$. Thus we prove, assuming from now on $\theta \in \overline{\mathbb{C}}_a^+$, because specular arguments hold for $\theta \in \overline{\mathbb{C}}_a^-$:

<u>2.2. Proposition.</u> Let $(F,\theta) \in \mathbb{C} \times \overline{\mathbb{C}}_a^+$, and $\dot{K}(F,\theta)$ be the operator family in $\mathcal{K}$ defined as the action of the l.h.s. of (2.9) on $\mathcal{D}$. Then $\dot{K}(F,\theta)$ is closable, and its closure $K(F,\theta)$ enjoys the following properties:

(1) There is $M > 0$ independent of $\theta \in \overline{\mathbb{C}}_a^+$ and F in the compacts of $\mathbb{R}$ such that $iK(F,\theta) + M$ is m-accretive and $iK(F,\theta)$ is the generator of the C_0-semigroup $\{\mathcal{U}(\sigma;F,\theta): \sigma \geq 0$ for $\theta \in \mathbb{C}_a^+$; $\sigma \in \mathbb{R}$ for $\theta \in \mathbb{R}$. In particular, $K(F,\theta)$ is self-adjoint for $\theta \in \mathbb{R}$.
(2) For $\theta \in \mathbb{C}_a^+$, $K(F,\theta)$ has domain $L^2(\mathbb{T}_\omega) \otimes H^2 \cap H^1(\mathbb{T}_\omega) \otimes \mathcal{H}$ and represents a type-A holomorphic family in $(F,\theta) \in \mathbb{C} \times \mathbb{C}_a^+$.
(3) For $\theta \in \overline{\mathbb{C}}_a^+$:

$$\mathcal{S}(\phi)\ K(F,\theta)\ \mathcal{S}(\phi)^{-1} = K(F,\theta+\phi), \qquad \phi \in \mathbb{R} \tag{2.10}$$

$\mathcal{S}(\phi) = I_{L^2(\mathbb{T}_\omega)} \otimes \tilde{S}(\phi)$. In particular for $\theta \in \mathbb{R}$ $K(F,\theta) = \mathcal{S}(\theta)K(F)\,\mathcal{S}(\theta)^{-1}$, $K(F) \equiv K(F,0)$.
(4) $K(F,\theta)$ is strongly continuous in the generalized sense as $\operatorname{Im}\theta \downarrow 0$, uniformly on compacts, in $(F,z) \in \mathbb{C} \times \{z: \operatorname{Im} z > M\}$.
(5) If $\lambda(F,\theta)$ is an isolated eigenvalue of $K(F,\theta)$, then λ is locally independent of $\theta \in \mathbb{C}_a^+$: $\lambda=\lambda(F)$. If $F \in \mathbb{R}$, $\operatorname{Im}\lambda(F) \leq 0$.
(6) Let $F \in \mathbb{R}$, $\theta \in \overline{\mathbb{C}}_a^+$, $\lambda = \lambda(F) \in \sigma_d(K(F,\theta))$: $K(F,\theta)f = \lambda(F)f$, $f \in D(K,\theta))$. Then $\psi(\cdot,t) = e^{-i\lambda(F)t}\ f(\cdot,t)$ solves the Schrödinger equation

$H(F,\theta,t)\psi = i\partial\psi/\partial t$: $f(\cdot,t) = e^{i\lambda(F)(t-s)}U(t,s;\ F,\theta)f(\cdot,s)$. In particular $e^{-(2\pi i/\omega)\lambda(F)}$ is eigenvalue of the Floquet operator $U(s+2\pi/\omega,s;\ F,\theta)$: $U(s+2\pi/\omega,s;\ F,\theta)\ f(\cdot,s) = e^{-(2\pi i/\omega)\lambda(F)}f(\cdot,s)$. Conversely if $U(s+2\pi/\omega,s;\ F,\theta)\phi(\cdot,s) = e^{-(2\pi i/\omega)\lambda(F)}\phi(\cdot,s)$, then $f(\cdot,t) \equiv$ $\equiv e^{i\lambda(F)(t-s)}U(t,s;\ F,\theta)\phi(\cdot,s)\in D(K(F,\theta))$ and $K(F,\theta)f = \lambda(F)f$.

Proof. We first prove (2). Remark that the operator family $K(0,\theta)\equiv K(\theta)$ defined as the action of $T(\theta) - i\partial/\partial t$ on $L^2(\mathbb{T}_\omega)\otimes H^2\cap h^1(\mathbb{T}_\omega)\otimes H$ is obviously type-A holomorphic in $\theta\in\mathbb{C}_a^+$. Since $\|A(t)\|^2$ is bounded, (2) holds if we can prove that for any $b>0$ there is $a>0$ independent of θ in the compacts of $\mathbb{C}_a^+$ such that

$$\|\sum_{k=1}^{N}<\vec{A}(t),-i\vec{\nabla}_k u>\| \le \|\sum_{k=1}^{N}-i\vec{\nabla}_k u\| \le b\,\|K(\theta)u\| + a\,\|u\| \tag{2.11}$$

for any $u\in\mathcal{D}$. Since $\mathrm{Im}\,\theta>0$, an elementary computation yields:

$$\begin{aligned}&\lim_{\mathrm{Im}z\to+\infty}\ \sup_{n\in\mathbb{Z}}\ \|\sum_{k=1}^{N}(-i\vec{\nabla}_k)(T_0(\theta)+n\omega-z)^{-1}\| = \\ &\lim_{\mathrm{Im}z\to+\infty}\ \sup_{n\in\mathbb{Z}}\ \|W(\theta)(T_0(\theta)+n\omega-z)^{-1}\| = 0\end{aligned} \tag{2.12}$$

uniformly on compacts in $\theta\in\mathbb{C}_a^+$, $T_0(\theta)$ being the action of $-\frac{1}{2}e^{-2\theta}\sum_{k=1}^{N}\Delta_k$ on H^2. Denote now by F_t the Fourier transform in $L^2(\mathbb{T}_\omega)$:

$$g(t) = \sum_{n=-\infty}^{+\infty}(F_t g)(n)e^{in\omega t},\ g\in L^2(\mathbb{T}_\omega) \tag{2.13}$$

Then $K(\theta)\cong\bigoplus_{n=-\infty}^{+\infty}(T(\theta)+n\omega)$ under the unitary equivalence $K\cong\ell^2(\mathbb{Z})\otimes H\cong\bigoplus_{n=-\infty}^{+\infty}H$, whence:

$$F_t\left(\sum_{n=1}^{N}-i\vec{\nabla}_k\right)(K(\theta)-z)^{-1}F_t^{-1} = \bigoplus_{n=-\infty}^{+\infty}\left(\sum_{k=1}^{N}-i\vec{\nabla}_k\right)(T(\theta)+n\omega-z)^{-1} \tag{2.14}$$

By (2.12) we see that given $\varepsilon>0$ there is $z(\varepsilon)\in\mathbb{C}^+$ independent of (n,θ), $n\in\mathbb{Z}$, θ in the compacts of $\mathbb{C}_a^+$, such that

$$\begin{aligned}&\|\left(\sum_{k=1}^{N}-i\vec{\nabla}_k\right)(T(\theta)+n\omega-z)^{-1}\|_H \le \\ &\le \|\left(\sum_{k=1}^{N}-i\vec{\nabla}_k\right)(T_0(\theta)+n\omega-z)^{-1}\{1+W(\theta)(T_0(\theta)+n\omega-z)^{-1}\}^{-1}\|_H < \varepsilon\end{aligned}$$

Hence $\|(\sum_{k=1}^{N} - i\vec{\nabla}_k)(K(\theta) - z)^{-1}\| < \varepsilon$ with the stated uniformities for some $z = z(\varepsilon) \in \mathbb{C}^+$. This proves (2.11) and hence (2). Next, remark that by the relative boundedness of Prop.2.1 the union $\overline{\Theta}$ over $|F| < +\infty, \theta \in \mathbb{C}_a^+$ of the numerical ranges $\Theta(F,\theta,t)$ of $H(F,\theta,t)$ is contained in the half-plane $\{z: \mathrm{Im} z < M\}$ for some $M > 0$. Since $\Theta(-i\partial/\partial t) = \mathbb{R}$, we see that the union over $|F| < +\infty$, $\theta \in \overline{\mathbb{C}}_a^+$ of $\Theta(iK(F,\theta)) \subset \{z: \mathrm{Re}\, z > -M\}$. Therefore $iH(F,\theta)$ is maximal accretive for $\theta \in \mathbb{C}_a^+$ because, by (2.11), $\rho(iK(F,\theta)) \subset \{z: \mathrm{Re}\, z > -M\}$. Now by Prop.2.1(5) $\mathcal{D}$ is invariant under the C_o-semigroup $\{\mathcal{U}(\sigma,F,\theta):\sigma \geq 0\}$, and thus is a core for its generator by a well known result (see e.g. Reed-Simon[17], Thm.X.49). On the other hand the restriction of the generator to $\mathcal{D}$ is $\dot{K}(F,\theta)$ by (2.9), and hence (1) is proved for $\theta \in \mathbb{C}_a^+$. For $\theta \in \mathbb{R}$, $\dot{K}(F,\theta)$ is obviously symmetric, and by (2.9) $L(F,\theta)|\mathcal{D} = \dot{K}(F,\theta)$ if $L(F,\theta)$ denotes the generator of $\{\mathcal{U}(\sigma;F,\theta):\sigma \in \mathbb{R}\}$ which leaves $\mathcal{D}$ invariant once more by (5) of Prop.2.1. Hence the assertion follows the corresponding result for one-parameter unitary groups (see e.g. Reed-Simon[17], Thm.VIII.10). Assertion (3) is obvious. To see (4), remark that $\|(K(F,\theta) - z)^{-1}\|$ is bounded uniformly with respect to $\theta \in \overline{\mathbb{C}}_a^+$ and z in $\{z: \mathrm{Im}\, z > M\}$ because $\|(K(F,\theta) - z)^{-1}\| \leq \mathrm{dist}(z,\overline{\Theta})$, $\overline{\Theta}$ as above. Since the K-valued function $\theta \longrightarrow K(F,\theta)u$ is continuous as $\mathrm{Im}\theta \downarrow 0$ for any $u \in \mathcal{D}$, (4) follows from a known result (see e.g. Kato[18], Thm.VIII.1.5). If $\lambda(F,\theta) \in \sigma_d(K(F,\theta))$, then $\lambda = \lambda(F)$ by (2.10) and standard dilation analyticity arguments; furthermore, if ϕ,ψ are $S(\theta)$-analytic vectors, by (2.10) and (4) we have, for $\mathrm{Im}\, z > M$:

$$\langle \phi, (K(F) - z)^{-1}\psi \rangle = \langle S(\overline{\theta})\phi, (K(F,\theta) - z)^{-1} S(\theta)\psi \rangle \qquad (2.15)$$

which extends to all $z \in \mathbb{C}^+$ by the self-adjointness of $K(F)$. Since the $S(\theta)$-analytic vectors are dense in K, (2.15) yields assertion (5). To see (6), we proceed as in Yajima[7]: if $K(F,\theta)f = \lambda(F)f$, $\theta \in \overline{C}_a^+$, $F \in \mathbb{R}$, then $\exp(-i\sigma K(F,\theta))f(\cdot,t) = e^{-i\lambda\sigma} f(\cdot,t) = U(t,t-\sigma;F,\theta)f(\cdot,t)$ by definition of $\mathcal{U}(\cdot)$. Hence $U(t+\sigma,t;F,\theta)f(\cdot) = e^{-i\lambda(F)\sigma} f(\cdot,t+\sigma)$ for all $\sigma > 0$ and almost any t. Hence the assertion is proved in one direction by the strong continuity of the propagator. Conversely, if $U(s+2\pi/\omega,s,F,\theta)\phi(\cdot,s) = e^{-(2\pi i/\omega)\lambda}\phi(\cdot,s)$, we have for $\sigma \geq 0$, $t \in \mathbb{T}_\omega$

$$e^{i\lambda(t-\sigma-s)}U(t,t-\sigma;F,\theta)U(t-\sigma,s;F,\theta)\phi = e^{-i\lambda\sigma}f(\cdot,t) = \\ = \exp(-i\sigma K(F,\theta))f(\cdot,t) \tag{2.16}$$ ■

The actual existence of resonances and the convergence of perturbation theory is now an immediate consequence of field strength analyticity taking place for $\theta \in \mathbb{C}_a^+$ fixed, i.e. of Prop.2.2(2): the isolated eigenvalues of $K(\theta)$ turn into resonances of $K(F,\theta)$ determined by regular perturbation theory. We first recall that if λ is an isolated eigenvalues of $T(\theta)$ (i.e. any non-threshold bound state or resonance of T), then λ is independent of θ , $\mathrm{Im}\,\lambda \leq 0$ and (for more detail see Ref.(16)) $\lambda + n\omega$ is an isolated eigenvalue of $K(\theta)$ for all $n \in \mathbb{Z}$. The perturbation expansion is generated by ordinary Rayleigh-Schrödinger perturbation theory in which the unperturbed operator is $K(\theta)$ and the perturbation is $F \sum_{k=1}^{N} \langle \vec{A}(t), -i\vec{\nabla}_k\rangle + A\,\frac{1}{2}\,NF^2\|A(t)\|^2$. More precisely:

<u>2.3. Proposition</u>. Let λ be an isolated eigenvalue of $T(\theta)$, $\theta \in \mathbb{C}_a^+$, of (algebraic) multiplicity $m_0(\lambda)$. Then there is $\bar{F}(\lambda) > 0$ such that for $F \in \mathbb{C}$, $|F| < \bar{F}(\lambda)$:

(1) Let $\lambda + n_j\omega$, $n_j \in \mathbb{Z}$, $j = 1,\ldots,\ell$, be the eigenvalues of $T(\theta)$ which differ from λ by integer multiples of ω, $m_j(\lambda)$ their (algebraic) multiplicities, $N(\lambda) = m_1(\lambda)+\ldots+m_\ell(\lambda)$. Then there exist exactly $N(\lambda)$ eigenvalues $\lambda_1(F),\ldots,\lambda_N(F)$ (counting multiplicity) of $K(F,\theta)$ such that $\lambda_i(F) \to \lambda$ as $|F| \to 0$, given by holomorphic functions of $F^{1/N}$ near $F = 0$. If in particular $N(\lambda) = 1$ (which occurs for almost every ω if $m_0(\lambda) = 1$) the unique eigenvalue $\lambda(F)$ near λ is holomorphic near $F = 0$ and its perturbation expansion has therefore a positive convergence radius.

(2) Each eigenvalue $\lambda(F)$ of $K(F,\theta)$ near $\lambda \in \sigma_d(K(\theta))$ is a second sheet pole of $(K(F) - z)^{-1}$, i.e. there exist a neighbourhood $\Omega(\lambda)$ of λ and $S(\theta)$-analytic vectors (ϕ,ψ) such that the function:

$$S_{\phi,\psi}(z) = \langle \phi, (K(F) - z)^{-1}\psi \rangle \tag{2.17}$$

a priori holomorphic for $z \in \mathbb{C}^+$ has a meromorphic continuation to $\bar{\mathbb{C}}^- \cap \Omega(\lambda)$ with poles exactly at $\lambda(F)$.

Proof. Statement (1) is a direct consequence of analytic perturbation theory (see e.g. Kato[18], VII.1.2). Assertion (2) follows from (2.10), Prop.2.2(5) and once more by regular perturbation theory which ensures the existence of a neighboourhood $\Omega(\lambda)$ in which the spectrum of $K(F,\theta)$ consists exactly in $N(\lambda)$ eigenvalues (counting multiplicity). ■

To obtain statements 2.(v), 2(vi) of Sect.1, we consider, as in Refs. (7) and (16), the standard particular case $\vec{A}(t) = (\omega^{-1}F \sin\omega t, 0, 0)$. We denote by $\Sigma = \inf \sigma_{ess}(T)$ the lowest threshold of T, and consider the following two cases in Prop.2.3.:

Case A. $\lambda < \Sigma$, $m_o(\lambda) = 1$, $\ell = 0$, $n = 0$

Case B. $\lambda < \Sigma$, $m_o(\lambda) = 1$, $\ell = 1$, $n_1 = \pm 1$, $m_1(\lambda) = 1$.

Then by the same argument of Yajima[7], Thm.3.5 (see also Ref.(16) for more details on the simplifications occurring in this case) we have:

2.4. Proposition. Let case A hold, and let $\sum_{i=0}^{\infty} C_i(\omega)F^i$, $C_o(\omega) = \lambda$, be the perturbation series of $\lambda(F)$. Then:

(1) $C_i(\omega)$ is θ-independent, and $C_{2i+1} = 0$, $i = 0,1,\ldots$

(2) Let $\lambda + n\omega < \Sigma$. Then $\operatorname{Im} C_{2i}(\omega) = 0$, $0 \leq i < n$.

(3) Let $\lambda + \omega > \Sigma$. Then:

$$\operatorname{Im} C_2(\omega) = - \left\langle \frac{dE(\mu+\omega)}{d\mu}\Big|_{\mu=\lambda} (\Sigma\vec{p}_k)_x\phi, (\Sigma\vec{p}_k)_x\phi \right\rangle \tag{2.18}$$

where ϕ is the eigenvector of λ, $T\phi = \lambda\phi$, and $\mu \to E(\mu)$ the spectral measure of T.

Formula (2.18) is of course the Fermi Golden Rule. Its version to the smallest order n in perturbation theory such that $\lambda + n\omega > \Sigma$ can be found in Ref.(16).

We conclude this exposition by stating, without proof, that the existence of forced oscillations, i.e. 2-(vi) of Sect.1, holds true without change, always in the two-body case $N = 1$, for the equation $(-i\vec{\nabla} - F\vec{A}(t))^2\psi + V\psi = i\partial\psi/\partial t$ under the following conditions on the dilation analytic potentials which are more general than those assumed by Yajima[7]:

(1) $(A(\theta), B(\theta)) \in L^p(\mathbb{R}^3)$, $\frac{3}{2} - \varepsilon \leq p \leq 6 + \varepsilon$

(2) $(\hat{A}(\theta), \hat{B}(\theta)) \in L^q(\mathbb{R}^3)$, $\frac{3}{2} - \varepsilon \leq q \leq \frac{3}{2} + \varepsilon$

Here $\theta \in \mathbb{C}_a^+$, $V(\theta) = V(e^{\theta}\vec{r})$, $A(\theta) = |V(\theta)|^{1/2}$, $B(\theta) = |V(\theta)|^{-1/2}$, $\hat{A}(\theta)$ and $\hat{B}(\theta)$ are the Fourier transforms of $A(\theta)$, $B(\theta)$, respectively. For the proof the reader is referred to Ref. (16). An example of potential fulfilling (1),(2) above is $V = e^{-\beta|\vec{r}|}|\vec{r}|^{-\alpha}$, $\beta>0$, $\alpha<1$.

REFERENCES

(1) W.P.Reinhardt: Complex coordinates in the theory of atomic and molecular structure and dynamics. Ann. Rev. Phys. Chem. 33, 223-255 (1982)

(2) S.Graffi, V.Grecchi: Resonances in Stark effect and perturbation theory. Commun. Math. Phys. 62, 83-96 (1978)

(3) I.Herbst: Dilation analyticity in constant electric field,I: The two-body problem. Commun. Math. Phys. 64, 279-298 (1979)

(4) S.Graffi, V.Grecchi: Resonances in the Stark effect of atomic systems and perturbation theory. Commun. Math. Phys. 79, 91-109 (1981)

(5) I.Herbst, B.Simon: Dilation analyticity in constant electric field, II: The N-body problem, Borel summability. Commun. Math. Phys. 80, 181-216 (1981)

(6) K.Yajima: Scattering theory for Schrödinger operators with potentials periodic in time. J. Math. Soc. Japan 29, 729-743 (1977)

(7) K.Yajima: Resonances for AC-Stark effect. Commun. Math. Phys. 87, 331-352 (1982)

(8) H.Kitada, K.Yajima: A scattering theory for time-dependent long-range potentials. Duke Math. J. 49, 341-376 (1981)

(9) E.Merzbacher, Quantum Mechanics, John Wiley, New York 1961

(10) S.Graffi, K.Yajima: Exterior complex scaling and the AC-Stark effect in a Coulomb field. Commun. Math. Phys. 89, 277-301 (1983)

(11) B.Simon: The definition of molecular resonance curves by the method of exterior complex scaling. Phys. Lett. 71A, 211-214 (1979)

(12) N.L.Manakov, A.G.Feinshtein: Quasi-stationary quasi-energy states and convergence of the perturbation series in a monochromatic field. Theor. Math. Phys. 48, 815-822 (1982)

(13) N.L.Manakov, M.A.Preobrazhznskii, L.P.Rapoport, A.G.Feinshtein: Higher order perturbation theory effects in the shift and width

of atomic levels in an optical field. Soviet Physics JETP 48, 626-635 (1978)

(14) N.L.Manakov, A.G.Feinshtein: Decay of a weakly bound state in a monochromatic field. Soviet Physics JETP 52, 382-388 (1980)

(15) J.S.Howland: Complex scaling of ac Stark Hamiltonians. J.Math. Phys. 24, 1240-1244 (1983)

(16) S.Graffi, V.Grecchi, H.J.Silverstone: Resonances and convergence of perturbation theory for N-body atomic systems in external AC-electric field. Bologna preprint, January 1984.

(17) M.Reed, B.Simon: Methods of Modern Mathematical Physics, Voll. I-IV, Academic Press, New York 1975-79

(18) T.Kato: Perturbation Theory for Linear Operators, Springer-Verlag Berlin-Heidelberg-New York 1966 (Second Edition 1976)

FERMI PSEUDOPOTENTIALS AND RESONANCES IN ARRAYS

Tai Tsun Wu*

Gordon McKay Laboratory, Harvard University
Cambridge, Massachusetts 02138, U.S.A.

Abstract

The method of Fermi pseudopotential for the Schrödinger equation is generalized to the case of the Maxwell's equations. Infinite, linear, uniform arrays of such pseudopotentials are analyzed, and infinitely narrow resonances are found. A list is given of related problems that are not yet solved but are guaranteed to be solvable.

1. Introduction

One of the most useful antennas is the Yagi-Uda array [1], invented over half a century ago. This antenna array, shown schematically in Fig. 1, can be seen on top of many houses for purposes of television or fm reception. The elements of the array, usually supported by a non-metallic structure not schown in Fig. 1, may consist of one or two reflectors, a folded dipole connected to the receiver, and a number of directors. Typically, the length of the folded dipole is about $\frac{1}{2}\lambda$, half the wavelength, the reflectors somewhat longer and the directors shorter.

It is interesting to inquire why the Yagi-Uda antenna array works so well. Here we shall study a much simplified version of this array. This analysis suggests many other arrays with exciting properties, but we shall not attempt to describe these novel arrays. The simplified version of the Yagi-Uda antenna array is obtained as follows:

(1) The use of the folded dipole is motivated by impedance matching to the transmission line connecting the antenna to the receiver. Roughly, the input impedance of a folded dipole is four times that of a dipole. Since impedance matching is not an important aspect of the antenna array, we replace the folded dipole by an ordinary dipole (which

*Work supported in part by the United States Department of Energy under Grant No. DE-FG02-84-ER40158 and in part by the United States Joint Services Electronics Program contract No. N00014-84-K-0465.

Fig. 1: A typical Yagi-Uda antenna for reception

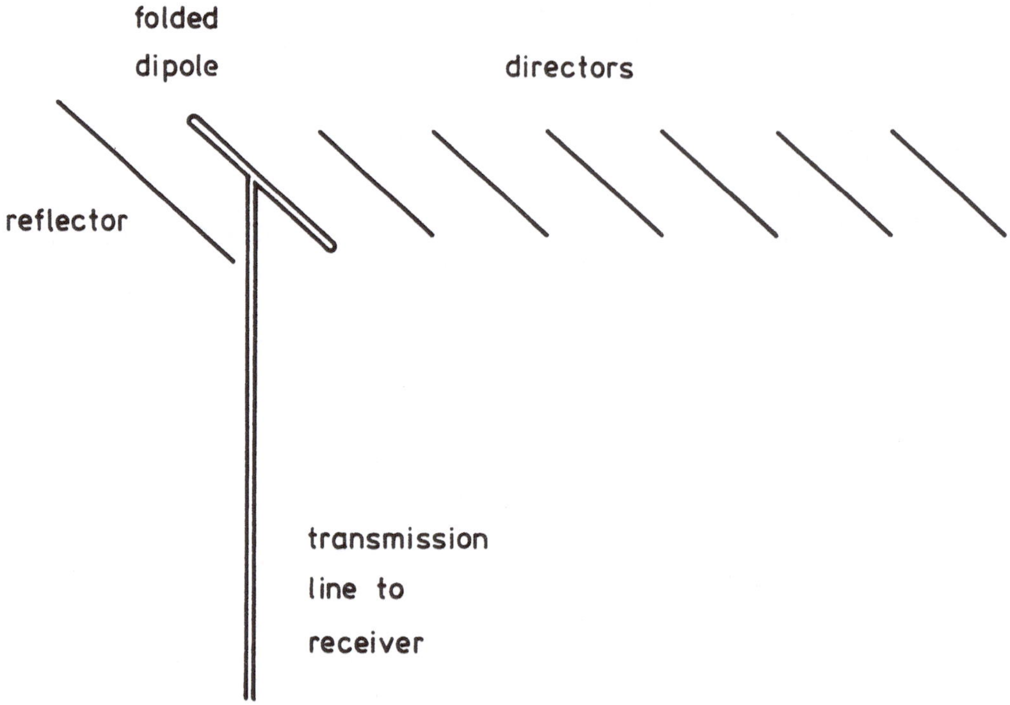

is in fact often the case for the Yagi-Uda array).

(2) We ignore the differences in lengths between the folded dipole, the reflectors, and the directors. Thus the antenna array is considered to be one with uniform spacing and lengths.

(3) The major step here is to replace each dipole antenna by a simpler object. This new object is to be studied in detail in Sec. 3.

2. Fermi Pseudopotential in Three Dimensions

The single dipole antenna of finite length is an extremely complicated object [2]. The basic method of analysis is the King-Middleton iterative solution of an integral equation of the first kind. Whether this iterative procedure can be made to converge is still an unsolved problem. Therefore, it is imperative to replace the dipole antennas by simpler objects.

We are thus led to pose the following question: What is the simplest non-trivial scatterer in the context of Maxwell's equations? For the non-relativistic Schrödinger equation, the answer is the Fermi pseudopotential [3], invented for nuclear physics about half a century ago. In this section, we study the basic properties of the Fermi pseudopotential. Let the Fermi pseudopotential be located at the origin, then the Hamiltonian is

$$H = -\nabla^2 + 4\pi a\ \delta^3(\vec{r})\ \frac{\partial}{\partial r}\, r \tag{2.1}$$

as given by Blatt and Weisskopf [4], where ∇^2 is the three-dimensional Laplacian, a is a real constant, and $r = |\vec{r}|$. Intuitively, the operator $\frac{\partial}{\partial r} r$ is introduced to remove the $\frac{1}{r}$ singularity of the free-space Green's function, i.e., the $\vec{r}$ representation of the resolvent $(-\nabla^2 - k^2)^{-1}$.

Superficially, the H of (2.1) does not seem to be self-adjoint. In the late fifties, the Fermi pseudopotential was used extensively to study the dilute hard-sphere gas [5-8]. It was always found that the troubles usually associated with non-self-adjoint operators miraculously did not appear. It was realized only a few years ago that this H of (2.1) is actually self-adjoint [9].

Consider the Green's function defined through the partial differential equation

$$[-\nabla^2 - k^2 + 4\pi a\ \delta^3(\vec{r})\ \frac{\partial}{\partial r} r\,]G(\vec{r},\vec{r}_o;k) = \delta^3(\vec{r}-\vec{r}_o). \tag{2.2}$$

This Green's function can be found explicitly. Let

$$A = 4\pi a\ \frac{\partial}{\partial r} r\ \ G(\vec{r},\vec{r}_o;k)\Big|_{r=o}\ , \tag{2.3}$$

then

$$(\nabla^2 + k^2)G(\vec{r},\vec{r}_o;k) = -\delta^3(\vec{r}-\vec{r}_o) + A\delta^3(\vec{r}). \tag{2.4}$$

Let

$$G_o(\vec{r},\vec{r}_o;k) = \frac{e^{ik|\vec{r}-\vec{r}_o|}}{4\pi|\vec{r}-\vec{r}_o|}\ , \tag{2.5}$$

which is the free-space Green's function, then the solution of (2.4) is

$$G(\vec{r},\vec{r}_o;k) = G_o(\vec{r},\vec{r}_o;k) - A\ G_o(\vec{r},0;k). \tag{2.6}$$

With A determined by (2.3) and (2.6), the required Green's function is explicitly

$$G(\vec{r},\vec{r}_o;k) = G_o(\vec{r},\vec{r}_o;k) - \frac{4\pi a}{1+ika} G_o(\vec{r},0;k)G_o(\vec{r}_o,0;k). \tag{2.7}$$

This Green's function has all the properties of that of a self-adjoint operator. Indeed, in the sixties and seventies, the Fermi pseudopotential was rediscovered and renamed point interaction [10-13]. Much of this work was based on the alternative description of the Fermi-pseudopotential as $\alpha\delta^3(\vec{r})$ with an infinitesimal α. The simplest way to show the equivalence of these two descriptions is to verify that the Green's functions are identical.

3. Pseudodipole

Guided by the Fermi pseudopotential, we proceed to find the simplest non-trivial electromagnetic scatterer, which will be called a pseudodipole. Maxwell's equations are, with suitable choice of units so that $\mu = \varepsilon = c = 1$,

$$\begin{cases} \nabla \times \vec{E} = i\,k\,\vec{B}, \\ \nabla \times \vec{B} = -\,i\,k\,\vec{E} + \vec{J}. \end{cases} \tag{3.1}$$

For a linear system, $\vec{J}$ is proportional to $\vec{E}$. Therefore, analogous to (2.1), we characterize the pseudodipole by

$$\vec{J} = \overleftrightarrow{V} \cdot \vec{E}\ , \tag{3.2}$$

with

$$\overleftrightarrow{V} = \delta^3(\vec{r})\ \overleftrightarrow{a}\ \mathcal{D}\ , \tag{3.3}$$

where $\overleftrightarrow{a}$ is a hermitian constant dyadic, and $\mathcal{D}$ is a generalization of $\frac{\partial}{\partial r}r$ for the Fermi pseudopotential.

In order to determine $\mathcal{D}$, we need the free-space electromagnetic Green's function $\overleftrightarrow{G}_o(\vec{r},\vec{r}_o;k)$, which may be found by

$$i\,k\,\overleftrightarrow{G}_o(\vec{r},0;k)\cdot\hat{z} = \vec{E}_o\ . \tag{3.4}$$

with

$$\begin{cases} \nabla \times \vec{E}_o = i\,k\,\vec{B}_o\ , \\ \nabla \times \vec{B}_o = -i\,k\,\vec{E}_o + \hat{z}\,\delta^3(\vec{r}). \end{cases} \tag{3.5}$$

Let the vector and scalar potentials $\vec{A}$ and ϕ be introduced in the usual way: $\vec{B} = \nabla\times\vec{A}$, $\vec{E} = -\nabla\phi + i\,k\,\vec{A}$ with the Lorentz condition $\nabla\cdot\vec{A} - ik\phi = 0$. Then it follows from (3.5) that

$$\vec{A}_o = \hat{z}\,\frac{e^{ikr}}{4\pi r}\,, \tag{3.6}$$

and accordingly

$$\begin{aligned}
E_{ox} &= \frac{i\,x\,z}{4\pi k r^5}\; e^{ikr}\,[3(1-ikr) - k^2r^2],\\
E_{oy} &= \frac{i\,y\,z}{4\pi k r^5}\; e^{ikr}\,[3(1-ikr) - k^2r^2],\\
E_{oz} &= \frac{i}{4\pi k r^3}\; e^{ikr}\,\{(-1+ikr+k^2r^2) + \frac{z^2}{r^2}[3(1-ikr)-k^2r^2]\}.
\end{aligned} \tag{3.7}$$

Eq. (3.4) together with rotational and translational invariances then give

$$\overleftrightarrow{G}_o(\vec{r},\vec{r}_o;k) = \overleftrightarrow{G}_o(\vec{r}-\vec{r}_o,0;k) \tag{3.8}$$

with

$$\overleftrightarrow{G}_o(\vec{r},0;k) = \frac{e^{ikr}}{4\pi k^2 r^3}\{\overleftrightarrow{1}\,(-1+ikr+k^2r^2) + \hat{r}\,\hat{r}\,[3(1-ikr) - k^2r^2\,\}], \tag{3.9}$$

where $\overleftrightarrow{1}$ is the unit dyadic.

It is seen from (3.9) that the singularity of $\overleftrightarrow{G}_o(\vec{r},\vec{r}_o;k)$ is a triple pole $|\vec{r}-\vec{r}_o|^{-3}$. Furthermore, the $|\vec{r}-\vec{r}_o|^{-2}$ term is absent. A natural choice for $\mathcal{D}$ is therefore

$$\mathcal{D} = \frac{\partial^3}{\partial r^3}\,r^3\,, \tag{3.10}$$

and hence

$$\overleftrightarrow{V} = \frac{\pi}{ik}\;\overleftrightarrow{a}\;\delta^3(\vec{r})\,\frac{\partial^3}{\partial r^3}\,r^3\,, \tag{3.11}$$

where the choice of the factor π is for later convenience. The constant dyadic $\overleftrightarrow{a}$ depends on the orientation of the dipole. For example, $\overleftrightarrow{a} = a\,\hat{z}\,\hat{z}$ for a pseudodipole in the z direction. It is $\overleftrightarrow{a} = a_x\hat{x}\,\hat{x} + a_y\hat{y}\,\hat{y}$ for a crossed pseudodipole that consists of two dipoles in the x and y directions.

The major difference between the present case and that of last section is the appearance of a higher derivative. That (3.10) contains a third derivative is less important. An alternative choice of $\mathcal{D}$ is $2\,\frac{\partial}{\partial r}\,\frac{1}{r}\,\frac{\partial}{\partial r}\,r^3$, which contains only a second derivative. This necessity of a higher derivative may be attributed to the Maxwell's equations being of first order.

4. Fermi Pseudopotential in Five Dimensions

In order to have a better understanding of this $\frac{\partial^3}{\partial r^3} r^3$ vs. $\frac{\partial}{\partial r} r$, we raise the question: Is there a situation in connection with the non-relativistic Schrödinger equation where the operator $\frac{\partial^3}{\partial r^3} r^3$ is needed for the Fermi pseudopotential? The answer is yes, for quantum mechanics in five dimensions [14].

In five dimensions, the free-space Green's function for the Schrödinger equation is

$$G_o(\vec{r},\vec{r}_o;k) = \frac{e^{ik|\vec{r}-\vec{r}_o|}(1-ik|\vec{r}-\vec{r}_o|)}{8\pi^2|\vec{r}-\vec{r}_o|^3} . \qquad (4.1)$$

Thus the singularity is again a triple pole $|\vec{r}-\vec{r}_o|^{-3}$, with the $|\vec{r}-\vec{r}_o|^{-2}$ term missing. The five-dimensional analog of (2.1) is therefore [14]

$$H = -\nabla^2 + 4\pi^2 a^3 \delta^5(\vec{r}) \frac{\partial^3}{\partial r^3} r^3 . \qquad (4.2)$$

Here ∇^2 is the five-dimensional Laplacian.

The procedure of Sec. 2 can be followed step by step. The partial differential equation for the Green's function corresponding to (4.2) is

$$[-\nabla^2 - k^2 + 4\pi^2 a^3 \delta^5(\vec{r}) \frac{\partial^3}{\partial r^3} r^3] G(\vec{r},\vec{r}_o;k) = \delta^5(\vec{r}-\vec{r}_o). \qquad (4.3)$$

Once again this Green's function can be found explicitly. Analogous to (2.3), let

$$A = 4\pi^2 a^3 \frac{\partial^3}{\partial r^3} r^3 G(\vec{r},\vec{r}_o;k)\Big|_{r=o} , \qquad (4,4)$$

then

$$(\nabla^2 + k^2) G(\vec{r},\vec{r}_o;k) = -\delta^5(\vec{r}-\vec{r}_o) + A\delta^5(\vec{r}). \qquad (4.5)$$

In terms of the free-space Green's function (4.1), the solution of (4.5) is

$$G(\vec{r},\vec{r}_o;k) = G_o(\vec{r},\vec{r}_o;k) - A\, G_o(\vec{r},0;k). \qquad (4.6)$$

Finally, with A determined by (4.4), G is explicitly

$$G(\vec{r},\vec{r}_o;k) = G_o(\vec{r},\vec{r}_o;k) - \frac{24\pi^2 a^3}{1+ik^3a^3} G_o(\vec{r},0;k) G_o(\vec{r}_o,0;k). \qquad (4.7)$$

In spite of the formal similarity, there are qualitative differences between (2.7) and (4.7), as discussed in detail in ref. [14]. The salient points are the following.

(1) If we limit ourselves to L_2 spaces, then Fermi pseudopotential cannot exist in five dimensions. Consider all self-adjoint extensions of the symmetric operator obtained by restricting $-\nabla^2$ to the space of smooth L_2 functions that vanish at the origin [15]. In five dimensions, the only self-adjoint extension is $-\nabla^2$ itself. Note that $G(\vec{r},\vec{r}_o;k)$ of (2.7) is square integrable, but that of (4.7) is not.

(2) The underlying space is a Pontrjagin space [16] of rank 1 instead of a Hilbert space. The indefinite inner product is

$$(f_1, f_2) = \int d^5\vec{r}\,[f_1^*(\vec{r})\, f_2(\vec{r}) - \frac{c_1^*}{r^3}\frac{c_2}{r^3}], \qquad (4.8)$$

where

$$c_j = \lim_{r\to 0} r^3 f_j(\vec{r}). \qquad (4.9)$$

With (4.8), the resolvent equation is satisfied.

5. Green's Function for Pseudodipole

Since the operator $\mathcal{D}$ of (3.10) appears in both (3.3) and (4.2), we follow the procedure of Sec. 4 to determine the Green's function for the pseudodipole. Since the Maxwell's equations (3.1) imply

$$\nabla \times \nabla \times \vec{E} - k^2 \vec{E} = i k \vec{J}\,, \qquad (5.1)$$

the partial differential equation for the Green's function is, in this case,

$$[\nabla\times\nabla\times - k^2 + \pi\, \overleftrightarrow{a}\cdot \delta^3(\vec{r})\, \frac{\partial^3}{\partial r^3} r^3]\, \overleftrightarrow{G}(\vec{r},\vec{r}_o;k) = \overleftrightarrow{1}\, \delta^3(\vec{r}-\vec{r}_o). \qquad (5.2)$$

By the way, the appearance of the factor ik in (3.4) is due to the ik on the right-hand side of (5.1). Analogous to (2.3) and (4.4), let

$$\overleftrightarrow{A} = \pi\, \overleftrightarrow{a} \cdot \frac{\partial^3}{\partial r^3} r^3\, \overleftrightarrow{G}(\vec{r},\vec{r}_o;k)\Big|_{r=o}\,, \qquad (5.3)$$

then

$$(-\nabla\times\nabla\times + k^2)\, \overleftrightarrow{G}(\vec{r}, \vec{r}_o;k) = -\overleftrightarrow{1}\, \delta^3(\vec{r}-\vec{r}_o) + \overleftrightarrow{A}\, \delta^3(\vec{r}), \qquad (5.4)$$

and hence

$$\overleftrightarrow{G}(\vec{r},\vec{r}_o;k) = \overleftrightarrow{G}_o(\vec{r},\vec{r}_o;k) - \overleftrightarrow{G}_o(\vec{r},0;k)\cdot \overleftrightarrow{A}\,. \qquad (5.5)$$

Since it follows from the explicit form (3.9) that

$$\frac{\partial^3}{\partial r^3} r^3 \overleftrightarrow{G}_o(\vec{r},0;k)\Big|_{r=0} = \frac{ik}{\pi} \overleftrightarrow{1} \tag{5.6}$$

with the $\hat{r}\hat{r}$ term not contributing, $\overleftrightarrow{A}$ of (5.3) satisfies

$$\overleftrightarrow{A} = 6\pi \overleftrightarrow{a} \cdot \overleftrightarrow{G}_o(\vec{r}_o,0;k) - ik \overleftrightarrow{a} \cdot \overleftrightarrow{A}. \tag{5.7}$$

Therefore

$$\overleftrightarrow{A} = \frac{6\pi \overleftrightarrow{a}}{1+ik\overleftrightarrow{a}} \cdot \overleftrightarrow{G}_o(\vec{r}_o,0;k), \tag{5.7}$$

and hence

$$\overleftrightarrow{G}(\vec{r},\vec{r}_o;k) = \overleftrightarrow{G}_o(\vec{r},\vec{r}_o;k) - \overleftrightarrow{G}_o(\vec{r},0;k) \cdot \overleftrightarrow{B}(k) \cdot \overleftrightarrow{G}_o(\vec{r}_o,0;k), \tag{5.9}$$

with

$$\overleftrightarrow{B}(k) = \frac{6\pi \overleftrightarrow{a}}{1+ik\overleftrightarrow{a}} . \tag{5.10}$$

This is the desired Green' function.

This Green's function shows features of the Fermi pseudopotential in both three and five dimensions, see (2.7) and (4.7).

First, (5.10) is very close to the corresponding factor $\frac{4\pi a}{1+ika}$ for the three-dimensional Fermi pseudopotential. In particular, in both cases the spectrum is always real. This is to be contrasted with the five-dimensional pseudopotential, where, when $a > 0$, the point spectrum consists of a complex conjugate pair $a^{-2}e^{\pm i\pi/3}$. Such a complex conjugate pair is allowed for a Pontrjagin space of rank 1 [16].

Secondly, the free-space Green's functions (3.9) and (4.1) share the property of not being square integrable. In this respect, the pseudo-dipole is much closer to the Fermi pseudopotential in five dimensions. It is this property that has presumably retarded the discovery of the Fermi pseudopotential in five dimensions and the pseudodipole. In both cases, as already discussed, a Pontrjagin space must be used instead of a Hilbert space.

6. Array of Fermi Pseudopotentials in Three Dimensions

Attention is now turned to the second topic, i.e., resonances in arrays. We shall consider here only one-dimensional uniform infinite arrays. Let b be the spacing between consecutive elements of the array, which are located at $\vec{r}_j$. Let $R_j = |\vec{r} - \vec{r}_j|$.

In three dimensions, the Schrödinger equation for such an array of identical Fermi pseudopotentials is

$$[-\nabla^2 - k^2 + 4\pi a \sum_j \delta^3(\vec{r}-\vec{r}_j)\frac{\partial}{\partial R_j}R_j]\psi = 0. \qquad (6.1)$$

The method of solution is virtually identical to that used in Sec. 2 to determine the Green's function. Let

$$A_j = 4\pi a \frac{\partial}{\partial R_j} R_j \psi \Big|_{R_j=0}, \qquad (6.2)$$

then, in the absence of an incident field, ψ is given by

$$\psi = -\sum_j A_j \frac{e^{ikR_j}}{4\pi R_j}. \qquad (6.3)$$

Substitution into (6.2) then yield the equations for A_j

$$(1+ika)A_j + a \sum_{n\neq j} r_{jn}^{-1} e^{ikr_{jn}} A_n = 0, \qquad (6.4)$$

where $r_{jn} = |\vec{r}_j - \vec{r}_n|$.

For the uniform infinite array under consideration, (6.4) simplifies greatly. Translational invariance gives

$$A_j = e^{ij\beta} A_o \qquad (6.5)$$

where β specifies the reduced Hamiltonian. Therefore resonances are determined by the transcendental equation

$$1 + ika + \frac{a}{b} \sum_{j\neq 0} |j|^{-1} e^{ik|j|b} e^{ij\beta} = 0. \qquad (6.6)$$

This is a complex equation. The important point is that it can be transformed into a real equation, because, for $kb < \pi$,

$$\begin{aligned} &\sum_{j\neq 0} |j|^{-1} e^{ik|j|b} e^{ij\beta} \\ &= -\ln\{[1-e^{i(kb+\beta)}][1-e^{i(kb-\beta)}]\} \\ &= -\ln[2(\cos kb - \cos\beta)] - ikb. \end{aligned} \qquad (6.7)$$

Thus (6.6) reduces simply to

$$1 - \frac{a}{b}\ln[2(\cos kb - \cos\beta)] = 0, \qquad (6.8)$$

which is a real equation provided that the argument of the logarithm is positive.

Since (6.8) is a real equation, resonances can occur for real k and real β. In other words, there are in this case infinitely narrow resonances [13]. Such infinitely narrow resonances are allowed because of the infinite size of the array. If the array is bent into a circle so that the size is no longer infinite, then these resonances acquire widths which are exponentially small in the radius of the circle [9].

7. Array of Fermi Pseudopotentials in Five Dimensions

Just as Sec. 4 is parallel to Sec. 2, the development of the preceding section will now be generalized step by step to the case of five dimensions [14].

The five-dimensional Schrödinger equation for an infinite linear array of uniformly spaced identical Fermi pseudopotentials is

$$[-\nabla^2 - k^2 + 4\pi^2 a^3 \sum_j \delta^5(\vec{r}-\vec{r}_j) \frac{\partial^3}{\partial R_j^3} R_j^3]\psi = 0. \qquad (7.1)$$

Let

$$A_j = 4\pi^2 a^3 \frac{\partial^3}{\partial R_j^3} R_j^3 \psi \Big|_{R_j=0} , \qquad (7.2)$$

then, in the absence of an incident field, ψ is given by

$$\psi = -\sum_j A_j \frac{e^{ikR_j}(1-ikR_j)}{8\pi^3 R_j^3}. \qquad (7.3)$$

The equations for A_j are then

$$(1+ik^3a^3)A_j + 3a^3 \sum_{n\neq j} r_{jn}^{-3} e^{ikr_{jn}} (1-ikr_{jn})A_n = 0 , \qquad (7.4)$$

analogous to (6.4). The use of (6.5) reduces (7.4) to one transcendental equation:

$$1 + ik^3a^3 + 3(\tfrac{a}{b})^3 \sum_{j\neq 0} |j|^{-3} e^{ik|j|b} (1-ik|j|b)e^{ij\beta} = 0. \qquad (7.5)$$

The task is therefore to rewrite (7.5) in a real form. This involves carrying out the sum

$$\sum_{n=1}^{\infty} \frac{z^n}{n^3} = \int_0^z \frac{dz'}{z'} \ln(1-z') \ln\frac{z'}{z} . \qquad (7.6)$$

After a suitable deformation of the contour, (7.5) is found to be equivalent to [14]

$$1+3\left(\frac{a}{b}\right)^3\left\{2\zeta(3)-\left[\int_0^{\beta+kb}+\int_0^{\beta-kb}\right]d\theta\,(\theta-\beta)\ln\left(2\sin\frac{\theta}{2}\right)\right\}=0, \quad (7.7)$$

the imaginary term ik^3a^3 having been cancelled.

Although (7.7) is more complicated, the similarity to (6.8) is striking. In (6.8), for kb and β less than π, β must be larger than kb in order for the logarithm to be real. Similarly in (7.7), $\beta \geq kb$ is needed to get a real equation.

The discussion in the last paragraph of the preceding section applies equally well here.

8. Array of Pseudodipoles

The development of the last two sections can also be applied to the corresponding array of pseudodipoles. However, this electromagnetic case is richer in structure.

Maxwell's equations for an infinite linear array of uniformly spaced identical pseudodipoles can be written in the form [see eq. (5.2)]:

$$[\nabla\times\nabla\times - k^2 + \pi\,\overleftrightarrow{a}\cdot\sum_j \delta^3(\vec{r}-\vec{r}_j)\,\frac{\partial^3}{\partial R_j^3}\,R_j^3]\vec{E} = 0. \quad (8.1)$$

Let

$$\vec{V}_j = \pi\,\overleftrightarrow{a}\cdot\frac{\partial^3}{\partial R_j^3}\,R_j^3\,\vec{E}\Big|_{R_j=0}, \quad (8.2)$$

then, in the absence of an incident electromagnetic field, the electric field $\vec{E}$ is given by

$$\vec{E}(\vec{r}) = -\sum_j \overleftrightarrow{G}_o(\vec{r},\vec{r}_j;k)\cdot\vec{A}_j, \quad (8.3)$$

where $\overleftrightarrow{G}_o$ is given by (3.9). The equations for $\vec{A}_j$ are then

$$(\overleftrightarrow{1}+ika\,\overleftrightarrow{V}_o)\cdot\vec{A}_j + 6\pi\,\overleftrightarrow{a}\cdot\sum_{n\neq j}\overleftrightarrow{G}_o(\vec{r}_j,\vec{r}_n;k)\cdot\vec{A}_n = 0. \quad (8.4)$$

This is analogous to (6.4) and (7.4) of the Schrödinger cases. The vector form of (6.5)

$$\vec{A}_j = e^{ij\beta}\,\vec{A}_o \quad (8.5)$$

can be used to reduce (8.4) to

$$(\overleftrightarrow{1} + ik\overleftrightarrow{a}) \cdot \overleftrightarrow{A}_o$$

$$+ \frac{3\overleftrightarrow{a}}{2k^2b^3} \cdot \sum_{j\neq 0} |j|^{-3} e^{ik|j|b} e^{ij\beta} \{\overleftrightarrow{1}[-1 + ik|j|b + (kjb)^2]$$

$$+ \hat{x}\,\hat{x}\,[3(1 - ik|j|b) - (kjb)^2]\} \cdot \vec{A}_o = 0\ , \qquad (8.6)$$

where the direction of the pseudodipole array has been taken as the x-axis.

The similarity of (8.6) to (6.6) and (7.5) is striking. Indeed the left-hand side of (8.6) is essentially a linear combination of these of (6.6) and (7.5):

$$\text{LHS of } (8.6) = \vec{A}_o + [\frac{3}{2a}\overleftrightarrow{a} \cdot (1 - \hat{x}\hat{x}) \cdot \vec{A}_o][\text{LHS of } (6.6) - 1] \qquad (8.7)$$

$$- [\frac{1}{2k^2a^3}\overleftrightarrow{a} \cdot (1 - 3\hat{x}\hat{x}) \cdot \vec{A}_o][\text{LHS of } (7.5) - 1].$$

It therefore follows from (6.8) and (7.7) that (8.6) can be put in the real form

$$\vec{A}_o - \frac{3}{2b}[\overleftrightarrow{a} \cdot (1 - \hat{x}\hat{x}) \cdot \vec{A}_o] \ln[2(\cos kb - \cos\beta)]$$

$$- \frac{3}{2k^2b^3}[\overleftrightarrow{a} \cdot (1 - 3\hat{x}\hat{x}) \cdot \vec{A}_o]$$

$$\{2\zeta(3) - [\int_0^{\beta+kb} + \int_0^{\beta-kb}] d\theta(\theta-\beta)\ln(2\sin\frac{\theta}{2})\} = 0. \qquad (8.8)$$

Resonances are determined by the existence of a non-trivial $\vec{A}_o$ that satisfies (8.8). The existence of infinitely narrow resonance for this infinite array is the underlying reason for the excellent properties of the finite array, including the Yagi-Uda array.

9. List of Problems

We conclude by presenting a partial list of related problems that are not yet solved but are guaranteed to be solvable without excessive work. Most of these problems have at least three versions, one for the non-relativistic Schrödinger equation in three dimensions, one for that in five dimensions, and one for Maxwell's equations. The corresponding problems for the Schrödinger equation in two, four, and perhaps six dimensions are somewhat more complicated, involving the appropriate

Bessel functions.

(A) The simplest generalization is to replace each element of the infinite, linear, uniform array by more than one pseudopotential or pseudodipole. An example involving two elements is shown in Fig. 2.

Fig. 2: An example of the infinite, linear, uniform array where each element consist of two Fermi pseudopotentials or two pseudodipoles.

(B) Another simple problem is that of a uniform, circular array, already mentioned in Sec. 6. The Schrödinger case in three dimensions has been treated [9], but not the other cases. Generalizations of type (A) can also be applied here.

(C) The Fermi pseudopotentials or pseudodipoles may be located uniformly on a helix rather than a straight line. The important point is that the array must be invariant under an appropriate element of the group of translation and rotation. For the case of the helix, this element is a translation together with a rotation.

For the electromagnetic problem, there is an especially simple case, where the pseudodipoles are located uniformly on a straight line as before, but their directions are rotated.

(D) We now go on to give a few more complicated cases. The infinite array studied in Secs. 6-8 may be replaced by a semi-infinite array. In principle, this semi-infinite problem, at least for the Schrödinger cases and suitable choices of $\overleftrightarrow{a}$ in the electromagnetic case, can be solved by the Wiener-Hopf technique. However, the kernel is sufficiently complicated so that the Wiener-Hopf factorization cannot be carried out explicitly.

(E) All the above cases can be solved exactly, even though the exact solution may be too complicated to be informative. In cases where the semi-infinite problem can be solved by the Wiener-Hopf technique, the corresponding problem of the finite, long array can be studied asymptotically. The technique used may be adopted from the one previously developed for the long dipole antenna [17].

Some of these problems can clearly be combined, an example being (AB) as already mentioned. Another example is (AC). However, since in general coupled Wiener-Hopf sum equations cannot be solved, the combinations (AD) and (AE) can be treated only in certain especially symmetrical cases.

Acknowledgments

I am grateful to Professor Sergio Albeverio and Professor Ludwig Streit for inviting me to this interdisciplinary conference on resonances. Much of the work on the Schrödinger equation was carried out in collaboration with Professor Alexander Grossmann, and I thank him and Professor S. Graffi for helpful discussions.

References

[1] R.W.P. King, R.B. Mack, and S.S. Sandler, Arrays of Cylindrical Dipoles, Cambridge University Press (1968). See especially Chapter 6.

[2] R.W.P. King, Theory of Linear Antennas, Harvard University Press (1956).

[3] E. Fermi, Ricerca Sci. 7, 13 (1936).

[4] J.M. Blatt and V.F. Weisskopf, Theoretical Nuclear Physics, John Wiley and Sons, Inc. (1952), pp. 74-75.

[5] K. Huang and C.N. Yang, Phys. Rev. 105, 767 (1957),

[6] K. Huang, C.N. Yang, and J.M. Luttinger, Phys. Rev. 105, 776 (1957).

[7] T.D. Lee, K. Huang, and C.N. Yang, Phys. REv. 106, 1134 (1957).

[8] T.T. Wu, Phys. Rev. 115, 1390 (1959).

[9] A. Grossmann and T.T. Wu, Preprint (1981).

[10] G. Flamand, Cargèse Lectures in Theoretical Physics, Gordon and Breach (1969), pp. 247-287.

[11] L.E. Thomas, J. Math. Phys. 20, 1848 (1979).

[12] S. Albeverio, J.E. Fenstad, and R. Høegh-Krohn, Trans. Amer. Math. Soc. 252, 275 (1979).

[13] A. Grossmann, R. Høegh-Krohn, and M. Mebkhout, J. Math. Phys. 21, 2376 (1980); Comm. Math. Phys. 77, 87 (1980).

[14] A. Grossmann and T.T. Wu, J. Math. Phys. 25, 1742 (1984).

[15] N. Dunford and J.T. Schwartz, Linear Operators, Part II, Interscience Publishers (1963).

[16] J. Bognár, Indefinite Inner Product Spaces (Springer-Verlag, 1970) Chapter IX.

[17] T.T. Wu, J. Math. Phys. 2, 550 (1961).

Time Evolution of Chemical Systems Far from Equilibrium

Erkki J. Brändas
Quantum Chemistry Group for Research
in Atomic, Molecular, and Solid State Physics
Uppsala University, S-75120 Uppsala, Sweden

I. Introduction

This meeting embodies a large spectrum of interdisciplinary interests. The focus is on the resonance picture of unstable states.

Eversince the pioneering work of Balslev and Combes [1] proving the non-existence of singularly continuous spectra in manybody Hamiltonians with dilation analytic interactions, work on the complex scaling method have revitalized treatments of problems involving decay phenomena; see refs. [2-4] for reviews.

For the benefit of the mathematicians I will start this lecture by pointing out some general problems facing present day statistical mechanics. We will consider the problem of imbedding the dynamics of a chemical system in interaction with an environment (Fig. 1) within a vaster formalism, of deriving generalized master equations and of viewing the irreversible process in terms of the resonance picture of unstable states mentioned above.

In section III resonance states are studied in the Hamiltonian formulation with the aim of obtaining rigorous results as existence and bounding theorems as well as expansion formulas.

Applications in scattering theory with particular reference to time resolved spectroscopy is the theme in section IV. Here the interest is focused on the spectral density concept, analytic continuation of the S-matrix and evaluation of associated Gamov waves and cross sections.

In the final section, we indicate some anomalous behaviour in the second order reduced density matrix associated with fundamental coherence phenomena in macromolecular systems. By interpreting this process in terms of the Bohm-Aharonov effect [5] complex scaling in an appropriate gauge is possible, see e.g. Howland [6].

Under each section we list the appropriate co-workers presently engaged. Although the research reviewed under each subsection represents separate research efforts as well as different supporting agencies, the common denominator, however, is the focus on resonances of unstable quantum states.

A more detailed account of the present development has been given during a lecture series on time evolution of chemical systems far from equilibrium at the Institute for Physical Chemistry, Technische Hochschule Darmstadt. I thank its chairman, Professor J. Brickmann, for generous hospitality and interest. This work was partly supported by the Deutsche Forschungsgemeinschaft, Bonn, under a contract with the Technische Hochschule Darmstadt.

II. Theory of Subdynamics

Together with C.H. Obcemea [7] . Supported in part by the National Foundation of Cancer Research under a contract with the University of Florida.

The present day problem of statistical mechanics is in fact more than a century old. It derives from the source problem: How to derive an irreversible dissipative macrodynamics of few degrees of freedom from a reversible microdynamics with many degrees of freedom. Many usually means $\sim 10^{23}$ and few less than 10^3. Consider a chemical system interacting with some environment depicted as reservoir in Fig. 1.

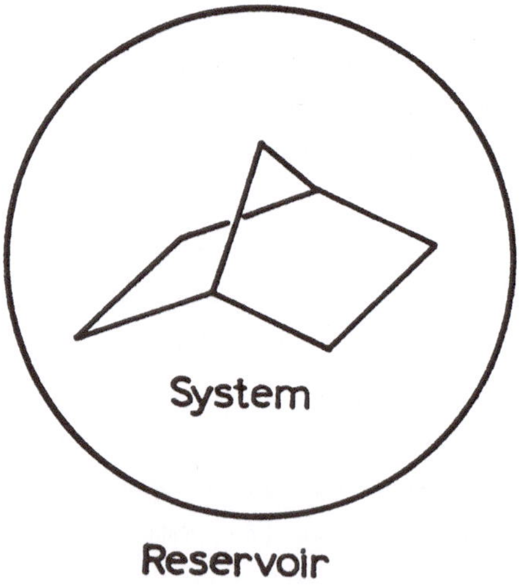

Fig. 1: System and reservoir

Table of some examples:

system		reservoir
atoms in a gas	colliding with	other atoms
light in a cavity	interacting with	walls
spin variables in NMR experiment	interacting with	other degrees of freedom
atoms or molecules	"sitting on a"	surface
chromophore	"inside a"	protein "overcoat"

The source problem can be recast into the following mathematical formulation, see ref. [8] and references therein for more details. For the system plus reservoir the Liouville equation describes the (reversible) time evolution of the density matrix $\rho = \rho_S \otimes \rho_R$, where the tensor product of the system and reservoir density operators is taken. However, in the thermodynamic formulation of Boltzmann both reversible and irrevesible parts occur in the equations of evolution. Hence the problem becomes that of going from left to right below.

$$i \frac{\partial \rho}{\partial t} = L\rho \qquad \overset{?}{} \qquad i \frac{\partial \tilde{\rho}}{\partial t} = (\Phi_r + \Phi_{ir})\tilde{\rho} \quad . \tag{1}$$

$$L^+ = L$$

In eq. (1) the left hand side is invariant under $t \to -t$ and $L \to -L$, while the right hand side contains a reversible part Φ_r and an irreversible one Φ_{ir}. $\tilde{\rho}$ is a transformed density operator that, in the terminology of the Brussels school [8], provides the equivalence map between the original density distribution and the manifold of decaying densities.

As is well known, the work of Boltzmann (to derive a kinetic theory) was heavily critizised by leading physicists and chemists as the basic laws of physics are reversible and not irreversible. Although a quantum version of the Boltzmann analysis was attempted by Pauli [9] leading to a Markovian master equation for the probability density distribution

$$\dot{\rho}^S_{mm}(t) = \sum_{n \neq m} (\rho^S_{nn}(t) W_{mn}) - \rho^S_{mm}(t) \sum_{n \neq m} W_{nm} \tag{2}$$

the source problem still existed. In eq. (2) $\rho_{mm}(t)$ gives the probability of finding the state m of the system S at time t and W_{mn} is the probability per unit time that a transition between levels n and m is induced by interaction with the environment R .

A rigorous derivation of eq. (2) from the left hand side of eq. (1) entails four basic approximations,

1. (Fano [10]) The density matrix corresponding to R is constant in time, i.e. $\rho(t) \simeq \rho_S(t) \otimes \rho_R(0)$.
2. (Markoff [11]) The time rate of change $\dot{\rho}_S(t)$ depends on $\rho_S(t)$ only.
3. (Redfield [12]) The correlation time τ of R is much smaller than the macroscopic time γ^{-1} for which $\rho_S(t)$ changes appreciably.
4. (Secular approximation) Survival of secular terms, compare for instance rotating wave approximation in scattering theory.

The employment of 1-4, each approximation alone destroying time reversal invariance, leads up to the PME (Pauli master equation) for the diagonal part of $\rho_S(t)$ subject to a semigroup time evolution generated by a non-selfadjoint, non-normal operator. In the two level case one obtains for instance

$$\rho_{11}(t) = e^{-W_{12}t}\rho_{11}(0) + \int_0^t dt' \, e^{-W_{21}(t-t')} W_{12}\rho_{22}(t') \tag{3}$$

showing that PME necessarily implies exponential decay. This is naturally in contrast with the group evolution of $S \oplus R$ governed by the self-adjoint Liouvillian $L = \frac{1}{\hbar}[H,\cdot]$, where $h = 2\pi\hbar$ is Planck's constant,

$$\rho(t) = e^{-iLt}\rho(0) \; . \tag{4}$$

The questions arising at this stage are crucial. First, is it possible to imbed the dynamics of the system within a vaster formalism? Second, is it possible to derive a corresponding GME, generalized master equation, in which the semigroup evolution, c.f. eq. (3) is a <u>generic</u> property inherent in the dynamical conditions rather than being brought in via approximations of type I-IV?

A positive answer to the source problem is given by the Brussels school, see e.g. Prigogine et al. [13]. In condensed form their answer is:

a. <u>Irreversible</u> processes are as <u>real</u> as reversible ones.

b. <u>Irreversible</u> processes play a fundamental constructive role in the physical world.

c. <u>Irreversibility</u> is deeply rooted in dynamics.

d. There exists an <u>exact subdynamics</u> of S such that the corresponding GME leads to a semigroup evolution $\tilde{\rho}$, see eq. (1) without invoking <u>any</u> approximation.

We are now at the point where precise mathematical questions can be asked. To do this we elaborate on point d. Noting that a diagonal density matrix ρ in an appropriate representation stays diagonal if there is no interaction between S and R, interaction by our choice appear in the off-diagonal part of ρ. Using projection operators P and $Q = J - P$ for the respective components $\rho_P = P\rho$ and $\rho_Q = Q\rho$ the following identity can be proven, (J is identity operator and z complex)

$$(L - zJ)^{-1} = (P + C(z))(PLP + \Psi(z) - zJ)^{-1}(P + \mathcal{D}(z)) + Q(QLQ - zJ)^{-1}Q \tag{5}$$

where the creation, destruction and collision operators are given by

$$\begin{aligned} C(z) &= -(QLQ - zJ)^{-1}QLP \\ \mathcal{D}(z) &= -\ PLQ(QLQ - zJ)^{-1} \\ \Psi(z) &= -\ PLQ(QLQ - zJ)^{-1}QLP \quad . \end{aligned} \tag{6}$$

Identities of the type above have occurred frequently in the theoretical development, see e.g. Löwdin [14], C. George et al. [15] or reference [7].

Using eq. (5) in eq. (4) via a Fourier-Laplace transform one obtains, $C^{\pm}$ is a path in the complex plane going from $-\infty \pm i0$, +(-)sign corresponds to $t > 0$ $(t < 0)$

$$\begin{aligned} \rho(t) &= e^{-iLt}\rho(0) = \pm\frac{1}{2\pi i}\int_{C^{\pm}} dz\ e^{-izt}\ (L - zJ)^{-1}\rho(0) \\ &= \pm\frac{1}{2\pi i}\int_{C^{\pm}} dz\ e^{-izt}\Big[(P + C(z))(PLP + \Psi(z) - zJ)^{-1}(P + \mathcal{D}(z)) + \\ &\qquad Q(QLQ - zJ)^{-1}Q\Big]\rho(0) \end{aligned} \tag{7}$$

and the retarded advanced GME

$$\begin{aligned} i\,\dot{\rho}_P^{\pm}(t) &= \pm i\{\delta(t)\rho_P(0) + PLQ\ G_Q^{\pm}(t)\rho_Q(0)\} + PLP\rho_P^{\pm}(t) \\ &\quad -i\ PLQ\int_0^t (\pm\, i\, G_Q(t-t')QLP\rho_P(t'))^{\pm}dt' \ . \end{aligned} \tag{8}$$

In eq. (8) $\delta(t)$ is the Dirac delta function and $G_Q^{\pm}(t)$ the propagator (θ is the Heaviside function)

$$G^{\pm}_{Q}(t) = \mp i\theta(t)\, e^{-iQLQt} = \frac{1}{2\pi} \int_{C^{\pm}} (zJ - QLQ)^{-1}\, e^{-izt} dz \; . \tag{9}$$

By adding the retarded and advanced parts in (8), we get the Nakajima [16] - Zwanzig [17] GME

$$i\, \dot{\rho}_P(t) = PLP\rho_P(t) + PLQ\, e^{-iQLQt} \rho_Q(0) - i\, LPQ \int_0^t e^{-iQLQ(t-t')} QLP\rho_P(t')dt' \tag{10}$$

with $\rho_P = \rho_P^{+} + \rho_P^{-}$ and $\rho_Q = \rho_Q^{+} + \rho_Q^{-}$. At this point one should note that eqs. (8) and (10) are identical with the full dynamics since no approximations have been made.

The main issue is now to find means to fully decouple the dynamics of the system S from the reservoir R such that the resulting dynamics would obey a semigroup (decay) law. Before proceeding, one should mention that the realization often associated with the treatments of spin systems coupled to a radiation field is a projection P defined as partial traces over reservoir variables [18], see also approximations 1-4. The Prigogine subdynamics [13] on the other hand, realizes P as a projection onto the diagonal part of the full density matrix.

Following the Brussels school we assume that complex poles of the reduced resolvent in eq. (7) exist via some analytic continuation procedure to be specified later. in Fig. 2 we display the spectral situaticn for the simple case that the Hamiltonian has a number of discrete eigenvalues and a continuum exhibiting complex resonances in the lower half-plane. The right hand side of Fig. 2 shows the corresponding spectra in the Liouville formulation, with all the bound states being mapped into the zero point and the resonances appearing on the negative imaginary axis.

From the integral theorem the residue contribution of the integrand in eq. (7), due to singularities of the reduced superoperator resolvent $P(PLP + \Psi(z) - zJ)^{-1}P$, yields the projection operator $\pi(0)$ onto the kinetic space of exponentially decaying density distributions evolving according to a semigroup law given by $\pi(t)$, $t > 0$

$$\pi(t) = \sum_k (P + C(z_k))(1 - \mu_k'(z_k))^{-1}\, O_k(P + D(z_k)\, e^{-iz_k t} \; . \tag{11}$$

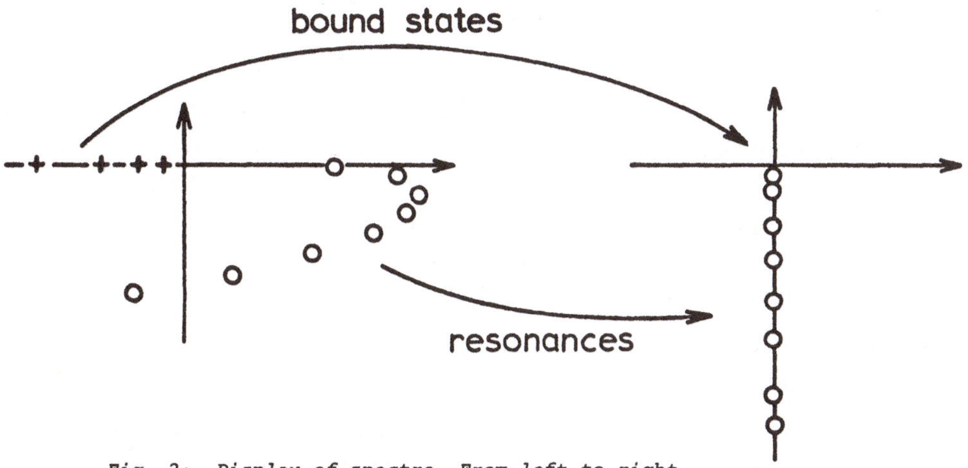

Fig. 2: Display of spectra. From left to right, Hamiltonian and Liouvillian picture, respectively. See text for details.

In eq. (11) $z_k = -i\Gamma_k$ is the solution of

$$\{PLP + \psi(z)\}\rho_{Pk} = \mu_k(z)\rho_{Pk} \tag{12}$$

$$z = \mu_k(z) = z_k$$

and

$$O_k(1 - \mu_k'(z_k))^{-1} = |\rho_{Pk}^+ >(1 - \mu_k'(z_k))^{-1}<\rho_{Pk}| \tag{13}$$

Note that there is no problem in associating the branch μ_k with an eigenvalue z_k unless eigenvalues of eq. (12) with Segré characteristics larger than one occur. The latter is possible in the complex plane. It can be handled straight forwardly by considering the appropriate Jordan blocks in the resolvent of eq. (7) before the residue evaluation. Rather than going into this slightly technical problem, we will instead concentrate on the analytic continuation procedure. This is obviously of outmost importance, since the existence of complex solutions of eq. (12) yields the decay factor $e^{-\Gamma_k t}$ in eq. (11). It is here the question mark in eq. (1) may be disposed of.

At first sight this seems to be an insoluble dilemma. A suggestion out of it was offered by Courbage [19]. He stated in a theorem that the creation operator $C(z)$ cannot have an analytic continuation similar to the destruction and collision operators. The proof that $C(z)$ cannot possibly have an analytic continuation is based on a contradiction resulting from the relation between the components ρ_P and ρ_Q

$$C(z_o)\rho_P = \rho_Q \tag{14}$$

for $z_o^* \neq z_o$. The complex scaling transformation, an unbounded similitude, must be carefully exercised as associated domain characteristics are vital. The Courbage analysis of this situation in terms of abstract domain and range considerations leading up to the contradiction, that the self-adjoint operator $L = L^+$ would have complex point eigenvalues did eq. (14) exist, is correct. However, the remainder of the theorem leads to the following questions:

i) in what sense do $\mathcal{D}(z)$ and $\Psi(z)$ admit analytic continuation in contrast to $C(z)$?

ii) what is the precise mathematical characterization of the appropriate domains appearing in the complex deformation formalism?

iii) in what sense should the whole subdynamics edifice be interpreted (see e.g. point d above)?

Note that we do not address the problem of defining the entropy here as it goes beyond the scope of the present lecture.

In answering i - iii we will depart from the content of the theorem referring to the impossibility of simultaneous continuability of $C(z)$, $\mathcal{D}(z)$ and $\Psi(z)$. The precise statement requires some detailed analysis of spectral deformations in the Hamiltonian formulation to be treated in the next section, see also reference [7] and [20].

In the Hamiltonian case a complex deformation of a dilation analytic operator [1], see also Simon [21], is defined as

$$A_d = \begin{cases} U\,A\,U^{-1} & D(A) = N \subset H \\ A(\eta) & D(A) \to H \end{cases} \tag{15}$$

where U is the operator of complex scaling and N is the Nelson class of entire vectors associated with the dilation group. N is a dense subset of Hilbert space, see Nelson [22], and $D(A)$ denotes the domain of A. For a superoperator a similar spectral deformation can be defined and we take A_d to mean the bounded extension of $(U A U^{-1})$. We then let, see ref. [7], $L \to L_d$ and

$$\begin{aligned} C(z) &\to C_d(z) \\ \mathcal{D}(z) &\to \mathcal{D}_d(z) \\ \Psi(z) &\to \Psi_d(z)\,. \end{aligned} \tag{16}$$

Complex poles $z_0 \neq z_0^*$ are given by

$$\{(PLP)_d + \Psi_d(z_0)\}\rho_P^d = z_0\rho_P^d$$

$$C_d(z_0)\rho_P^d = \rho_Q^d \quad ; \tag{17}$$

$$\Psi_d(z_0) = \mathcal{D}_d(z_0)(QLP)_d = (PLQ)_d\, C_d(z_0) \ .$$

In analogy with the Hamiltonian case [20], one can prove that the deformed density matrix $\rho^d = \rho_P^d + \rho_Q^d$ corresponding to $z_0^* \neq z_0$ does not belong to the domain of unbounded similarity transformations, i.e. there does not exist a ρ_P and ρ_Q (of Hilbert-Schmidt type) such that

$$U(\rho_P + \rho_Q) = (\rho_P^d + \rho_Q^d) \ . \tag{18}$$

If (18) were true, one would simply deduce that $z_0^* \neq z_0$ is in the spectrum of $L = L^+$, which is a contradiction. In contrast to the theorem of Courbage $C_d(z)$, $\mathcal{D}_d(z)$ and $\Psi_d(z)$ exist simultaneously as defined in eq, (16). However, the space of decaying distributions as given by the projection operator $\pi_d(0)$ derived from

$$\pi_d(t) = \sum_k (P_d + C_d(z_k))(1 - \mu_k'(z_k))^{-1}\, O_k^d(P_d + \mathcal{D}_d(z))e^{-\Gamma_k t} \tag{19}$$

c.f. eq. (11) cannot be obtained as an equivalence map between the physical manifold described by the projection operator P and the $\pi_d(0)$ space. Furthermore, the subdynamics in the quantum version refers to L_d rather than L. Irreversibility is manifested in the dilated equations as solutions appear that do no longer belong to the class of dilation analytic vectors (or the corresponding generalization in the Liouville formulation). In a way coarse-graining and time-smoothing, compare approximations 2 and 3, come back in "disguise", as the present formulation of subdynamics in terms of the resonance picture of unstable quantum states necessitates some kind of spectral deformation. Once the latter is performed, the resulting subdynamics cannot be transformed back to the original dynamics! Nevertheless, the temporal behaviour of our chemical system is precisely governed by the deformed projection operator π_d via eqs. (7) and (8). For more details we refer to Obcemea and Brändas [7].

III. Resonance States in the Hamiltonian Formulation

Together with P. Froelich. Supported in part by the Swedish Natural Science Research Council.

The unsolved problems of complex scaling has been well summarized by Simon [23] and we will here review and comment on some of them. First an abbreviated list of problems and difficulties associated with the general concept of a resonance:

1. The unique definition
2. Existence proofs and error bounds
3. Expansion formulas
4. Channel decompositions in scattering theory.

The pitfalls and contradictions that one may fall into if one were to introduce complex resonance states without sufficient carefulness, termed Howland's [24] razor, have been vividly described by Simon [23]. With the appropriate conditions given for a unique definition of a resonance; it could for instance be certain analytic properties on the potential like dilatation analyticity [1, 21] or Weyl's limit point, see e.g. Coddington and Levinson [25], we would view it as a complex eigenvalue of an analytically continued Hamiltonian or simultaneously as a pole of an analytically continued (associated) resolvent. Note that this definition of a resonance is different from the one in terms of its physical effects, see e.g. Lavine [26], although for narrow resonances (small widths) the two may be intimately related.

To demonstrate some of the unexpected features connected with the resonance concept as well as comment on 1-4, we introduce an analytic extension of the scalar product which in one dimension becomes

$$\langle f|g\rangle_{C^\pm} = \int_{C^\pm} f^*(z^*)\; g(z)dz \tag{20}$$

where $C^\pm$ are complex paths in the upper or lower halfplane. One realization would be $z = \eta x$ where η is a complex scale factor, and integration interval is the positive real axis. If f and g belong to the domain of complex rotations, i.e. the set N of entire vectors associated with the dilation group [22], then we can write

$$\begin{aligned}\langle f|g\rangle_\eta &= \int_{\eta R} f^*(\eta^* x)\; g(\eta x)\; d\eta x = \int_R f^*(x)\; g(x)\; dx \\ &= \langle f|g\rangle = \int_R \left[(\eta^*)^{\frac{1}{2}}\; f(\eta^* x)\right]^*\; \eta^{\frac{1}{2}}\; g(\eta x)dx\end{aligned} \tag{21}$$

$$= \langle u(\eta^*)f \mid u(\eta)g\rangle = \langle f \mid u^+(\eta^*)\, u(\eta) \mid g\rangle \,.$$

From (21) it is obtained for all $f,g \in N$, the "star unitary" property,

$$\begin{aligned} u^+(\eta^*) &= u^{-1}(\eta) \\ u^+(\eta) &= u^{-1}(\eta^*) \,. \end{aligned} \tag{22}$$

For simplicity we denote $u(\eta)\varphi(x) = \eta^{3N/2}\varphi(\eta x) = \varphi(\eta)$ where $3N$ are the number if coordinates to be scaled. Hence for all $\varphi,\phi \in N$

$$\langle\varphi\mid\phi\rangle = \langle\varphi\mid u^+(\eta^*)u(\eta)\mid\phi\rangle = \langle\varphi(\eta^*)\mid\phi(\eta)\rangle \quad . \tag{23}$$

The scalar product (23) was introduced in ref. [27] in connection with the derivation of a bi-variational principle and the associated complex virial theorem. Consider now the matrix element

$$w(z) = \langle\varphi\mid(z-H)^{-1}\mid\phi\rangle = \langle\varphi(\eta^*)\mid(z-H(\eta))^{-1}\mid\phi(\eta)\rangle = W_\eta(z) \tag{24}$$

in which $H(\eta) = u(\eta)H(1)u^{-1}(\eta)$ and $z \in \rho(H)$ (the resolvent set of $H(1)$). Because of relation (24) the matrix element appearing to the right is finite for all z, $\text{im}z \neq 0$, still the resolvent $(z-H(\eta))^{-1}$ has poles associated with

$$H(\eta)\;\psi(\eta) = \varepsilon\Psi(\eta) \;;\; \varepsilon = E - i\,\frac{\Gamma}{2} \;;\; \eta = e^{i\theta}. \tag{25}$$

In Fig. 3a-c) the situation is depicted for a single continuum, rotated the angle $-2\arg\eta = -2\theta$, displaying the bound states, some exposed resonances and the rotated cut. The contour is chosen in order to calculate the survival amplitude [28] $(t > 0)$

$$A(t) = \langle\psi_o\mid e^{-iHt}\mid\psi_o\rangle = \frac{i}{2\pi}\int_{C^+} \langle\psi_o\mid(z-H)^{-1}\mid\psi_o\rangle e^{-izt}dz \tag{26}$$

where ψ_o is the wave packet at time $t = 0$, which we assume to be in the domain of $u(\eta)$. We will come back to this calculation as well as Figure 3 during the discussion of point 3 in the list above.

1. We then proceed to point 1 on the list.

With due consideration of Howland's razor, see below, we will take the view employed in ref. [20], see also [30]. The resonance "eigenvalue and "eigenfunction" is based on the existence of the Cauchy sequence

$$\| f_n^\eta - f_m^\eta \| \xrightarrow[m\to\infty]{n\to\infty} 0 \tag{27}$$

with the bounded sequence $\{f_n^\eta\}$ satisfying $\{z_o \notin \sigma(H(1))\}$

$$\|(H(\eta) - z_o)f_n^\eta\| \xrightarrow{n\to\infty} 0 \tag{28}$$

for some sufficiently large rotation angle, or in general some necessary spectral deformation. The domain $\mathcal{D}_r(H)$ is here defined so that

$$(H - z_o)\ \mathcal{D}_r(H) = \mathcal{N}\ , \tag{29}$$

for more details see ref. [20]. The actual structure needed to dispose of the Howland's razor is showing up in the spectral deformation needed for the existence of (28), and for the choice of the associated set $\mathcal{N}$. In the next section we will discuss another structure based on the Titchmarsh-Weyl theory. In general one must be aware of the fact that different polestrings for different rotation angles may appear as certain critical angles are bypassed, see refs. [31-34] for examples.

2. Recently an attempt to derive an error formula based on

$$\sigma_H(\varepsilon) = \langle(H(\eta)-\varepsilon)\varphi|(H(\eta)-\varepsilon)\varphi\rangle = \rho_W^2 \tag{30}$$

$$\varepsilon \in \mathbb{C}\ ,\ \varphi \in \mathcal{D}(H(\eta))\ ; \qquad \langle\varphi|\varphi\rangle = 1$$

presented [35] via the introduction of a special matrix representation. The error estimate was obtained from a special matrix norm, equal to the bound given by the Lévy-Hadamard-Gerschgorin theorem and also equal to the Weinstein radius ρ_W. Although the circle theorem applies to non-normal matrices, one must be certain that no other circles may overlap with the one obtained from the extremum variation principle derived from $\sigma_H(\varepsilon)$. To be rigorous such an analysis should be made, but in practice the non-intersecting situation may appear provided $\rho_W = \sigma^{1/2}(\bar{\varepsilon})$, $\bar{\varepsilon} = \langle\varphi|H|\varphi\rangle$ is sufficiently small and $\sigma^{1/2}(\bar{\varepsilon}^*)$ is not too large. Several attempts [36,37] to provide counter examples have been made, mostly related to the break-down of the Weinstein estimate in the case of arbitrary matrices. It is true that we have not used any particular property of the analytic family of operators except that we know the associated spectrum to be <u>non empty</u>. For more work on error bounds see N. Moiseyev [38] these proceedings.

Before leaving point 2, it is important to emphasize the consequence of (29). Since every member of the sequence $(H - z_o)f_n$ belongs to $\mathcal{N}$ it can be rotated at will. However, as shown in ref. [20] it is <u>impossible</u> to find a sequence (27) with $\eta = 1$, corresponding to

$$\| (H(1) - z_o) f_n^1 \| \xrightarrow{n \to \infty} 0 \qquad (31)$$

since the existence of (31) implies that z_o is in the spectrum of the self-adjoint operator $H(1)$. As a consequence the resonance "eigenfunction" f_o^η defined by (27) and (28) does not belong to N . If it did the relation

$$H(\eta) f_o^\eta = z_o f_o^\eta = UH(1)U^{-1} f_o^\eta = z_o f_o^\eta \qquad (32)$$

or

$$H(1)U^{-1} f_o^\eta = z_o \, U^{-1} f_o^\eta \qquad (33)$$

again would imply that z_o is in the spectrum of $H(1)$. The latter conclusion is the one used in the analysis and reinterpretation of the Courbage theorem, see section II.

3. A general expansion theorem for non-normal operators is still an unsolved problem although a few results for special cases have been obtained, see Simon [23] for some rigorous completeness results. However, a rigorous expansion formula can still be obtained as presented recently [29]. Using general inner projection techniques [14] it is possible to make a decomposition of the matrix element appearing in eq. (24) (and then indirectly to $A(t)$, eq, (26)) into the bound state part, the resonance contribution and the remainder associated primarily with the rotated cut.

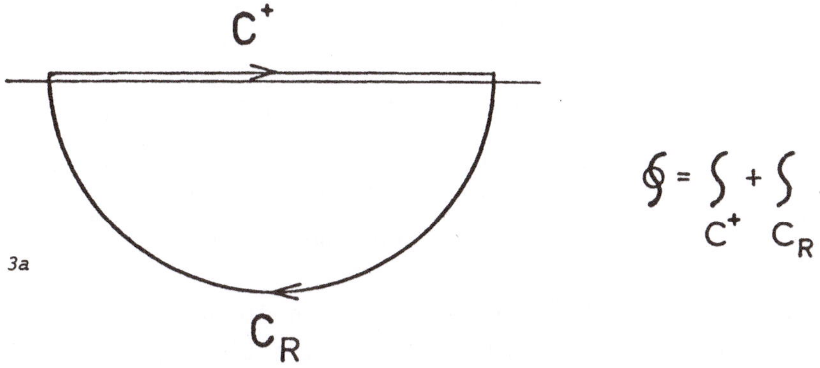

Fig. 3a

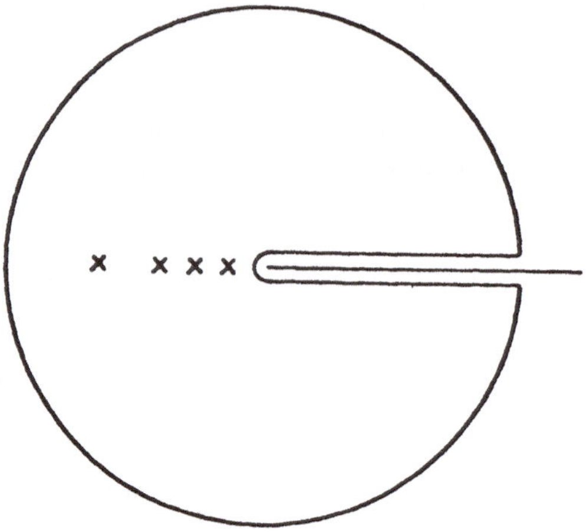

Fig. 3b

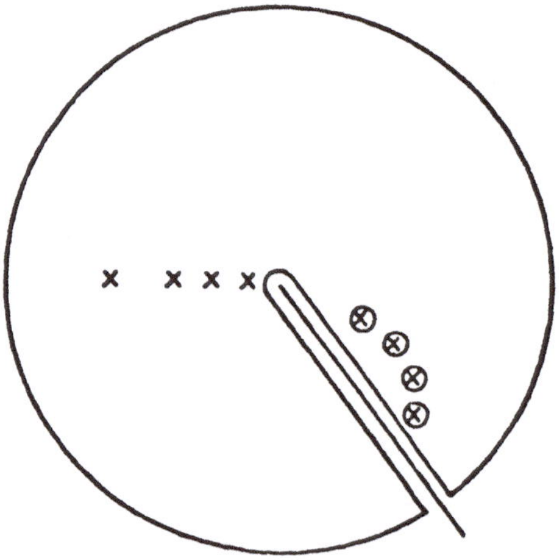

Fig. 3c

<u>*Fig. 3:*</u> *The contour shown in a) displays the various parts before $R \to \infty$. No spectra of H is indicated, as they are supposed to be below C^+ . In b) and c) the contours are formally distorted so as to account for the cut.*

Note that the rigour stressed in the discussion of ref. [29] is due to the simplifying assumption that φ , ϕ (or ψ_0) is in D(u). This does not restrict the physics of the problem and hence an expansion formula valid in a dense subset of Hilbert space, rather than the whole of it, can be rigorously derived and used for applicatory purposes.

It is interesting to note that the matrix element W(z) defined in eq. (24) is finite even if z happens to be a resonance eigenvalue

z_o of $H(\eta)$. In the latter case, however, we will find that W_{res} in the decomposition

$$W_\eta(z) = W_{bound} + W_{res} + W_{cut} \qquad (24')$$

must have a pole at $z = z_o$. Since $W_\eta(z)$ is finite this singularity must be cancelled by contributions in W_{cut} . For a definition of W_{cut} see reference [29].

4. The problem of a satisfactory channel decomposition is in fact very closely related to the situation encountered in the subdynamics formulation. Again, one can use inner projection techniques to estimate the various contributions. In this way one may find projection operators associated with bound states, resonances and the rotated cut, being the ingrediences in the spectrally deformed Prigogine subdynamics [7]. Working in reference manifolds constructed from well-defined dense sets of Hilbert space appropriate limits are guaranteed. Like the subdynamics formulation the channel decomposition is based on a biorthogonal construction, which does not allow rotation back to the real axis.

In the next section, we will exemplify these aspects from the view-point of scattering theory and the Titchmarsh-Weyl formulation.

IV. Scattering Theory and Analytic Continuation

Together with Magnus Rittby and Nils Elander. Supported in part by the Swedish Natural Science Research Council.

We have previously given a presentation of the Titchmarsh-Weyl theory for a singular second-order differential equation (or set of equations) and its connections with scattering theory [39-43]. For a comprehensive review of the current status of the theory, see Everitt and Bennewitz [44].

The Titchmarsh-Weyl theory focus its interest on the m-function (or m-matrix). In the limit point case, provided the potential satisfies certain conditions, see e.g. ref. [25] for simple cases, m is uniquely given by the condition that χ below should be square integrable

$$\chi = \varphi + \psi m \ . \qquad (34)$$

In (34) φ and ψ are initial value solutions of

$$Hu = \left(-\frac{d^2}{dr^2} + V(r)\right)u = (E + i\varepsilon)u \ ; \quad \varepsilon \neq 0 \qquad (35)$$

with ψ being regular at the origin and φ is the irregular one, usually chosen so that $W(\varphi\psi) = \varphi\psi' - \varphi'\psi = 1$. The limit point classification, which is different from the conditions required in scattering theory, quarantees a unique square integrable solution χ of (35) for an infinite interval. The key property of m follows by its intimate relation with the spectral function ρ associated with H, i.e.

$$\rho(E_\beta) - \rho(E_\alpha) = \lim_{\varepsilon\to 0+0} \frac{1}{\pi} \int_{E_\alpha}^{E_\beta} \mathrm{Im}[m(E+i\varepsilon)]dE\ . \tag{36}$$

The inversion formula (36) implies that m can be written

$$m(\lambda) = -\mathrm{tg}\alpha + \int_{-\infty}^{+\infty} \frac{d\rho(\omega)}{\omega-\lambda}\quad ;\quad \alpha \neq \pi/2\ . \tag{37}$$

The angle α is chosen so that the regular solution ψ has a logarithmic derivative equal to $\mathrm{tg}\alpha$ at origin. If origin is a singular point one chooses a point $a > 0$. In the latter case α becomes dependent on λ although the dependence of m on α is not explicitly denoted. A simple Wronskian formula for m can be proven to yield

$$m = -\{W(f\psi)\}^{-1}\ W(f\psi) = \frac{iz\ \tan\alpha + 1}{\tan\alpha - iz}\quad ;\quad iz = f'f^{-1} \tag{38}$$

where we have made use of the fact that χ is proportional to the Jost solution (when the potential also satisfies the requirements of scattering theory), i.e.

$$\chi = fc \tag{39}$$

with

$$c = -\{W(\psi,f)\}^{-1} = J^{-1}\ . \tag{40}$$

Observing that (40) exhibits two branches the connections with scattering theory follow directly. J is the Jost-function, i.e.

$$J^{\pm} = -W(\psi f^{\pm}) = |J|e^{\mp i\delta} \tag{41}$$

with δ the phase shift. Finally the S-matrix (and cross sections) can be obtained from

$$S = \{J^{+}\}^{-1}\ J^{-}\ . \tag{42}$$

Note that the equations involving wronkians are written so that they are immediately applicable in a matrix formulation. It is also easy to deduce that the m-function is entirely dependent on the logarithmic derivatives of the solution ψ and χ, while the S-matrix need more information [41,43]. The α-dependence of m leads to oscillations in the real and imaginary parts of m, see Figs. 5 and 6. The peaks reveal that resonances may exist (but not always [20]!), and one can prove that the actual position of the pole is independent of α. Since most potentials are accompanied by a centrifugal term $\frac{\ell(\ell+1)}{r^2}$ origin is a singular point. Hence the matching point a where the logarithmic derivative of ψ was $\mathrm{tg}\,\alpha$ cannot approach zero, unless $\ell = 0$. In the latter case one obtains $(E \in \sigma_C(H))$

$$\mathrm{im}\{m(E)\} = \frac{k}{|J(k)|^2} \xrightarrow{E \to \infty} k = \sqrt{E} \tag{43}$$

see Fig. 4. In this case (37) assumes a more general Nevanlinna form [44] including also $\alpha = \frac{\pi}{2}$. For $\ell \neq 0$ m cannot be constructed according to (37) or its Nevanlinna generalization, since

$$\pi\left(\frac{d\rho}{d\omega}\right)_{W=E} = \frac{k^{2\ell+1}}{|J_\ell(k)|^2} \tag{44}$$

grows too fast when $k \to \infty$. Fore more details see ref. [45].

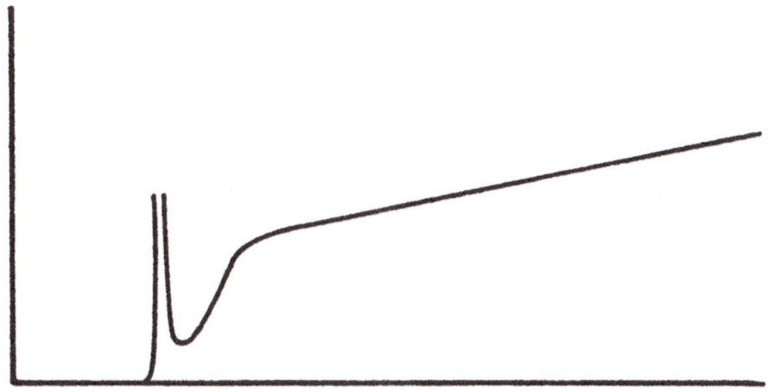

Fig. 4: *Spectral density, see eq. (43), for $V = V_o r^2 e^{-r}$ exhibiting one primary resonance.*

In order to explore the complex pole string associated with the differential equation (or sets of equations) defined in eq. (35) we perform the scaling or exterior scaling transformation [41,46,47].

$$r \to \begin{cases} r, & r < R_o \\ R_o + \eta(r - R_o), & r \geq R_o \quad , \quad \eta = e^{i\theta} \quad , \quad \theta > 0 \end{cases} \tag{45}$$

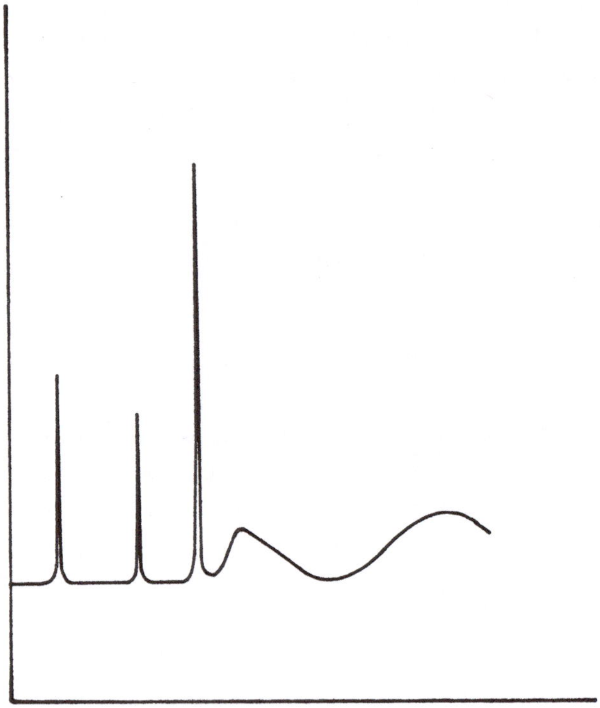

Fig. 5: Imaginary part of $m(E)$ for a numerical potential, see ref. [43], with $a \neq 0$ in eq. (38). Note the onset of oscillations after the occurrence of the shoulder-like behaviour. See also Fig. 6.

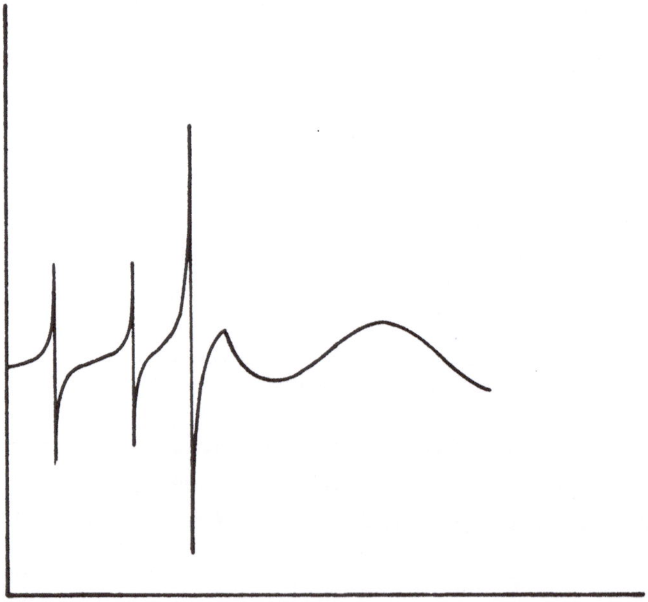

Fig. 6: Display of the real part of $m(E)$ in the application to the $A^1\Pi$ state of CH^+ also referred to in Fig. 5.

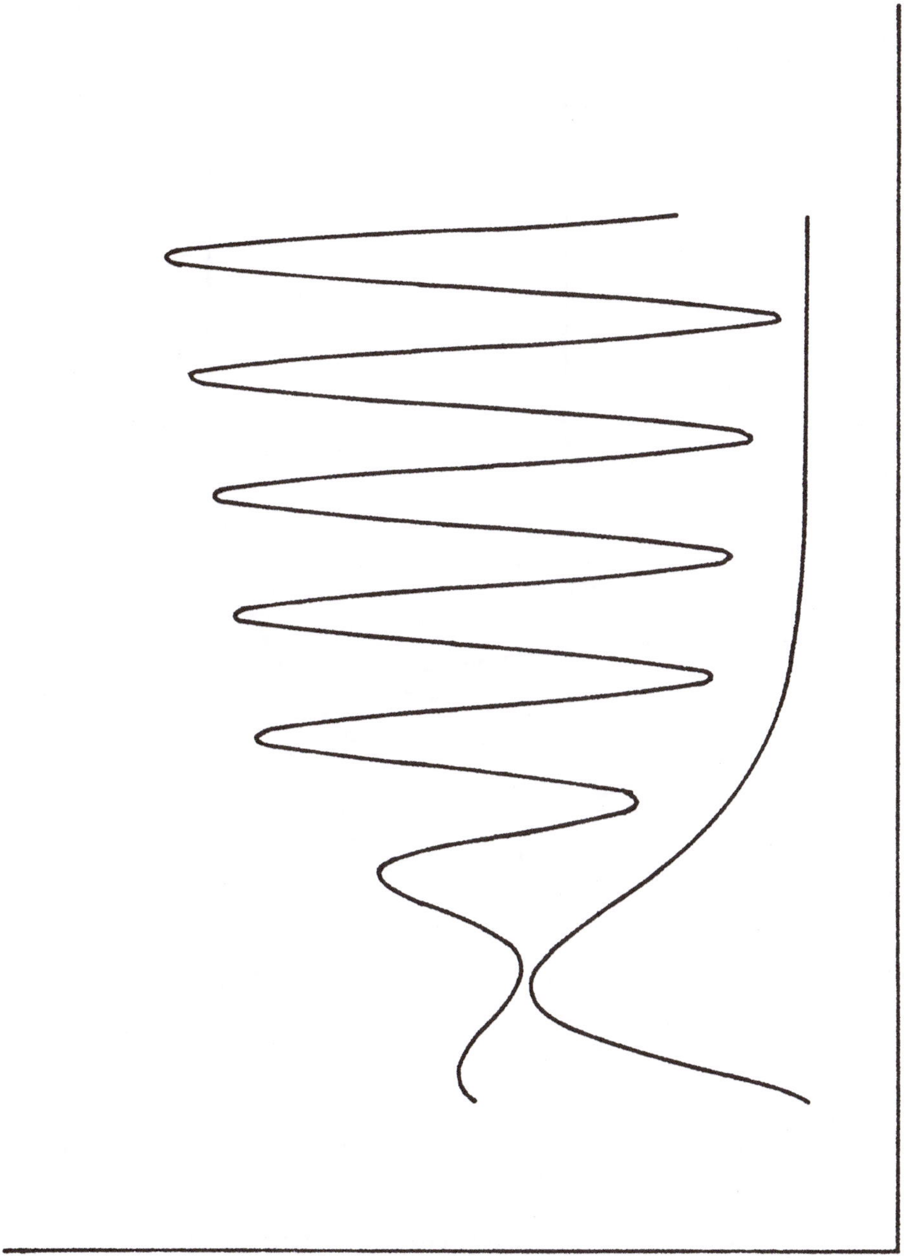

Fig. 7: Real part of Gamow wave for the potential defined in Fig. 4.

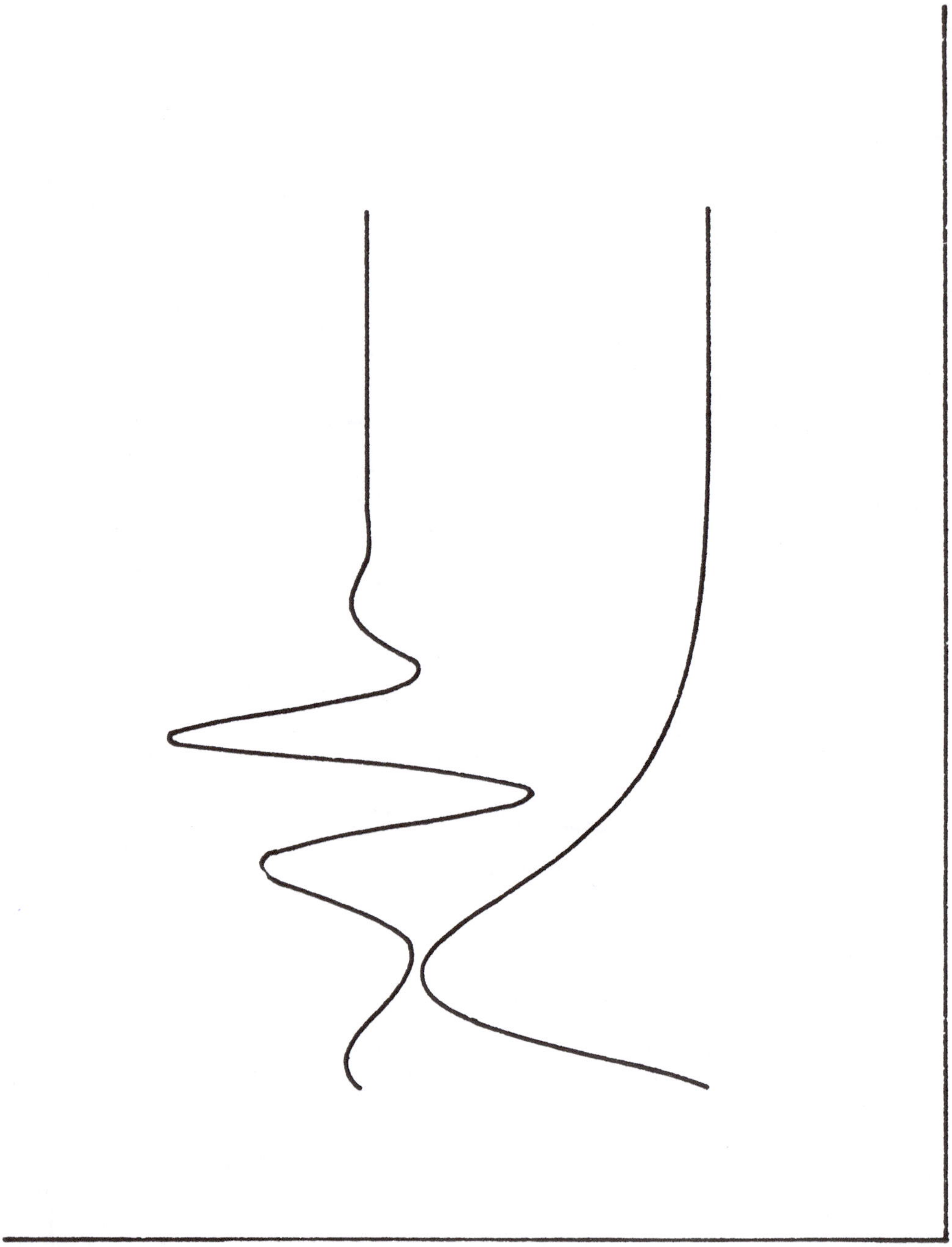

Fig. 8: *Real part of the solution corresponding to the Gamow wave in Fig. 7 after exterior scaling has been performed*

From the view-point of application complex scaling is a device to make an instable procedure stable [20,43,45], compare also the self-correction effect mentioned by Atabek and Lefebvre [33]. Once transformation (45) is performed $J^{\pm}$ and S as given by (41) and (42) can be calculated. Note also that J^{+} and J^{-} can be scaled independently.

In Fig. 7 the real part of a Gamow wave is displayed. Here the Gamow vector is a formal solution to the Schrödinger equation, regular at origin and with a pure exponential increase at infinity. In Fig. 8 the same wave (its real part) is shown but with the exterior scaling transformation employed. The point R_o is chosen just before the third maximum in the wave function. The potential shown in the figure is the same as the one in Fig. 4.

In this context one should not forget that Weyl's limit circle - limit point classification may incorporate potentials that are not dilation analytic. Moreover, actual potentials may change their asymptotic behaviour when certain critical angles are attained [32]. Different pole-strings may then occur [33,34]. Despite the fact that two or more totally different sets of poles appear, they are nevertheless part of the total analytic information intrinsic to the potential. Irrespective of all the complications that are possible at the present level, we surmize that an analysis of the type displayed in the last two sections is sufficiently rich in concrete details and rigorous formulations to motivate an abstraction and generalization initially aimed at in the development of a non-contradictory subdynamics. At all levels there are tremendous amounts of work left to be done, in terms of obtaining more rigorous results, increased familiarity with fundamental concepts like Gamow vectors, Jost functions, spectral densities, equilibrium and non-equilibrium situations and finally in obtaining more applications to concrete problems.

V. Some Concluding Comments

In the table of examples on chemical systems and their environment all but the last one concern the situation when the number of particles in the reservoir are many many orders of magnitude larger. In the last case mentioned, i.e. in the study of macromolecular behaviour of bacteriorhodopsin [48], the protein overcoat may consist of $10^4 - 10^6$ electrons, while the polyene system may contain $\sim 10^2$. As a consequence system and reservoir, here building up the macromolecule, may interact in a very complex way, with wave mechanical behaviour spreading out over macroscopic dimensions.

To study such complicated phenomena one need to analyze the spectrum of the second order reduced density matrix

$$\Gamma^{(2)}(x_1x_2|x_1'x_2') = \binom{N}{2} \int \Psi(x_1x_2x_3...x_N)\Psi^*(x_1'x_2'x_3...x_N)dx_3\ dx_N \quad (46)$$

which in extreme situations may exhibit off-diagonal-long-range-order (ODLRO) [49].

ODLRO implies the existence of a macro-wavefunction [48], which can be utilized to explain a manifold of anomalous macromolecular behaviour. Although an interpretation via the Bohm-Aharonov effect [5] seems far from the topics described in this review it is perhaps surprising to note that rigorous spectral deformation theories can be applied to the ac Stark Hamiltonian provided an appropriate gauge is chosen [6].

Although this connection is rather vague it illustrates the necessary ingrediences in a general formulation of time evolution of quantum chemical systems, namely the interplay between general spectral classifications, spectral deformations (or unbounded similarity transformations) and the precise temporal behaviour of wave-packets or, in the Liouville formulation, initial density distributions.

A detailed analysis of the limiting procedures leading to infinite systems or infinite boxes that surround the system (and reservoir) has not been explicitly emphasized. On the other hand, it should be stressed that the existence of the essential spectrum means that one must go beyond confining the particles to a finite volume in space. The existence of a continuous spectrum therefore assumes a boundary value problem where all the appropriate degrees of freedom are defined over infinite intervals. The present interpretation of subdynamics is a consequence of the general spectral properties associated with such a formulation.

References

[1] E. Balslev and J. Combes, Commun. Math. Phys. 22, 280 (1971)

[2] W.P. Reinhardt, Ann. Rev. Phys. Chem. 33, 223 (1982)

[3] B.R. Junker, Advan. At. Mol. Phys. 18, 207 (1982)

[4] Proceedings of the International Workshop of the 1978 Sanibel Symposium, Int. J. Quant. Chem. 14, (1978)

[5] Y. Aharonov and D. Bohm, Phys. Rev. 115, 483 (1959)

[6] J.S. Howland, J. Math. Phys. 24, 1240 (1983)

[7] C.H. Obcemea and E. Brändas, Ann. Phys. 151, 383 (1983)

[8] I. Prigogine, From Being to Becoming. Time and Complexity in the Physical Sciences. W.H. Freeman and Co. San Francisco (1980)

[9] W. Pauli, in "Probleme der Modernen Physik, Festschrift zum Geburtstage A. Sommerfelds", Hirzel, Leipzig, 1928

[10] U. Fano, Rev. Mod. Phys. 29, 74 (1957)

[11] W.H. Loisell, Quantum Statistical Properties of Radiation, Wiley, New York (1973)

[12] A.G. Redfield, IBM J. Res. Develop. 1, 19 (1957), Adv. Magn. Res. 1, 1 (1965)

[13] I. Prigogine, C. George, F. Henin, and L. Rosenfeld, Chem. Scripta 4, 5 (1973)

[14] P.O. Löwdin, Phys. Rev. 139, A357 (1965)

[15] C. George, I. Prigogine and L. Rosenfeld, Kungl. Dansk. Vedensk. Selsk. Mat. Fys. Meddel. 38, 1 (1972)

[16] S. Nakajima, Progr. Theoret. Phys. 20, 948 (1958)

[17] R. Zwanzig, J. Chem. Phys. 33, 1338 (1980)

[18] P. Argyres and P. Kelley, Phys. Rev. 134, 89 (1964)

[19] M. Courbage, J. Math. Phys. 23, 646 (1982)

[20] E. Brändas, P. Froelich, Ch. Obcemea, N. Elander and M. Rittby, Phys. Rev. A26, 3656 (1982)

[21] B. Simon, Ann. Math. 97, 247 (1973)

[22] E. Nelson, Ann. Math. 70, 572 (1959)

[23] B. Simon, Int. J. Quant. Chem. 14, 529 (1978)

[24] J. Howland, Pac. J. Math. 55, 157 (1974)

[25] E.A. Coddington and N. Levinson, Theory of Ordinary Differential Equations, (McGraw-Hill, New York, 1955)

[26] R. Lavine, in Proceedings of London, Ont. Conf. on Atomic Scattering, June 1978

[27] E. Brändas and P. Froelich, Phys. Rev. A16, 2207 (1977)

[28] C.H. Obcemea and E. Brändas, On the Anderson Localization Model, Preprint June 1982 (unpublished)

[29] E. Brändas and P. Froelich, Int. J. Quant. Chem. 17S, 113 (1983)

[30] E. Brändas, P. Froelich and M. Hehenberger, Int. J. Quant. Chem. 14, 419 (1978)

[31] H.J. Korsch, H. Laurent and R. Möhlenkamp, Phys. Rev. A26, 1799 (1982)

[32] M. Rittby, N. Elander and E. Brändas, Phys. Rev. A26, 1804 (1982)

[33] O. Atabek and R. Lefebvre, Nuovo Cimento 76B, 176 (1983)

[34] N. Moiseyev and J. Katriel, Chem. Phys. Letters 105, 194 (1984)

[35] P. Froelich, E.R. Davidson and E. Brändas, Phys. Rev. 28A 2641 (1983)

[36] H. Siedentop, Priv. communication

[37] R. Ahlrichs, Priv. communication

[38] N. Moiseyev, see these proceedings

[39] E. Brändas, M. Hehenberger and H.V. McIntosh, Int. J. Quant. Chem. 9, 103 (1975)

[40] M. Hehenberger, B. Laskowski and E. Brändas, J. Chem. Phys. 45, 4559 (1976)

[41] M. Rittby, N. Elander and E. Brändas, Int. J. Quant. Chem. 23 865 (1983)

[42] E. Brändas and N. Elander, Proceedings Bielefeld Symposium on Energy Storage and Redistribution in Molecules. Ed. J. Hinze Plenum Press (1983) p. 543

[43] M. Rittby, N. Elander and E. Brändas, Chem. Phys. (in press)(1984)

[44] W.N. Everitt and G. Bennewitz, in Tribute to Åke Pleijel, Proceedings of the conference, Sept. 1979, Dept. of Mathematics, University of Uppsala, Uppsala, Sweden, p. 49

[45] E. Brändas, M. Rittby and N. Elander (to be submitted (1984)

[46] B. Simon, Phys. Letters. A71, 211 (1979)

[47] J.D. Morgan and B. Simon, J. Phys. B14, L167 (1981)

[48] L.J. Dunne, A.D. Clark and E.J. Brändas, Chem. Rev. (accepted for publication) (1984)

[49] C.N. Yang, Rev. Mod. Phys. 34, 694 (1962)

GEOMETRICAL QUARK CONFINEMENT AND HADRONIC RESONANCES

E. van Beveren, T.A. Rijken, and C. Dullemond
Institute for Theoretical Physics, University of Nijmegen
NL-6525 ED NIJMEGEN, The Netherlands

and

G. Rupp
Zentrum für Interdisziplinäre Forschung
Universität Bielefeld
D-4800 BIELEFELD 1, FR Germany

(presented by E. van Beveren)

1. Introduction

In this talk I would like to introduce a model for the description of hadronic resonances and to discuss its connection to the field theory for strong interactions.

If we neglect all interactions but strong interactions, hadronic resonances and bound states are determined by two in a sense competing properties of strong interactions: permanent confinement and pair creation and annihilation. The principle of permanent confinement dictates that quarks can never be isolated and has been formulated after it turned out that quarks could not be isolated experimentally, but always form colorsinglets with other quarks and/or antiquarks (see however [1]). Although in principle all colorsinglet combinations of quarks and antiquarks are allowed [2], uptill now hadronic resonances could be explained by assuming that they exist of either three valence quarks (the baryons) or of one valence quark and one valence antiquark (the mesons) [3]. In the literature other colorsinglet quark configurations are proposed for the explanation of some hadronic resonances [4], but whether it is necessary to open this Pandora's box is still a point of discussion (see e.g. [5]).

In this talk I restrict myself to mesons which I will consider to be described by a quark + antiquark pair. Examples are the π^+ meson which consists of an up quark and a down antiquark ($u\bar{d}$), its antiparticle the π^- meson ($\bar{u}d$) with lifetimes in the order of 10^{-8} seconds and the π^0 meson ($n\bar{n}$, where n = nonstrange stands for up or down) with a lifetime in the order of 10^{-16} seconds. These lifetimes are large compared to the internal motion, the period of which is in the order of 10^{-23} seconds (or a frequency of about 0.2 GeV in units $\hbar = c = 1$). Other examples are the K mesons consisting of one nonstrange quark (or antiquark) and one strange antiquark (or quark) ($n\bar{s}$ or $\bar{n}s$), with lifetimes ranging from about 10^{-10} up to 10^{-8} seconds; the J/ψ meson and the Υ meson consisting respectively of a charm and a bottom $q\bar{q}$ pair ($c\bar{c}$ and $b\bar{b}$) and with lifetimes of the order of 10^{-20} seconds. These examples (and some others) are called stable particles in hadronic physics [6]. Consequently in an S-matrix description of some scattering process these particles appear as poles on the real axis in the complex energy plane.

The principle that quarks and antiquarks are permanently confined suggests that all hadrons are stable. However, the phenomenon of pair creation allows most of the hadronic states to decay rapidly into other hadrons without violating the principle of permanent confinement. As an example let us study the ψ (3770) resonance ($c\bar{c}$) which meson decays with a mean lifetime of 3.10^{-23} seconds into a D meson ($n\bar{c}$) and an anti-D meson ($c\bar{n}$) [6]. Suppose we had at our disposal a beam of D mesons and another beam of $\bar{D}$ mesons in order to study the elastic scattering process $D+\bar{D} \rightarrow D+\bar{D}$. Then at a centre of mass energy of about 3.77 GeV the resonance ψ (3770) could be formed due to the annihilation of the nonstrange partons of the D and the $\bar{D}$ mesons. The remaining charmed partons have to form a hadron in order not to violate the principle of permanent confinement. In terms of strings: the two strings keeping confined to each other the two partons in each D meson are glued together in the annihilation process to form one string between the two charmed partons.

The next step in the scattering process is the creation of another nonstrange $q\bar{q}$ pair which breaks up again the string between the two charmed partons under the formation of new charmed mesons; in this case a new $D\bar{D}$ pair. The process is depicted in figure 1.1.

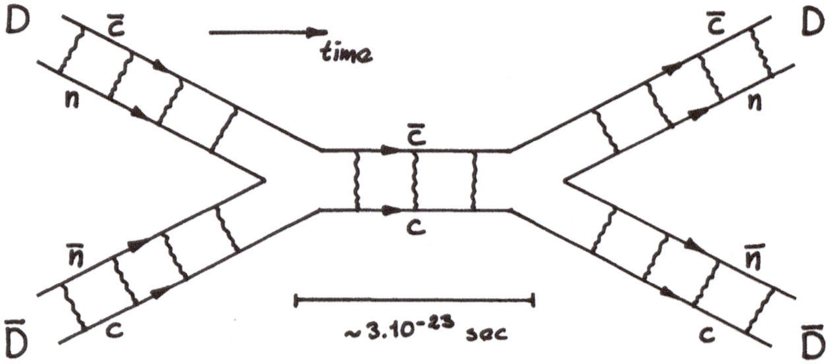

Figure 1.1: $D\bar{D} \rightarrow \psi$ (3770) $\rightarrow D\bar{D}$.

In Schrödinger scattering theory one could try to find a potential which yields as a spectrum the resonances in $D\bar{D}$ scattering. Such a description needs at least two different channels: one channel describing the $D\bar{D}$ state and another channel for describing the $c\bar{c}$ state. The two channels communicate via the annihilation and creation of $n\bar{n}$ pairs [7, 8].

In the absence of pair creation or annihilation the $c\bar{c}$ pair would be stable due to the phenomenon of permanent confinement. This suggests that a potential which describes the interaction in the $c\bar{c}$ channel is infinitely rising with distance. Although the precise form of this potential is not known and could not be found starting from the theory of strong interactions (which is supposed to be QCD) uptill now, many proposals exist in the literature [9, 10].

2. A simplified model of a hadron

In a series of papers a model for the description of a hadron is elaborated which is on one hand not too simple as not to contain some general features of hadrons, but which on the other hand is simple enough to allow easy and transparant calculations and which is based on the idea that the dominant interaction in the scattering channel is the communication of the two (or more) channels which are introduced in the previous section [11, 12].

Let me for a short introduction of the model restrict myself to the case of one permanently closed channel for the flavor quarks ($c\bar{c}$ in the case of ψ resonances) and one decay channel (e.g. $D\bar{D}$ in the case of ψ resonances). For this purpose we make use of the framework of the non-relativistic Schrödinger equation. We describe a hadron by a two-component wave function. One component describes the confinement sector of valence quarks which are permanently bound. The other component describes the decay sector of decay products, in the scattering process of which the hadron under consideration would appear as a resonance or a bound state.

For instance a meson would in this picture adopt the form

$$\begin{pmatrix} \psi_{q\bar{q}} \\ \psi_{MM} \end{pmatrix} \tag{2.1}$$

where $\psi_{q\bar{q}}$ is the wave function of the confined partons and where ψ_{MM} is the wave functio for the decay products, in this case consisting of two mesons. For the moment we leave out of consideration decay channels with more than two particles, but we will come back to this point.

If hadronic decay is forbidden, the two channels in the wave function (2.1) have no communication. The wave equation for this case would be

$$\begin{pmatrix} H_{q\bar{q}} - E & 0 \\ 0 & H_{MM} - E \end{pmatrix} \begin{pmatrix} \psi_{q\bar{q}} \\ \psi_{MM} \end{pmatrix} = 0 \tag{2.2}$$

where $H_{q\bar{q}}$ contains the confinement potential of the valence quarks and where H_{MM} contains the scattering potential of the two decay products. Equation (2.2) gives rise to two disconnected spectra: the bound state spectrum of the confinement sector (see figure 2.1)

Figure 2.1: The bound state spectrum of $H_{q\bar{q}}$.

and the continuum spectrum of the scattering sector (possible bound states of H_{MM} are not treated here) (see figure 2.2).

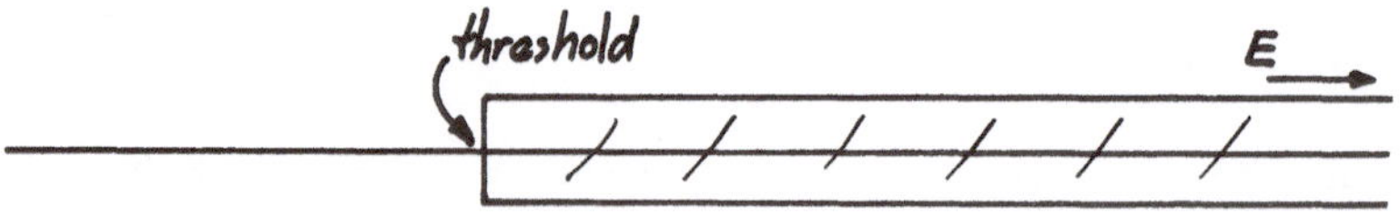

Figure 2.2: The continuum spectrum of H_{MM}.

Hadronic decay can in this model be simulated by introducing a transition potential H_{decay} at the off-diagonal positions of the potential matrix in eq. (2.2), leading to:

$$\begin{pmatrix} H_{q\bar{q}} - E & H_{decay} \\ H^{+}_{decay} & H_{MM} - E \end{pmatrix} \begin{pmatrix} \psi_{q\bar{q}} \\ \psi_{MM} \end{pmatrix} = 0 \quad . \tag{2.3}$$

An object of interest might in this case be the scattering matrix S(E) in the scattering sector. Bound states and resonances appear then as poles in S(E) in the complex variable E.

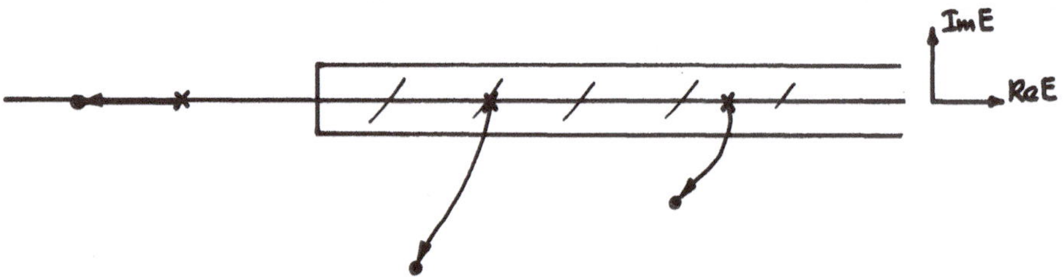

Figure 2.3: Bound states and resonances of the Hamiltonian (2.3).

In figure 2.3 we have superimposed figures 2.1 and 2.2; the arrows indicate how the poles (dots) in S(E) move as the strength of H_{decay} is increased. This way we obtain a model which simulates the effects of hadronic decay on the properties of hadrons. In this model we can switch on and off hadronic decay and study what happens to the pole positions in S(E). In the limit of no coupling all bound states and resonances are at the real axis (fig. 2.1), these positions correspond to the so-called bare hadron spectrum.

In this conference George Rupp presented the numerical solution of such a many coupled permanently closed and scattering channel Schrödinger equation [13]. Our main aim was to find the characteristics of the infinitely rising potential in the confinement sector.

Our project is now well defined: find for some $H_{q\bar{q}}$, H_{decay}, and H_{MM} such poles in S(E) which fit the physical resonances and bound states of the mesons. Then switch off the transition potential H_{decay} and see what happens to the pole positions. This way we obtain the bare spectrum of $H_{q\bar{q}}$ which should also be described by the gluonsector of the QCD Lagrangian if this sector is responsible for confinement.

As a result we found that a flavor mass dependent harmonic oscillator with universal frequency can perfectly serve for the description of the confinement force quarks feel at intermediate distances:

$$V(r) = \frac{1}{2} \mu_{q\bar{q}} \omega^2 r^2 \tag{2.4}$$

where r is the distance between the quarks, where $\mu_{q\bar{q}}$ is the reduced mass of the effective quark masses and where ω is the frequency of the harmonic oscillator force which turns out to be the same (about 0.2 GeV) for the light mesons (masses of about 1 GeV) as for the other mesons, even the very heavy mesons (masses of about 10 GeV).

3. Results

Let us take in $H_{q\bar{q}}$ the harmonic oscillator (2.4) for perfect confinement. Let us take no final state interactions in H_{MM} just for simplicity. For H_{decay} we use the $^3P_0$ pair creation model [14] with an overall strength parameter g.

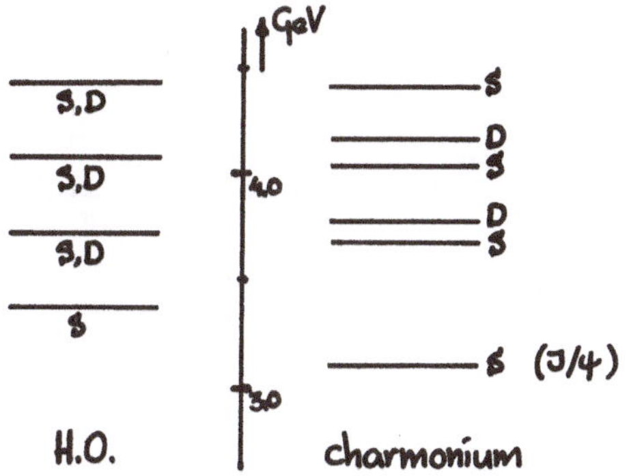

Figure 3.1: Harmonic oscillator spectrum versus charmonium states.

In figure 3.1 we compare a harmonic oscillator spectrum with the $J^{PC} = 1^{--}$ charmonium states [6]. We might conclude that there is not too much similarity. In [13] are depicted the real parts of the pole positions for several values of the strength of the transition potential. From that figure we conclude that it is very well possible to obtain the physical charmonium spectrum from a bare harmonic oscillator spectrum. Moreover we notice the remarkable feature of our model that the level spacing between the ground state and the first radial excitation in the "physical" spectrum is enlarged with respect to the same spacing in the bare spectrum. In figure 3.2 we see that many physical spectra of radial excitations exhibit this feature [6, 15].

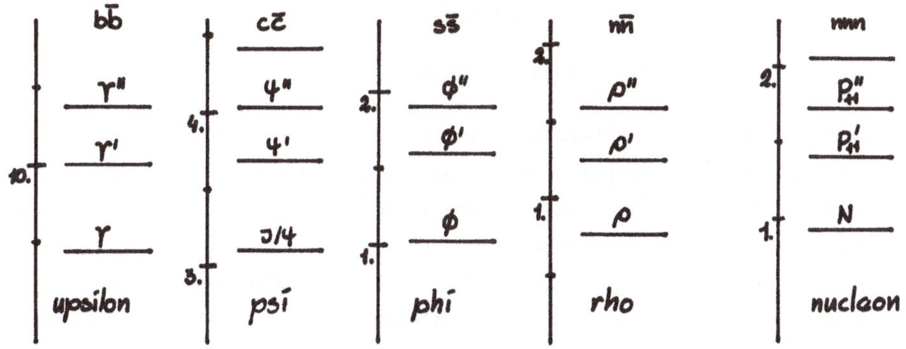

Figure 3.2: The experimentally available radial excitations of some hadrons.

A more extended form of the model of eqs (2.1 - 2.3) is described in [12]. The parameters of the model (quark masses, oscillator frequency ω, coupling strength g and one parameter for the so-called color splitting between pseudoscalar and vector mesons) are fixed in an overall fit to the central positions of bound states and resonances of pseudoscalar and vector mesons.

Let me briefly comment the results:

1. For bound states we can easily compute the wave functions and via the Van Royen-Weisskopf formula [16], the leptonic widths. Good results.

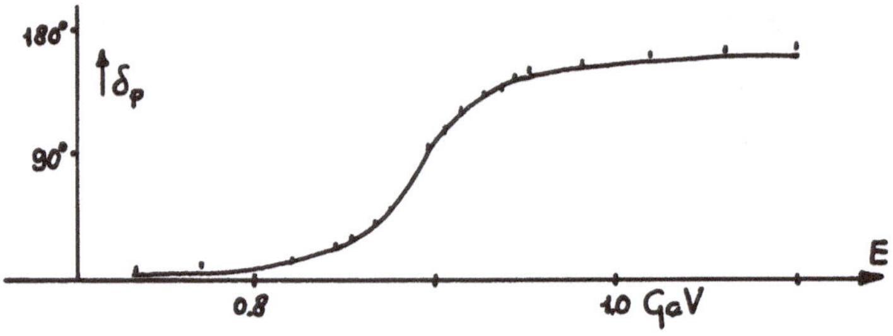

Figure 3.3: P-wave $K\pi$ elastic scattering phaseshifts.

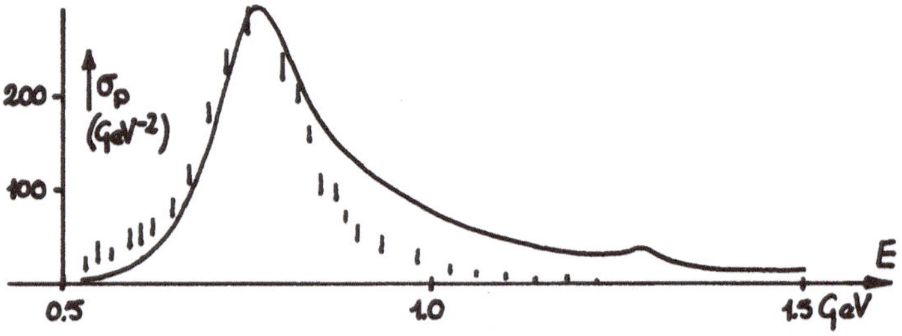

Figure 3.4: P-wave $\pi\pi$ elastic scattering cross sections.

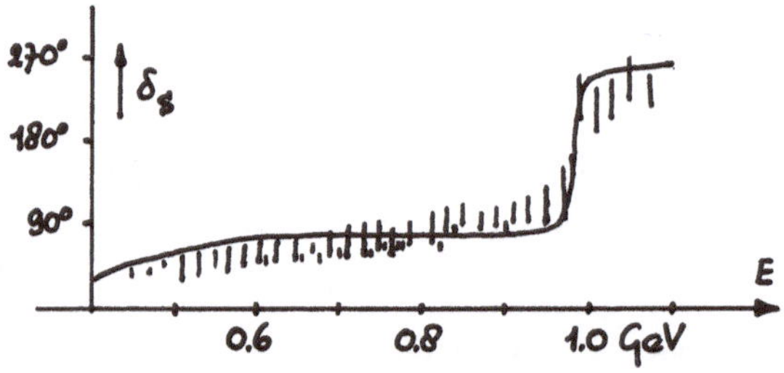

Figure 3.5: S-wave $\pi\pi$ elastic scattering phaseshifts.

2. For resonances we can compute scattering properties in the scattering channels. In the figures 3.3, 3.4 and 3.5 we have compared the results (remember: no final state interactions in the present version of our model!) with the available data. The solid lines are the results of our model calculations; the data are taken from [17].

3. On the influence of three or more particle decays, the following remarks can be made: Although up till now we have not done a complete calculation, we have some information available on this point, dealing with the ω (785) meson. This resonance can only decay into

three pions. This decay can be explained by treating the ω as a resonance in the virtual $\rho\pi$ scattering process. The cascade of $\rho\pi \rightarrow 3\pi$, can for this resonance be imitated by giving the ρ meson mass in the $\rho\pi$ scattering channel an imaginary part, which immediately amounts in the right width for ω meson. In this case we also found that turning on the width of the ρ mass leads only to a small shift in the resonance position of the ω meson, which demonstrates the point that the three particle decay channel has not much influence, albeit essential for the width of the ω meson. That it is reasonable to assume that the ω meson decays via the above described virtual process, can be checked at the ϕ (1020) meson, which contains a small admixture of a pure isosinglet $n\bar{n}$ state and therefore has a small width into three pion decay. Here it is known that more than 80% of this decay goes via the (in this case real) $\rho\pi$ two particle decay [6]. This example we generalize to all cases, concluding that it is safe to neglect three or more particle decay for a first rough description of a hadron.

4. There is another surprise which should be mentioned in this talk, namely that the bare spectrum not only deviates from the physical spectrum in the respect that they lead to different "masses", but even that the number of states might differ for both spectra. An example is the $I^GJ^{PC} = 0^+0^{++}$ resonance at .5 GeV, which as the σ meson disappeared from the particle properties tables already long ago [6]. This resonance can easily be explained as a pole in the S-matrix for $\pi\pi$ scattering in the model of [12]. The bare spectrum has a pole at about 1.3 GeV. If we turn on the couplings to the scattering channels to the values which give the best overall fit for the mesons, we do not find one but two poles in the complex energy plane, at about 0.5 GeV and 1.3 GeV central values. One has its origin at infinite negative imaginary part in the energy, the other comes from the bare spectrum. But which one is connected to which one of the two starting positions depends on the way we turn on the coupling constants. Unfortunately this is not a physical process, because only the end points represent something measurable and we can not turn on and off coupling constants in nature. So we have to conclude that both poles in the physical spectrum stem from one and the same pole in the bare spectrum.

4. The pole position as a function of the transition potential

In our model it is easy to study the behavior of poles in the complex energy plane. For instance in the neighborhood of the lowest threshold there is a difference in behavior of S-wave poles and P-, D-, and higher wave poles. If we increase the strength of the decay potential in the simplest version of the model, i.e. with one permanently closed and one scattering channel, the poles move as follows: for S-waves the resonance pole (R) and its compagnon (C) meet in the complex k-plane at the negative imaginary axis and continue along the imaginary axis, first both forming a virtual bound state but then one of them passes to the first Riemann sheet ($\mathrm{Im}\, k > 0$) to become a real bound state (B). The other pole continues to $\mathrm{Im}\, k = -\infty$ remaining a virtual bound state (V). Their trajectories are shown in figure 4.1.

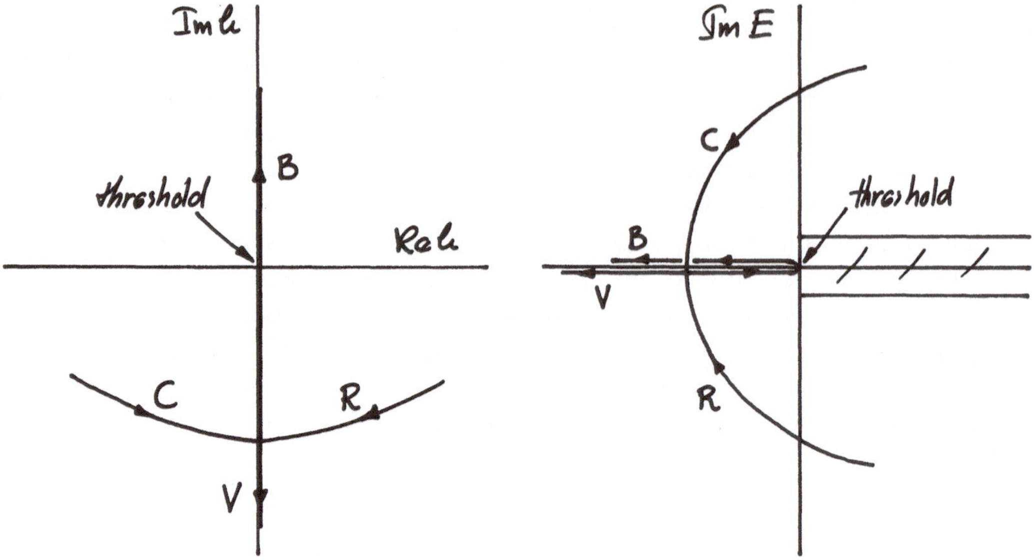

Figure 4.1: Trajectories of S-wave elastic scattering poles near threshold in the complex k- and E-planes.

For P-, D-, and higher waves the resonance pole (R) and its compagnon (C) meet at threshold and then continue along the imaginary k-axis, one pole forming a bound state (B), the other pole a virtual bound state (V). In figure 4.2 the situations in the complex k- and E-planes are shown.

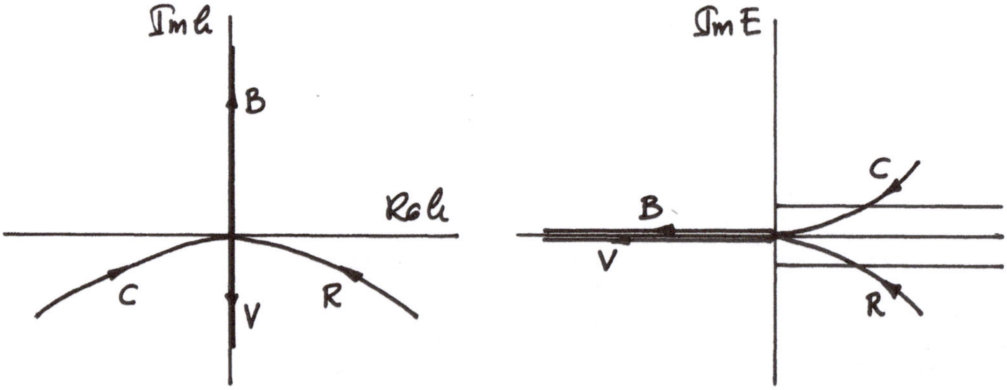

Figure 4.2: Trajectories of P-, D-, and higher wave elastic scattering poles near threshold in the complex k- and E-planes.

An example of the latter type of poles is the ψ (3680) state which is a bound state in $D\bar{D}$ P-wave scattering. For our model parameters (m_c = 1.6 GeV, ω = 0.19 GeV) the harmonic oscillator eigen-energy for the first radial excitation is at

$$E = \omega\,(2 + \frac{3}{2}) + m_c + m_{\bar{c}} = 3.83 \text{ GeV} \quad , \tag{4.1}$$

which is about 100 MeV above the $D\bar{D}$ threshold at $m_D + m_{\bar{D}} = 3.73$ GeV. For a very small decay potential (2.3) the pole in the lower half complex energy plane is very close to this value (see figure 4.3). But if we increase the transition potential (2.3) then the pole moves towards the $D\bar{D}$ threshold and becomes a bound state for the final model parameters.

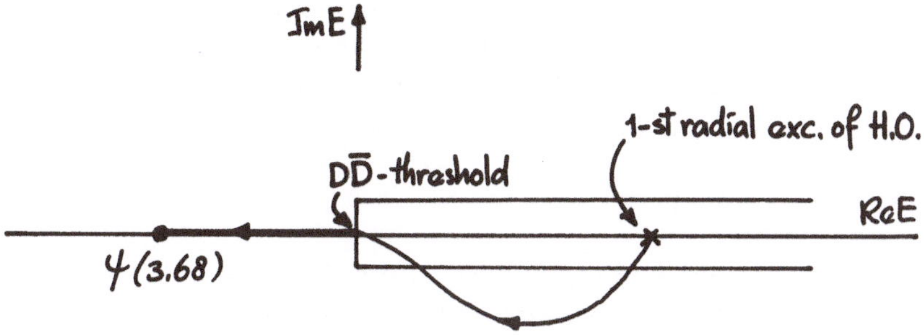

Figure 4.3: Relation between ψ (3.68) bound state pole and the bare $c\bar{c}$ spectrum.

A nice example of the S-wave poles near threshold is the scalar meson S* (990) which is just below the $K\bar{K}$ threshold with a small imaginary part. Unfortunately however, $K\bar{K}$ is not the lowest threshold in this case, since the S* (990) also couples to $\pi\pi$. In fact this resonance is found as a sharp peak in $\pi\pi$ scattering [6]. In figure 4.4 the trajectory of the S* pole is shown for the case that only the transition potential for decay to the $K\bar{K}$ channel is varied around the model parameters. We see that the pole changes Riemann sheet above threshold.

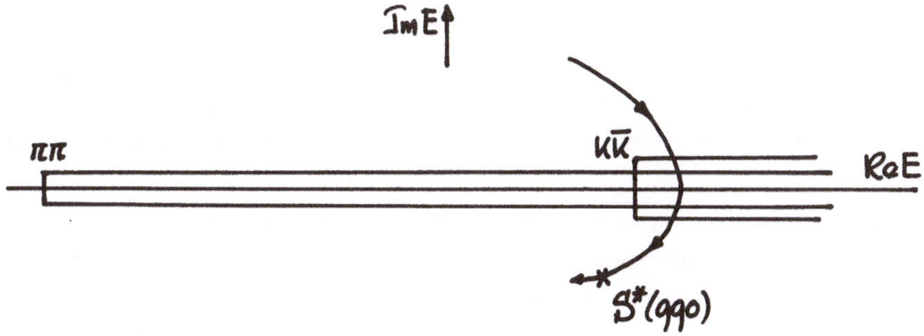

Figure 4.4: Behavior of the S* (990) resonance pole as a function of the strength of the transition potential to $K\bar{K}$.

Unfortunately, the strength of the transition potential (2.3) cannot (yet?) be varied in nature so none of the above results could be verified experimentally.

5. The confinement sector

It is nowadays generally accepted that bare hadrons are relativistic composite systen of quarks, antiquarks and gluons. The gluons and virtual quark-antiquark pairs form a kinc of medium in which the flavor quarks and/or antiquarks are permanently confined to each other. Can such a complicated system be described by simple equations of motion for the flavor partons? And when so, can these equations of motion be generated from a theory for the strong interacting fields?

First we would like to go beyond the non-relativistic equation of motion (2.4), whicł is an easy task because the relativistic analogue of (2.4) is the relativistic harmonic oscillator, which we for our purpose have written in the form [18]

$$H = \left\{ p^2 - \omega^2 (\vec{r}\cdot\vec{p})^2 + \frac{m^2}{1-\omega^2 r^2} \right\}^{1/2} \quad . \tag{5.1}$$

The Hamiltonian (5.1) describes a finite spherical system in space with radius $R = \omega^{-1}$ similar to the MIT-bag [19], but with the advantage that all angular excitations are allowed. Moreover the Hamiltonian (5.1) can be written in a more compact form

$$g_{\mu\nu}\, p^\mu\, p^\nu = m^2 \quad , \tag{5.2}$$

for $g_{\mu\nu}$ given by

$$g_{00} = \frac{1}{1-\omega^2 r^2} \quad , \qquad g_{0i} = g_{i0} = 0 \quad (i = 1,\, 2,\, 3) \quad ,$$

$$g_{ij} = \frac{1}{1-\omega^2 r^2} \left\{ \eta_{ij} - \frac{\omega^2 x^i x^j}{1-\omega^2 r^2} \right\} \qquad (i,j = 1,\, 2,\, 3) \quad . \tag{5.3}$$

Equation (5.2) is just the Klein-Gordon equation for a curved space.

So our picture is now that the medium of gluons and virtual $q\bar{q}$ pairs acts as a background metrical field, which only couples to the strong interacting fields and does not influence the motion of e.g. electrons and photons. In [20] we have written an equation similar to (5.1) for the fermions and solved it. The resulting spectrum is equidistant wit level spacings ω and the wave functions are confined to a finite region in space.

Let us next come to the second question. It is generally assumed that the interaction between the quarks and the gluons are described correctly by Quantum Chromo Dynamics (QCD) [21]. The QCD Lagrangian is supposed to give a good description of deep inelastic scatteri phenomena. Here the interaction takes place during a very short time interval and with ver high momentum transfer, so that one can use the asymptotic freedom property of QCD [22] in justifying perturbation theory calculations. In this realm the binding between the quarks can be neglected in first instance.

The description of quarks inside a hadron is much more complicated. The interactions responsible for the quark binding (and confinement) are likely to be dominated by small an moderate momentum transfer processes and clearly perturbation theory is not applicable.

The derivation of the forces between quarks for all momentum transfers or all distances in the framework of QCD is a formidable program and far from completed at presen

Although it seems safe to conclude that for constituent quarks at large distances the potential is linear, the form at intermediate and small distances is not established theoretically. In this situation it is opportune for phenomenological reasons to reformulate the theory in such a way that confinement is built in a priori by applying boundary conditions [19] and/or choosing a phenomenological confining potential [10, 23].

Another notable suggestion has been to describe the gluon sector, which is generally believed to be responsible for quark confinement, in terms of a color dielectricum. One assumes that due to its non-abelian character the QCD gauge field creates as a nonperturbative effect at low energies a very special chromodielectricum which has the property that it is only nonzero in a small domain of space around quarks, representing the hadron. The chromodielectricum function $\varepsilon(\vec{r})$ is nonzero inside this domain, tends to zero at the boundaries, and is zero outside. To describe the dielectricum one usually introduces a scalar field, say $\phi(x)$ [24, 25]. If $Z(\phi)$ denotes the renormalization factor in front of the kinetic term of the gluon sector in the QCD Lagrangian, then $\varepsilon(\vec{r})$ is proportional to $Z(\phi(\vec{r}))$ [24, 26, 27], leading to the condition that the scalar field should also vanish at the boundaries of the system.

The coupling of this scalar field to quarks and gluons is a delicate matter for which several proposals can be made. The most usual one is a Yukawa coupling. However, in a nonperturbative approach of QCD, Nielsen and Patkós [28] make it plausible that gauge invariance together with the introduction of an effective field for the description of the chromodielectricum gives rise to an interaction term of the form

$$\bar{\psi}\ \phi\ \gamma^{\mu}\ \partial_{\mu}\ \psi$$

in the Lagrangian. The resulting Dirac equation contains $\phi(x)$ acting as a tetrad field, similar to the one we have introduced some years ago [20]. There we assumed SO(3,2) symmetry for a baglike hadron model. The reason is that this symmetry is supported by nature because its associated spectrum is flavor mass independent as are the physical spectra, and is also equidistant. In the non-relativistic Schrödinger model (2.3) we have demonstrated that an equidistant bare spectrum in a natural way can lead to the physical spectra once the influence of hadronic decay has been taken into account.

If we restrict ourselves to the gluon sector we know that the Lagrangian is naively conformal invariant. It is however, well-known that conformal symmetry is severely broken in nature and in field theory as well. In the work of Fubini [29] it is shown that breaking to an SO(3,2) symmetric solution is possible, if one starts from a conformal invariant Lagrangian for a scalar field σ. This is not at variance with field theory as it has been shown that indeed SO(3,2) is the maximum allowed symmetry, which can survive radiation corrections [30].

In [31] we assume that the Lagrangian of the scalar field σ which is to be associated with the chromodielectricum, has the naive symmetry of the gluon sector. Furtheron, we assume that the SO(3,2) symmetric solution for σ is relevant for the description of hadrons. In conformal coordinates the Fubini solution has singularities in space and time. For the description of confinement we use the central projection coordinates [20] for which the singularities become spacelike and static, defining a closed localized domain in space. The spatial dependence of the SO(3,2) invariant scalar field σ, motivates us to make the Ansatz that $\sigma(x)$ is proportional to $1/\phi(x)$. Our study of σ leads to a geometrical picture of quark confinement.

We assume that broken conformal symmetry is a valuable framework to understand the hadronic spectra. In our choice we are left with anti-De Sitter symmetry which does not contain Poincaré symmetry. Note however, that the broken O(4,2) group describes part of the internal structure of a hadron, rather than the whole world. Our σ solution which represents the gluon sector of a localized system like a hadron, by itself does not need to preserve Poincaré symmetry.

In [31] we describe in detail how the coupling of photons to the scalar field σ via a tensor $t_{\mu\nu}$ leads to the above mentioned central projection. The tensor $t_{\mu\nu}$ can be considered as the photon metric and becomes Minkowskian (i.e. diagonal 1, -1, -1, -1) under the central projection. This way we obtain two metrical fields: one curved metric t which the strong interacting fields are coupled and a flat metric which couples to electromagnetism. In this talk I wish to discuss the connection between the strong curvature and the colordielectricum.

6. Curvature and the electromagnetic properties of a medium

In a curved space the inhomogeneous Maxwell equations read

$$[\sqrt{-g(x)}\, F^{\mu\nu}(x)]_{,\mu} = -\sqrt{-g(x)}\, J^{\nu}(x) \quad . \tag{6.1}$$

In flat inertial coordinates the similar equations read

$$\nabla\cdot\vec{D}(\vec{x},t) = J^{0}(\vec{x},t) \quad \text{and} \quad \nabla\times\vec{H}(\vec{x},t) - \frac{\partial\vec{D}(\vec{x},t)}{\partial t} = \vec{J}(\vec{x},t) \quad . \tag{6.1}$$

If we identify the conserved currents $J^{\mu}(\vec{x},t)$ from (6.1b) and $-\sqrt{-g(x)}\, J^{\mu}(x)$ from (6.1a) w obtain the following identities:

$$D_i(\vec{x},t) = -\sqrt{-g(x)}\, F^{0i}(x) \quad \text{and} \quad H_i(\vec{x},t) = -\frac{1}{2}\sqrt{-g(x)}\, \varepsilon_{ijk}\, F^{jk}(x) \quad . \tag{6.2}$$

The homogeneous Maxwell equations read

(i) in curvilinear coordinates

$$F_{\mu\nu,\sigma}(x) + F_{\nu\sigma,\mu}(x) + F_{\sigma\mu,\nu}(x) = 0 \quad , \tag{6.3}$$

(ii) in flat coordinates

$$\nabla\cdot\vec{B}(\vec{x},t) = 0 \quad \text{and} \quad \nabla\times\vec{E}(\vec{x},t) + \frac{\partial\vec{B}(\vec{x},t)}{\partial t} = 0 \quad . \tag{6.3}$$

Comparison of (6.3a) with (6.3b) leads to the identifications

$$E_i(\vec{x},t) = F_{0i}(x) \quad \text{and} \quad B_i(\vec{x},t) = -\frac{1}{2}\varepsilon_{ijk}\, F_{jk}(x) \quad . \tag{6.4}$$

One might easily check that the identifications (6.2) and (6.4) lead to the identificatio of the Lagrangians

(i) in curvilinear coordinates

$$\mathcal{L}(x) = -\frac{1}{4}\sqrt{-g(x)}\, F^{\mu\nu}(x)\, F_{\mu\nu}(x) \quad , \tag{6.5a}$$

(ii) in flat coordinates

$$\mathcal{L}(\vec{x},t) = \frac{1}{2}\{\vec{E}(\vec{x},t)\cdot\vec{D}(\vec{x},t) - \vec{H}(\vec{x},t)\cdot\vec{B}(\vec{x},t)\} \quad . \tag{6.5b}$$

With (6.4) and (6.2) we obtain a relation between $\vec{E}(\vec{x},t)$, $\vec{B}(\vec{x},t)$, $\vec{D}(\vec{x},t)$ and $\vec{H}(\vec{x},t)$:

$$D_i(\vec{x},t) = -\sqrt{-g(x)}\,\{g^{00}(x)\, g^{ij}(x)\, E_j(\vec{x},t) - g^{0j}(x)\, g^{i0}(x)\, E_j(\vec{x},t) - \frac{1}{2}\, g^{0j}(x)\, g^{ik}(x)\, \varepsilon_{jk\ell}\, B_\ell(\vec{x},t)\} \tag{6.6}$$

and similar for $\vec{H}(\vec{x},t)$.

In the lowest order approximation of the strong interaction Lagrangian which is given in [31], we have chosen a stationary metric $g(\vec{x},t) = g(\vec{x})$ for which $g_{i0}(x) = g_{0i}(x) = 0$. In that case (6.6) reduces to

$$D_i(\vec{x},t) = -\sqrt{-g(\vec{x})}\, g^{00}(\vec{x})\, g^{ij}(\vec{x})\, E_j(\vec{x},t) \quad , \tag{6.7a}$$

and for $\vec{H}(\vec{x},t)$ and $\vec{B}(\vec{x},t)$ we obtain similarly

$$B_i(\vec{x},t) = -\sqrt{-g(\vec{x})}\, g^{00}(\vec{x})\, g^{ij}(\vec{x})\, H_j(\vec{x},t) \quad . \tag{6.7b}$$

So if we interpret the metrical description of [31] in terms of the color electromagnetic properties of the hadronic medium, then we obtain for the color dielectric constant $\varepsilon_{ij}(\vec{x})$ and the magnetic susceptibility

$$\varepsilon_{ij}(\vec{x}) = \mu_{ij}(\vec{x}) = -\sqrt{g(\vec{x})}\, g^{00}(\vec{x})\, g^{ij}(\vec{x}) \quad . \tag{6.8}$$

The result (6.8) might not be very useful in gravity, because there we do not have an inertial frame, but in our description of a hadron where we do have an inertial frame defined by ordinary electromagnetism it might be very useful.

For the specific geometry (5.3) the color electromagnetic constants (6.8) read

$$\varepsilon_{ij}(\vec{x}) = \mu_{ij}(\vec{x}) = (1 - \omega^2 r^2)^{-1/2}\,\{\delta_{ij} - \omega^2 x_i x_j\} \quad . \tag{6.9}$$

Let me here briefly discuss the properties of the medium (6.9) for the classical abelian field.

7. The light velocity

In classical abelian electromagnetism the velocity of light is given by the Fresnel equation [32]. This is in general a very complicated equation, but we might use a local approximation, which for the medium (6.9) reads

$$\{1 - (\delta_{ij} - \omega^2 x_i x_j)\, n_i n_j\}^2 = 0 \quad . \tag{7.1}$$

For light which moves on the z-axis ($x_1 = x_2 = 0$) in the direction of the z-axis ($n_1 = n_2 = 0$) we obtain from (7.1) the equation

$$\{1 - (1-\omega^2 z^2) n_z^2\}^2 = 0 \quad . \tag{7.2}$$

Equation (7.2) is solved by

$$v_z = \frac{1}{n_z} = \sqrt{1-\omega^2 z^2} \quad , \tag{7.3}$$

which describes harmonically oscillating light along the z-axis, which just touches the surface of the sphere and with frequency 1/R.

For light which moves on the z-axis perpendicular to the z-axis (e.g. $n_2 = n_3 = 0$) (7.1) reduces to

$$\{1 - n_x^2\} = 0 \quad . \tag{7.4}$$

In this case we deal with light with the velocity 1, also with respect to the inertial coordinates.

In general the motion of light can be decomposed into three oscillations one of which just touches the surface of the sphere, which is precisely the motion found by the geometrical method [18].

8. The one "gluon" exchange potential

For a static abelian charge q at the centre of the hadron described by the dielectricum (6.9) we can use spherical symmetry to solve the Maxwell equations (6.1b and 6.3b). The solution in terms of a potential reads

$$\phi(r) = -(1-\omega^2 r^2)^{1/2} \frac{q}{4\pi r} \quad . \tag{8.1}$$

It is interesting to study the behavior of the potential (8.1) for small values of the distance r:

$$\phi(r) \xrightarrow[r\to 0]{} -\frac{q}{4\pi r} + \frac{q\omega^2}{8\pi} r \quad . \tag{8.2}$$

We obtain a Coulombic part and a linear part for this potential near the origin. This is precisely the form of the potential which is used in many non-relativistic quark models [33] and might explain why these models are so successful. In our description of a bare hadron this potential comes out as the first order correction to the harmonic oscillator force which still dominates at intermediate quark distances.

9. Conclusions and outlook

Hadron models which are derived from the gluon sector of the QCD Lagrangian should be tested at the bare hadrons rather than at the physical hadrons. This might for example lead to totally different types of quarkonium potentials.

Guided by the observation that bare hadron spectra are compatible with those of harmonic oscillators, we have constructed a geometrical model [18, 20, 31] for bare hadrons

In [31] we use spontaneous symmetry breaking of the naive conformal symmetry of the gluon sector of the QCD Lagrangian to SO(3) x SO(2), for the description of strong interactions in the energy domain of hadrons. Besides gluon and quark fields also two Higgs fields are introduced in an effective conformally invariant Lagrangian, which in an approximation can serve as the origin of the geometrical model [20] and its derived dielectricum. For heavy constituent quarks this boils down to a flavor dependent harmonic oscillator potential in a multi channel Schrödinger equation. This gives a fairly good description for the radial spectra of charmonium and beautonium, and to our surprise also for the light quarkonia and baryons.

Acknowledgements

I wish to thank Dr. L. Ferreira and Prof.dr. L. Streit for kindly inviting me to this conference and the Zentrum für Interdisziplinäre Forschung for the hospitality during my stay in Bielefeld.

References

1. G.S. Larue, W.M. Fairbank and A.F. Hebard, Phys.Rev. Lett. 38, 1011 (1977); G.S. Larue, W.M. Fairbank and J.D. Phillips, Phys.Rev.Lett. 42, 142, 1019 (E) (1979); G.S. Larue, J.D. Phillips and W.M. Fairbank, Phys.Rev.Lett. 46, 967 (1981).
2. P.J. Mulders, A.T. Aerts and J.J. de Swart, Phys.Rev. D 19, 2635 (1979); D 21, 1370, 2653 (1980).
3. M. Gell-Mann, Phys.Lett. 8, 214 (1964); G. Zweig, CERN-report no. TH-401 and TH-412 (1964); J.J.J. Kokkedee, The Quark Model (W.A. Benjamin Inc., New York, 1969).
4. R.L. Jaffe, Phys.Rev. D 15, 276, 287 (1977).
5. N.A. Törnqvist, Phys.Rev.Lett. 49, 624 (1982).
6. Particle Data Group, Phys.Lett. 111B (1982).
7. C. Dullemond and E. van Beveren, Ann.Phys. (NY) 105, 318 (1977).
8. R.F. Dashen, J.B. Healy and I.J. Muzinich, Ann.Phys. (NY) 102, 1 (1976); Phys.Rev. D 14, 2773 (1976).
9. For many references to the literature, see e.g. A. Martin, Phys.Lett. 93B, 338 (1980).
10. N. Isgur and G. Karl, Phys.Lett. 72B, 109 (1977); Phys.Rev. D 18, 4187 (1978); Phys. Lett. 74B, 353 (1978); Phys.Rev. D 19, 2653 (1979); Phys.Rev. D 20, 1191 (1979).
11. C. Dullemond, T.A. Rijken, E. van Beveren and G. Rupp, in: Proceedings of the VI Warsaw Symposium on Elementary Particle Physics, Kazimierz (Poland), 257 (1983); E. van Beveren, C. Dullemond and T.A. Rijken, Z.Phys. C 19, 275 (1983).
12. E. van Beveren, C. Dullemond and G. Rupp, Phys.Rev. D 21, 772 (1980); E. van Beveren, G. Rupp, T.A. Rijken and C. Dullemond, Phys.Rev. D 27, 1527 (1983); E. van Beveren, Z.Phys. C 17, 135 (1983); Z.Phys. C 21, 291 (1984); C. Dullemond, G. Rupp, T.A. Rijken and E. van Beveren, Comp.Phys.Comm. 27, 377 (1982); G. Rupp, Spectra and decay properties of pseudo-scalar and vector mesons in a multi-channel quark model (thesis, Nijmegen, 1982); E. van Beveren, The influence of strong decay on the spectra of hadrons (thesis, Nijmegen, 1983).
13. E. van Beveren, C. Dullemond, T.A. Rijken and G. Rupp, invited talk presented by G. Rupp in this conference, figure 3.

14. L. Micu, Nucl.Phys. B10, 521 (1969);
R. Carlitz, M. Kislinger, Phys.Rev. D 2, 336 (1970);
A. Le Yaouanc, L. Oliver, O. Pène, J.-C. Raynal, Phys.Rev. D 8, 2223 (1973);
M. Chaichian, R. Kögerler, Ann.Phys. (NY) 124, 61 (1980).
15. N.M. Budnev, V.M. Budnev and V.V. Serebryakov, Phys.Lett. 70B, 365 (1977);
D. Aston et al., Phys.Lett. 92B, 211, 215, 219 (1980);
F. Mané et al., Phys.Lett. 99B, 261 (1981);
G. Höhler, Handbook of pion-nucleon scattering, (Physics Data 12 - 1, Fachinformationzentrum, Karlsruhe, 1979).
16. R. Van Royen and V. Weisskopf, Nuov.Cim. 50A, 617 (1967).
17. P. Estabrooks and A.D. Martin, Nucl.Phys. B 79, 301 (1974);
P. Estabrooks et al., Nucl.Phys. B 133, 490 (1978);
N.N. Biswas et al., Phys.Rev.Lett. 47, 1378 (1981).
18. C. Dullemond and E. van Beveren, Phys.Rev. D 28, 1028 (1983).
19. A. Chodos, R.L. Jaffe, K. Johnson, C.B. Thorn, V.F. Weisskopf, Phys.Rev. D 9, 3471 (1974);
T.A. DeGrand, R.L. Jaffe, K. Johnson, J.J. Kiskis, Phys.Rev. D 12, 2060 (1975).
20. E. van Beveren, C. Dullemond and T.A. Rijken, Nijmegen report no. THEF-NYM-79.11 (1979) (unpublished).
21. H. Fritzsch and M. Gell-Mann, in: Proceedings of the VIth International Conference on High Energy Physics, Chicago, vol. II, 135 (1972).
22. H.D. Politzer, Phys.Rep. 14 C, 129 (1974);
D. Gross and F. Wilczek, Phys.Rev.Lett. 30, 1343 (1973);
S. Coleman and D. Gross, Phys.Rev.Lett. 31, 851 (1973).
23. W.A. Bardeen, M.S. Chanowitz, S.D. Drell, M. Weinstein and T.-M. Yan, Phys.Rev. D 11, 1094 (1975).
24. J.B. Kogut and L. Susskind, Phys.Rev. D 9, 3501 (1974).
25. T.D. Lee, Phys.Rev. D 19, 1802 (1979).
26. G. 't Hooft, in: Proceedings of the Marseille Conference on Recent Progress in Lagrangian Field Theory and Applications (C. Korthals-Altes et al., eds; CNRS, Marseille, 1974); Nucl.Phys. B 138, 1 (1978).
27. P. Hasenfratz and J. Kuti, Phys.Rep. 40 C, 75 (1978).
28. H.B. Nielsen and A. Patkós, Nucl.Phys. B 195, 137 (1982).
29. S. Fubini, Nuov.Cim. 34A, 521 (1976);
V. de Alfaro, S. Fubini and S. Furlan, Nuov.Cim. 34A, 569 (1976).
30. I.T. Drummond, Nucl.Phys. B 94, 115 (1975).
31. C. Dullemond, T.A. Rijken and E. van Beveren, Nijmegen report no. THEF-NYM-83.02 (1983) (to be published in Nuov.Cim.).
32. L.D. Landau and E.M. Lifshitz, Electrodynamics of continuous media, chapter XI (Pergamon Press).
33. T. Appelquist, A. De Rujula, S.L. Glashow and H.D. Politzer, Phys.Rev.Lett. 34, 365 (1975);
R. Barbieri, R. Gatto, R. Kögerler and Z. Kunzst, Phys.Lett. 57 B, 445 (1975);
E. Eichten, K. Gottfried, K.D. Lane, T. Kinoshita and T.-M. Yan, Phys.Rev. D 17, 3090 (1978); D 21, 313 (E) (1980); D 21, 203 (1980);
J.S. Kang and H.J. Schnitzer, Phys.Rev. D 12, 841 (1975);
J. Pumplin, W. Repko and A. Sato, Phys.Rev.Lett. 35, 1538 (1975);
J.L. Richardson, Phys.Lett. 82B, 272 (1979);
J. Rafelski and R.D. Viollier, in: Fundamental interactions and structure of matter, Proceedings of the Second Course of the International School of Physics of Exotic Atoms, Erice, Sicily, Italy (March 25 - April 5, 1979), 101 (N.Y., Plenum, 1980).

MODEL-INDEPENDENT DETERMINATION OF RESONANCE PARAMETERS FOR REACTIONS INVOLVING ONLY ZERO-SPIN PARTICLES

Z. Basrak*, F. Auger+, P. Charles+,
W. Tiereth** and H. Voit**

*Rudjer Bošković Institute, POB 1016, 41001 Zagreb, Yugoslavia
+DPhN, CEN Saclay, BP 2, 91190 Gif-sur-Yvette, France
**Physikalisches Institut der Universität Erlangen-Nürnberg, 8520 Erlangen, FR Germany

The identification of a resonance and the determination of its parameters is a rather difficult task if the resonant state interacts nonnegligibly with the background and/or when one faces strongly overlapping resonances. The phase-shift analysis is the usual approach to the search for resonances and to the determination of their quantum numbers, because it is solely based on the most general conservation laws and is intimately related to experimental data.

For a two-body reaction involving spinless particles, the reaction amplitude $f(\theta)$ has a particularly simple form when expanded in Legendre polynomials, and the expression for the reaction differential cross section reads

$$\frac{d\sigma}{d\Omega}(\theta) = |f(\theta)|^2 = \left|\frac{1}{k}\sum_\ell (2\ell+1) f_\ell P_\ell(\cos\theta)\right|^2$$

$$= \left|\sum_\ell (A_\ell + iB_\ell) P_\ell(\cos\theta)\right|^2 . \qquad (1)$$

The ℓth partial-wave amplitude f_ℓ is usually parametrized in terms of the phase shift δ_ℓ and the inelasticity η_ℓ:

$$f_\ell = \eta_\ell \exp(2i\delta_\ell)/(2i) . \qquad (2)$$

For a truncated expansion ($A_\ell = B_\ell = 0$, for $\ell > L$), the model-independent determination of amplitudes f_ℓ from experimental data would mean a χ^2 best-fit search in the space with dimension 2L+1 (the overall phase is free)[1]. However, this procedure is tedious and, moreover, there are in general a number of acceptable minima of χ^2.

As any function of polar angle, the differential cross section can be directly expanded in Legendre polynomials:

$$\frac{d\sigma}{d\Omega}(\theta) = \sum_\ell c_\ell P_\ell(\cos\theta) \approx \sigma_L(x) = \sum_\ell^{2L} c_\ell P_\ell(x) , \qquad (3)$$

where C_ℓ are real and $x = \cos\theta$. The expansion is truncated at $\ell = 2L$, so that there are again 2L+1 independent real quantities to be determined in the χ^2 fitting procedure. However, for the linear expansion (3), the χ^2 problem can be solved analytically and has a unique solution $\{C_\ell\}$ [2]). We shall show that, starting from a set $\{C_\ell\}$, one is able to generate in a straightforward way all of the 2^L equivalent sets of partial-wave amplitudes $\{f_\ell\}$, all of them giving exactly the same cross section $\sigma_L(x)$ via the relation (1) [1,3]).

The function $\sigma_L(x)$ is a polynomial of degree 2L in x and can be factorized in terms of its 2L complex zeros z_i, allowing us to write

$$\sigma_L(x) = \sum_{\ell=0}^{2L} C_\ell x^\ell = C_{2L} \prod_{i=1}^{L} (x-z_i)(x-z_i^*)$$
$$= \left| \sqrt{C_{2L}} \prod_{i=1}^{L} (x-z_i) \right|^2 = |\phi_L(x)|^2 \;. \tag{4}$$

The complex function $\phi_L(x)$ (polynomial of degree L in x with complex coefficients) is simply the reaction amplitude $f(\theta)$ where the partial-wave expansion is truncated at $\ell = L$. It is an easy exercise to rearrange the expansion coefficients in order to derive the partial-wave amplitudes f_ℓ.

Following the above procedure, we not only avoid a tedious search for f_ℓ's by a direct χ^2 fit with expression (1), but also obtain an elegant way of generating all possible 2^L solutions of the phase-shift analysis. These solutions are derived from the starting solution $\phi_L(x)$ by consecutive conjugation of zeros $z_i \to z_i^*$.

The phase-shift analysis is greatly simplified when the entrance or exit reaction channel consists of identical bosons, since many solutions violate the symmetry requirement, i.e. have nonvanishing contributions from odd partial waves. Generally, the number of physical solutions is reduced to $2^{L/2}$.

From the way solutions are generated it is obvious that f_ℓ^* is coupled to each partial-wave amplitude f_ℓ (i.e. $\phi_L^*(x)$ is coupled to $\phi_L(x)$) by mere conjugation of all zeros. It follows that the modulus $|f_\ell|$ has half the ambiguity of f_ℓ and is not affected by phase ambiguity.

On the basis of the analyticity of the reaction amplitude and the fact that solutions are unambiguous below the inelastic threshold, one expects that only one solution will be truly physical. To remove the phase-shift ambiguity, one must examine the measured differential cross section in sufficiently fine energy steps. The most widely used criterion to distinguish between different solutions is the continuity criterion; one can impose this criterion as smoothness in energy dependence of partial-wave amplitudes or/and zero trajectories[4]). We have adopted the "three-points" shortest-path method for the choice among different connections between solutions belonging to three neighbouring energies E_o and $E_{\pm} = E_o \pm \Delta E$. The path function reads

$$d_{ij} = \sum_{\ell} \left(|f_{\ell}^{o}(E_-)-f_{\ell}^{i}(E_o)| + |f_{\ell}^{i}(E_o)-f_{\ell}^{j}(E_+)| \right) .$$

The indices i,j run over all solutions. The solution $\{f_{\ell}^{i_o}(E_o)\}$ which corresponds to the smallest value of the function d_{ij} is adopted as a unique physical solution obtained by means of the shortest-path method and becomes a new starting point $\{f_{\ell}^{o}(E_-)\}$, etc. The smoothness of zero trajectories was inspected by eye and helped to define a unique solution when a few d_{ij} values were close to the minimal one.

The validity of this procedure was tested on synthetic data of four partial waves ℓ = 0,2,4 and 6, displaying overlapping resonances in the presence of nonresonant background. The shortest-path method turned out to be a powerful tool for finding the right set of partial-wave amplitudes $\{f_{\ell}(E)\}$ up to a phase.

The procedure described above was applied to the $^{12}C(^{12}C,\alpha_o)^{20}Ne_{g.s.}$ reaction. Angular distributions were measured at 14 equally spaced angles and in the energy range $E_{c.m.}$ = 5.035-6.133 MeV, with a step of 50 keV. Using expansion (3) truncated at 2L = 8, the acceptable χ^2 fits were obtained at all energies, yielding two sets of solutions for the moduli of partial-wave amplitudes $\{|f_{\ell}|\}$ (open and full circles in fig.1). The result of shortest-path method is given by the heavy solid line in fig.1. On the basis of this analysis we were able to identify two new resonances at $E_{c.m.} \approx$ 5.3 and 5.5 MeV which coincide with structures found already in the angle integrated cross section of the $^{12}C(^{12}C,\alpha_o)^{20}Ne$ reaction [5]. The J values of these resonances are 4(2) and 0, respectively. Besides this, we were able to confirm all resonances (including their J values) reported already in the literature for the energy range studied. The discovery of a new 0^+ resonance is of particular interest for our understanding of the process of fragmentation of nuclear-molecular resonances, since it provides new bandhead for resonant bands already observed at higher angular momenta.

This work was supported in part by the Internationales Büro der KFA Jülich, Jülich, FR. Germany.

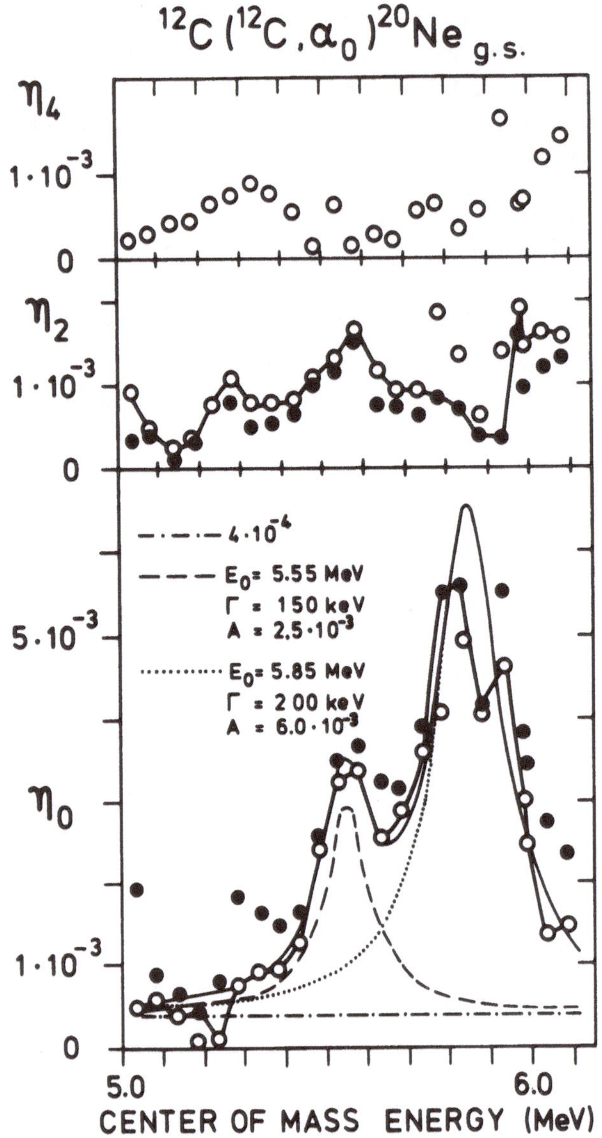

Fig.1 Inelasticity η_ℓ as a function of the energy for ℓ=0,2 and 4, respectively. The thin solid line is the result of a Lorentzian fit to the η_0 values (dashed and dotted curves;

$$\eta_0 = \frac{A\cdot(\Gamma/2)^2}{(E-E_0)^2+(\Gamma/2)^2}\ ),$$

and a small constant background (dashed-dotted line). The fitting parameters are given in the figure.

References

1. A. Gersten, Nucl.Phys. A219 (1974) 317.
2. G.E. Forsythe, J.Soc.Indust.Appl.Math. 5 (1957) 74.
3. N.P. Klepikov, Sov.Phys.JETP 14 (1962) 846.
4. F. Nichitiu, Sov.J.Part.Nucl. 12 (1981) 321.
5. W. Galster, W. Treu, P. Dück, H. Fröhlich and H. Voit, Phys,Rev. C15 (1977) 950.

SUMMARY OF THE CONFERENCE AND SOME OPEN PROBLEMS

Volker Enss
Institut für Mathematik I
Freie Universität Berlin
Arnimallee 2-6
D-1000 Berlin 33, W.-Germany

Resonance phenomena have been studied for several hundreds of years, but nevertheless they have not yet been fully understood. Rather than looking for a general explanation of "the" resonance it may well be more appropriate to distinguish between various kinds of resonance phenomena. The classical resonator in continuum mechanics has some proper frequencies. A small periodic exterior forcing with the same period will cause a very large response. A constructive phase relation between the driving force and the oscillations of the resonator allow to pump a lot of energy into the resonator. The energy per time is small but the available time is long. If the resonator is good i.e. the energy loss due to damping is small, then the effect may be stopped by spoiling the constructive phase relation: nonlinearities in the resonator typically cause an amplitude-dependence of the eigenfrequencies, or the system may disintegrate like the well-known example of crystal wine glasses broken by a violin sound. Also the possible instabilities of the approximately (quasi-)periodic motion in celestial mechanics are due to resonances (technically small denominators). Here no exterior perturbations are present, but almost independent subsystems influence each other.

In quantum mechanics the short time scale periodic behaviour is usually discussed in stationary terms of matching energies rather than frequencies. Here it should be possible to establish the correspondence between the different techniques and concepts. With regard to the long time behaviour, however, the attitudes are opposite. In classical mechanics a resonant perturbation helps to destroy an oscillating system, or to change its state drastically. A better match of the resonance parameter typically causes shorter times of approximate stability. On the other hand, in quantum physics a resonance has a long time of approximate stability compared to an otherwise unstable system. Here the "better" resonance is the one with longer lifetime before it decays eventually. It is an open question whether a further discussion of analogies and differences between classical and quantum physics will improve our understanding of resonances. From here on we will restrict ourselves to "quantum resonances", as it has been done at the conference.

We follow the talk of R.H. Dalitz with the programmatic title "What is a resonance" by starting with a phenomenological description of experimentally observed resonances. A first indication of a resonance is a "bump" of some observed quantity like a scattering cross section plotted against a parameter like the energy. The identification is simple if the bump is narrow, high, and well separated from other bumps, but often there are reasons for doubts: It may be difficult to distinguish a "bump"  (resonance) from a "step"  (threshold) if two of them are close (and of different height). If the energy resolution is not fine enough compared to the width and spacing of narrow resonances then the latter will not be seen and misinterpreted as a smooth background. It is a matter of feeling and experience where one draws the cutoff line between a (broad) bump and a variation of the background. In any event one needs some yardsticks which fix the scale for interpreting measured data in terms of resonances. This turns out to depend heavily on the type of interaction considered. This is particularly clear in lifetime measurements of metastable states. In high energy physics a system with a lifetime of 10^{-20} sec. can effectively be treated as stable. On the other hand, this lifetime is by several orders of magnitude too short for consideration as a resonance in atomic physics where the inverse Rydberg frequency sets the scale.

Therefore a couple of additional criteria were given in the talks of R.H. Dalitz and C. Mahaux to distinguish resonances: (a) they have to be visible in several scattering and reaction channels; (b) they should have definite values of the spin and possibly other quantum numbers; (c) in the Argand plot the data with increasing energy should be located on a circle counterclockwise. Often the available experimental data are not sufficient to perform all checks. In addition the reconstruction of scattering paramters from the data as needed for (c) is a difficult task which was treated in the seminar of A. Basrak.

It was impressive to see the myriads of very narrow densely spaced resonance peaks obtained from high energy resolution experiments in nuclear physics. This shows clearly that any model which approximately describes all these phenomena must be so complex that it is practically not solvable. As discussed by C. Mahaux one aims to construct a "dynamical", "microscopic" model using the available information from nuclear shell models etc. and use dynamical symmetries to eliminate unnecessary complexity. The evaluation of the model can be used to obtain statistical properties of resonances like the distributions of their energies, widths, etc.

In contrast to the latter the approach with so-called "structural methods" aims at the study of individual resonances. One then has to study simplified models where one or few particles move in a field obtained by averaging the effects of all other particles. G. Rupp and E. van Beveren presented the analysis of a model for interacting quarks and gluons. It is designed to explain both the hadron spectrum (which quasi bound states do exist?) and hadron decay on an equal footing.
A simple quasimolecule model of two touching balls (nuclei) which are glued together is amazingly successful to explain C^{12}-C^{12} scattering and reactions like $C^{12}(C^{12},\alpha)Ne^{20}$, as presented by N. Cindro. Starting from Faddeev equations with separable potentials A. Fonseca used methods similar to the Born-Oppenheimer approximation to construct an effective optical potential for the approximating two body model system.

The gap between high energy- and nuclear physics on one side and atomic and molecular physics on the other is bridged at several places. On the experimental side N. Cindro reported on observations of the Landau-Zener effect which is well-known in atomic and solid state physics. In addition, models, concepts, approximation schemes, and intuition for strongly interacting particles often are taken from nonrelativistic quantum mechanics. This was the starting point in Gamow's work in the early days of quantum mechanics.

For atoms and molecules, there is general agreement that a good description of the underlying dynamics is given by the Schrödinger equation. It is pleasant to have this solid foundation but it is of limited practical value. Most of the resonances in nature occur in multiparticle systems. Even if one knows the exact equations one cannot solve them. Only physical and mathematical insight lead to a reduction of the equations to a tractable size. A number of problems can be understood by dividing the system into two (or few) subsystems and studying the effective interaction between them. This enlarges the class of interesting pair potentials considerably.
Besides the basic potentials like the Coulomb- or Yukawa potential, etc. one encounters for instance multipole forces with power decay or the potential like

which leads to the celebrated "shape resonances".

But even these approximate models cannot be treated exactly and reliable approximations are necessary. H.J. Korsch showed us that a combination of the complex dilation methods with the semiclassical approximation allows to do calculations with phantastic precision. There are two very pleasing aspects of this: First, quantum system can be solved so precisely. Second, from the numerical point of view, we need not feel bad about the fact, that despite of 60 years of quantum mechanics we still often think of it as "classical mechanics + quantum corrections". Among the treated models are the usual (DC-)Stark effect and small perturbations of the quantum analogues of completely integrable systems. A complex procedure was presented by N. Moiseyev which gave good results for autoionization and other problems. He led us back to the "real" world by proposing a hermitian approach which is suitable for variational calculations.

The energy levels of shape resonances were analysed by R. Seiler with a "tunneling expansion". Inserting a Dirichlet boundary condition at the top of the "dividing hill" one obtains a bound state which is decaying in the original system due to tunneling. This yields a convergent power series expansion for the resonance energy in terms of a quantity which is exponentially small in $\hbar$ or the inverse mass. A family of Hamiltonians with potentials converging to point interactions was studied by F. Gesztesy. While the behaviour of bound states and of resonances is similar in various dimensions the methods of proof are different as a consequence of the various analyticity structures of the free resolvent. The existence or non-existence of resonances and bound states in atomic systems and properties of the wave functions were studied in the seminars of M. and T. Hoffmann-Ostenhoff.

The interaction of (bounded) electrons with photons were the topic of three talks. A. Maquet advocated to replace the hydrogen atom eigenfunctions by rescaled ones which give a complete set of eigenfunctions ("Sturmian expansion"). Multiphoton absorption of hydrogen in laser light has been calculated by evaluating the divergent series with Padé approximations; similarly for electrons which are scattered inelastically by charges in laser light. The dipole approximation for hydrogen-like atoms in a circularly polarized field was treated by A. Tip. The Hamiltonian is time independent in a rotating reference frame but then a term proportional to the angular momentum operator has to be added. Adaption of the complex dilation method to this situation yields good agreement with experimental data for multiphoton absorption. S. Graffi studied the genuinely time dependent problem of atoms in a linearly polarized electromagnetic field, the "AC-Stark effect". He managed to extend the efficient machinery of complex scaling to this case, including even Coulomb singularities. In particular he showed convergence of the expansion of resonance energies in terms of the field strength. It is known to be divergent in the DC-case.

A model for an idealized TV antenna was used by T.T. Wu to analyse resonances for electromagnetic waves. In three dimensions the Schrödinger operator can be constructed for an infinite linear periodic array of point interactions. It is an interesting question how the infinitely narrow resonances of the infinite array turn into usual resonances for the finite array. The three dimensional Maxwell equations, however, are analogous to a five dimensional Schrödinger equation where the construction of point interactions is not clear.

The analyticity which played a dominant rôle in concrete numerical calculations as well as in convergence proofs was the central object in the lecture of A. Grossmann. Besides the usual X- or P-space also representations related to certain two parameter subgroups of the canonical transformation group lend themselves naturally to complexification. Fascinating mathematical perspectives are the connection with the analysis of the hyperbolic half-plane, automorphic forms, etc.

Another impressive line of thought was in E. Brändas' lecture connecting numerical calculations of strings of resonances in the complex plane with the famous question of obtaining irreversible thermodynamic behaviour from reversible microscopic dynamics. The careful analysis of analytic continuation seems to shed some new light on this old problem.

Let us return to the question "what is a resonance", but this time from a mathematical point of view within the framework of quantum mechanics. R. Høegh-Krohn singled out the combination of the Hamiltonian with the position as a selected observable. If the integral kernel $(H-k^2)^{-1}(x,y)$ has an analytic continuation in k, then those k-values where the function has a pole for "all" x and y is defined to be a resonance. At a resonance value k_n the residue is expressed by a resonance function ψ_n as $\psi_n(x)\psi_n(y)$. For suitable functions of H and special values of x,y one obtains a completeness relation, e.g.

$$\exp(-i\ H_+t) = \Sigma \exp(-i\ k_n\ t)\ \psi_n(x)\ \psi_n(y) \quad \text{for} \quad |x-y| < t\ ,$$

where the sum runs over all resonance values k_n.

A different point of view has been stressed by E. Balslev. Two structures to be compared are the interacting and free Hamiltonians. For suitable potentials (exponentially decaying + dilation analytic) the resolvent as an operator has an analytic continuation, its poles are the resonances and there are resonance functions belonging to them. As a check that really resonances have been found one verifies that the analytically continued S-matrix has a pole as well. He has shown a reconstruction theorem: knowing a resonance function (even its modulus is sufficient) one can determine the potential!

Many *open problems* have been given in the lectures, they can be found in these proceedings. I restrict myself here to some problems related to the definition of a resonance. We begin by summing up a few of the lessons we have learned in the lectures.

a) Preferred objects of study are one-dimensional and rotationally symmetric systems. Certainly basic forces in nature like the Coulomb interaction are spherically symmetric, and resonances in nuclear and high energy physics seem to have definite spin values. The main reason for the restriction, however, seems to be technical: ordinary differential equations are simpler than PDE and one knows a lot more about them.

b) A key rôle is played by analyticity: analytic continuation of the resolvent or the S-matrix to the second ("unphysical") sheet and complex scaling methods. This approach provides the basis for most definitions of resonances and for most of the rigorous mathematical work about them within nonrelativistic quantum mechanics. In addition much of the numerical work on atomic and molecular systems is based on complex scaling. It gives very accurate results (compared to experiments) even where the mathematical foundation is not yet so firm. The complex dilation turns a resonance into a discrete eigenvalue problem for a non self-adjoint operator. For this case a good perturbation theory exists in contrast to the case of a continuous spectrum. A mathematically sound theory is available for potential scattering and is being developed for multiparticle system for potentials of the type "dilation analytic + exponentially decaying". For practical purposes this class seems to be large enough. Despite its overwhelming success we express the heretical view that a resonance is a much more basic phenomenon than the subtleties of analyticity. Therefore, we pose as an open question:

Define and study resonances using less analyticity (and less rotational symmetry).

Apparently related to this question is the following observation. Mathematicians tend to call any pole of the resolvent or S-matrix in the second sheet a resonance. Control of all these resonances may well be as subtle as analyticity. Physicists consider only poles "close" to the real axis as resonances, i.e. small width or long lifetime are required. A striking example is a positive Gaussian as a potential, $V(x) \sim \exp(-x^2)$; the Hamiltonian has infinitely many "mathematical" resonances but evidently it cannot have any "physical" ones [for references see Balslev]. If for a system the poles are known one may ask which ones lead to physically observable effects like bumps in measured curves. This class should be independent of subtleties. One cannot expect a sharp dividing line and nothing better than "small compared to" should be expected. The question is which quantities have to be compared. This is meant by:

Single out the physically relevant resonances among all possible ones.

A similar situation arises in statistical mechanics. For a large but finite system the interesting quantities depend analytically on the temperature and other parameters, but for some values the dependence is very "steep". For the infinite system analyticity may be lost and it is clear e.g. how to distinguish discontinuity (phase transition) from continuity. The thermodynamic limit which is independent of the particular system is the generally accepted procedure to turn quantitative differences into qualitative ones. We do not know of a counterpart to it in the study of resonances.

Many examples of resonances arise as "small" perturbations of systems with an eigenvalue embedded into the continuous spectrum. In the unperturbed system (found by physical insight or intuition) the transition between the eigenstate and the continuum is prohibited e.g. by a symmetry (Auger effect in atoms, strange elementary particles) or by a Dirichlet boundary condition (shape resonance). The perturbation (interaction between electrons, weak interaction; tunnel effect) turns the eigenvalue into a resonance. Instead of creating a family of operators by analytic continuation one "turns off" some interactions. In both cases one connects the Hamiltonian of interest by a one (or more) parameter family of operators to one with simpler properties. In the present situation the changes of the system are typically very drastic. The singular perturbations are small only on very special vectors. Thus a reliable perturbation theory is difficult. But a widespread conjecture is the positive answer to the following question:

Is there a natural procedure to find for a resonance a nearby system with embedded eigenvalue such that the perturbation can be controlled?

The problems in perturbation theory may be related to the observation that resonances are roughly speaking an "infinite dimensional phenomenon" in contrast to the one (or finite) dimensional eigenvalues reached by perturbation or continuation. Beyond the trivial statement that a continuous spectrum is possible only in infinite dimensional spaces we want to stress with this remark that it is hard to miss a resonance. If the energy resolution in a scattering experiment is good enough then the resonance is visible in quite different experimental setups and it shows up for various kinds of initial conditions. E.g. a resonance may be coupled to a whole subspace of states whith given angular momentum and energies in a suitable range.

It has been called "Howland's razor" by B. Simon that a resonance cannot be understood in terms of a single Hamiltonian operator alone on an abstract Hilbert space. Most often the additional structure used is the free Hamiltonian which generates a comparison dynamics and/or the position as a preferred observable. Also the behaviour in time is studied. For the case of bound states D. Ruelle gave a satisfactory characterization in terms of their space-time behaviour. A "geometric bound state" Ψ is localized uniformly in time, i.e. for any $\varepsilon > 0$ there is an $R(\varepsilon)$ such that

$$\sup_t \|F(|x| > R(\varepsilon)) \exp(-iHt)\Psi\| < \varepsilon .$$

Here $F(\cdot)$ denotes multiplication in x-space with the characteristic function of the indicated region. For almost any potential this characterization is equivalent to the traditional spectral theoretic one: bound states are superpositions of eigenvectors of H. The connection is by "local compactness" which couples the Hamiltonian to the space-structure: $F(|x| < R)\ (H+i)^{-1}$ is compact for any R.

Under the same condition scattering states observe local decay, i.e.

$$\lim_{T\to\infty} \frac{1}{T} \int_0^T dt\ \|F(|x| < R) \exp(-iHt)\Psi\| = 0$$

for any R. The long time average can be replaced by the time limit, if the Hamiltonian does not have a singular continuous spectrum as is usually the case. This does not distinguish between resonances and ordinary scattering states. We expect that a suitable refinement will involve the free time evolution for comparison:

Give a geometric definition of resonances in terms of their space-time behaviour and study its relation to traditional definitions.

A related problem is the study of exponential decay in time which has been observed with high precision for very long times compared to the lifetime of the metastable resonant state, the latter being long compared to intrinsic times of the model. On the other hand, semiboundedness of the Hamiltonian implies that exponential decay cannot hold forever. A hard problem of practical relevance is:

For specific systems give realistic estimates on the deviation from the exponential decay in time.

It is not yet clear whether one should study here a quantity like $(\Psi_0, \exp(-iHt)\Psi_0)$ where Ψ_0 is a square integrable approximation of a Gamow vector, or the probability $\|F(|x| < R)\exp(-iHt)\Psi_0\|$ to find the state after time t in a suitable area of radius R, or whether other quantities are more appropriate for the problem.

We conclude our list of some open problems in quantum dynamics. It is intended to be closely related but complementary to the long list of problems mentioned in the lectures of this conference. We are most indebted to the organizers of the meeting, Sergio Albeverio, Lidia Ferreira, and Ludwig Streit for exposing us to a wide range of resonance phenomena. We had the chance to learn a lot, and found ample opportunities for discussions.

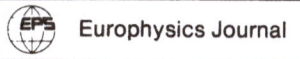

Lecture Notes in Physics

Vol. 173: Stochastic Processes in Quantum Theory and Statistical Physics. Proceedings, 1981. Edited by S. Albeverio, Ph. Combe, and M. Sirugue-Collin. VIII, 337 pages. 1982.

Vol. 174: A. Kadić, D.G.B. Edelen, A Gauge Theory of Dislocations and Disclinations. VII, 290 pages. 1983.

Vol. 175: Defect Complexes in Semiconductor Structures. Proceedings, 1982. Edited by J. Giber, F. Beleznay, J. C. Szép, and J. László. VI, 308 pages. 1983.

Vol. 176: Gauge Theory and Gravitation. Proceedings, 1982. Edited by K. Kikkawa, N. Nakanishi, and H. Nariai. X, 316 pages. 1983.

Vol. 177: Application of High Magnetic Fields in Semiconductor Physics. Proceedings, 1982. Edited by G. Landwehr. XII, 552 pages. 1983.

Vol. 178: Detectors in Heavy-Ion Reactions. Proceedings, 1982. Edited by W. von Oertzen. VIII, 258 pages. 1983.

Vol. 179: Dynamical Systems and Chaos. Proceedings, 1982. Edited by L. Garrido. XIV, 298 pages. 1983.

Vol. 180: Group Theoretical Methods in Physics. Proceedings, 1982. Edited by M. Serdaroğlu and E. İnönü. XI, 569 pages. 1983.

Vol. 181: Gauge Theories of the Eighties. Proceedings, 1982. Edited by R. Raitio and J. Lindfors. V, 644 pages. 1983.

Vol. 182: Laser Physics. Proceedings, 1983. Edited by J. D. Harvey and D. F. Walls. V, 263 pages. 1983.

Vol. 183: J. D. Gunton, M. Droz, Introduction to the Theory of Metastable and Unstable States. VI, 140 pages. 1983.

Vol. 184: Stochastic Processes – Formalism and Applications. Proceedings, 1982. Edited by G.S. Agarwal and S. Dattagupta. VI, 324 pages. 1983.

Vol. 185: H.N. Shirer, R. Wells, Mathematical Structure of the Singularities at the Transitions between Steady States in Hydrodynamic Systems. XI, 276 pages. 1983.

Vol. 186: Critical Phenomena. Proceedings, 1982. Edited by F.J.W. Hahne. VII, 353 pages. 1983.

Vol. 187: Density Functional Theory. Edited by J. Keller and J.L. Gázquez. V, 301 pages. 1983.

Vol. 188: A.P. Balachandran, G. Marmo, B.-S. Skagerstam, A. Stern, Gauge Symmetries and Fibre Bundles. IV, 140 pages. 1983.

Vol. 189: Nonlinear Phenomena. Proceedings, 1982. Edited by K. B. Wolf. XII, 453 pages. 1983.

Vol. 190: K. Kraus, States, Effects, and Operations. Edited by A. Böhm, J.W. Dollard and W.H. Wootters. IX, 151 pages. 1983.

Vol. 191: Photon Photon Collisions. Proceedings, 1983. Edited by Ch. Berger. V, 417 pages. 1983.

Vol. 192: Heidelberg Colloquium on Spin Glasses. Proceedings, 1983. Edited by J. L. van Hemmen and I. Morgenstern. VII, 356 pages. 1983.

Vol. 193: Cool Stars, Stellar Systems, and the Sun. Proceedings, 1983. Edited by S.L. Balliunas and L. Hartmann. VII, 364 pages. 1984.

Vol. 194: P. Pascual, R. Tarrach, QCD: Renormalization for the Practitioner. V, 277 pages. 1984.

Vol. 195: Trends and Applications of Pure Mathematics to Mechanics. Proceedings, 1983. Edited by P.G. Ciarlet and M. Roseau. V, 422 pages. 1984.

Vol. 196: WOPPLOT 83. Parallel Processing: Logic, Organization and Technology. Proceedings, 1983. Edited by J. Becker and I. Eisele. V, 189 pages. 1984.

Vol. 197: Quarks and Nuclear Structure. Proceedings, 1983. Edited by K. Bleuler. VIII, 414 pages. 1984.

Vol. 198: Recent Progress in Many-Body Theories. Proceedings, 1983. Edited by H. Kümmel and M. L. Ristig. IX, 422 pages. 1984.

Vol. 199: Recent Developments in Nonequilibrium Thermodynamics. Proceedings, 1983. Edited by J. Casas-Vázquez, D. Jou and G. Lebon. XIII, 485 pages. 1984.

Vol. 200: H.D. Zeh, Die Physik der Zeitrichtung. V, 86 Seiten. 1984.

Vol. 201: Group Theoretical Methods in Physics. Proceedings, 1983. Edited by G. Denardo, G. Ghirardi and T. Weber. XXXVII, 518 pages. 1984.

Vol. 202: Asymptotic Behavior of Mass and Spacetime Geometry. Proceedings, 1983. Edited by F. J. Flaherty. VI, 213 pages. 1984.

Vol. 203: C. Marchioro, M. Pulvirenti, Vortex Methods in Two-Dimensional Fluid Dynamics. III, 137 pages. 1984.

Vol. 204: Y. Waseda, Novel Application of Anomalous (Resonance) X-Ray Scattering for Structural Characterization of Disordered Materials. VI, 183 pages. 1984.

Vol. 205: Solutions of Einstein's Equations: Techniques and Results. Proceedings, 1983. Edited by C. Hoenselaers and W. Dietz. VI, 439 pages. 1984.

Vol. 206: Static Critical Phenomena in Inhomogeneous Systems. Edited by A. Pękalski and J. Sznajd. Proceedings, 1984. VIII, 358 pages. 1984.

Vol. 207: S. W. Koch, Dynamics of First-Order Phase Transitions in Equilibrium and Nonequilibrium Systems. III, 148 pages. 1984.

Vol. 208: Supersymmetry and Supergravity/Nonperturbative QCD. Proceedings, 1984. Edited by P. Roy and V. Singh. V, 389 pages. 1984.

Vol. 209: Mathematical and Computational Methods in Nuclear Physics. Proceedings, 1983. Edited by J. S. Dehesa, J.M.G. Gomez and A. Polls. V, 276 pages. 1984.

Vol. 210: Cellular Structures in Instabilities. Proceedings, 1983. Edited by J. E. Wesfreid and S. Zaleski. VI, 389 pages. 1984.

Vol. 211: Resonances – Models and Phenomena. Proceedings, 1984. Edited by S. Albeverio, L. S. Ferreira and L. Streit. VI, 369 pages. 1984.

Selected Issues from

Lecture Notes in Mathematics

Vol. 909: Numerical Analysis. Proceedings, 1981. Edited by J.P. Hennart. VII, 247 pages. 1982.

Vol. 912: Numerical Analysis. Proceedings, 1981. Edited by G. A. Watson. XIII, 245 pages. 1982.

Vol. 920: Séminaire de Probabilités XVI, 1980/81. Proceedings. Edité par J. Azéma et M. Yor. V, 622 pages. 1982.

Vol. 921: Séminaire de Probabilités XVI, 1980-81 Supplément: Géométrie Différentielle Stochastique. Proceedings. Edité par J. Azéma et M. Yor. III, 285 pages. 1982.

Vol. 922: B. Dacorogna, Weak Continuity and Weak Lower Semicontinuity of Non-Linear Functionals. V, 120 pages. 1982.

Vol. 923: Functional Analysis in Markov Processes. Proceedings, 1981. Edited by M. Fukushima. V, 307 pages. 1982.

Vol. 926: Geometric Techniques in Gauge Theories. Proceedings, 1981. Edited by R. Martini and E.M.de Jager. IX 219 pages. 1982.

Vol. 927: Y. Z. Flicker, The Trace Formula and Base Change for GL (3). XII, 204 pages. 1982.

Vol. 928: Probability Measures on Groups. Proceedings 1981. Edited by H. Heyer. X, 477 pages. 1982.

Vol. 929: Ecole d'Eté de Probabilités de Saint-Flour X - 1980. Proceedings, 1980. Edited by P.L. Hennequin. X, 313 pages. 1982.

Vol. 930: P. Berthelot, L. Breen, et W. Messing, Théorie de Dieudonné Cristalline II. XI, 261 pages. 1982.

Vol. 931: D.M. Arnold, Finite Rank Torsion Free Abelian Groups and Rings. VII, 191 pages. 1982.

Vol. 932: Analytic Theory of Continued Fractions. Proceedings, 1981. Edited by W.B. Jones, W.J. Thron, and H. Waadeland. VI, 240 pages. 1982.

Vol. 934: M. Sakai, Quadrature Domains. IV, 133 pages. 1982.

Vol. 935: R. Sot, Simple Morphisms in Algebraic Geometry. IV, 146 pages. 1982.

Vol. 936: S.M. Khaleelulla, Counterexamples in Topological Vector Spaces. XXI, 179 pages. 1982.

Vol. 937: E. Combet, Intégrales Exponentielles. VIII, 114 pages. 1982.

Vol. 938: Number Theory. Proceedings, 1981. Edited by K. Alladi. IX, 177 pages. 1982.

Vol. 942: Theory and Applications of Singular Perturbations. Proceedings, 1981. Edited by W. Eckhaus and E.M. de Jager. V, 363 pages. 1982.

Vol. 953: Iterative Solution of Nonlinear Systems of Equations. Proceedings, 1982. Edited by R. Ansorge, Th. Meis, and W. Törnig. VII, 202 pages. 1982.

Vol. 956: Group Actions and Vector Fields. Proceedings, 1981. Edited by J.B. Carrell. V, 144 pages. 1982.

Vol. 957: Differential Equations. Proceedings, 1981. Edited by D.G. de Figueiredo. VIII, 301 pages. 1982.

Vol. 963: R. Nottrot, Optimal Processes on Manifolds. VI, 124 pages. 1982.

Vol. 964: Ordinary and Partial Differential Equations. Proceedings, 1982. Edited by W.N. Everitt and B.D. Sleeman. XVIII, 726 pages. 1982.

Vol. 968: Numerical Integration of Differential Equations and Large Linear Systems. Proceedings, 1980. Edited by J. Hinze. VI, 412 pages. 1982.

Vol. 970: Twistor Geometry and Non-Linear Systems. Proceedings, 1980. Edited by H.-D. Doebner and T.D. Palev. V, 216 pages. 1982.

Vol. 972: Nonlinear Filtering and Stochastic Control. Proceedings, 1981. Edited by S.K. Mitter and A. Moro. VIII, 297 pages. 1983.

Vol. 978: J. Ławrynowicz, J. Krzyż, Quasiconformal Mappings in the Plane. VI, 177 pages. 1983.

Vol. 979: Mathematical Theories of Optimization. Proceedings, 1981. Edited by J.P. Cecconi and T. Zolezzi. V, 268 pages. 1983.

Vol. 982: Stability Problems for Stochastic Models. Proceedings, 1982. Edited by V. V. Kalashnikov and V. M. Zolotarev. XVII, 295 pages. 1983.

Vol. 989: A.B. Mingarelli, Volterra-Stieltjes Integral Equations and Generalized Ordinary Differential Expressions. XIV, 318 pages. 1983.

Vol. 994: J.-L. Journé, Calderón-Zygmund Operators, Pseudo-Differential Operators and the Cauchy Integral of Calderón. VI, 129 pages. 1983.

Vol. 999: C. Preston, Iterates of Maps on an Interval. VII, 205 pages. 1983.

Vol. 1000: H. Hopf, Differential Geometry in the Large, VII, 184 pages. 1983.

Vol. 1003: J. Schmets, Spaces of Vector-Valued Continuous Functions. VI, 117 pages. 1983.

Vol. 1005: Numerical Methods. Proceedings, 1982. Edited by V. Pereyra and A. Reinoza. V, 296 pages. 1983.

Vol. 1007: Geometric Dynamics. Proceedings, 1981. Edited by J. Palis Jr. IX, 827 pages. 1983.

Vol. 1015: Equations différentielles et systèmes de Pfaff dans le champ complexe - II. Seminar. Edited by R. Gérard et J.P. Ramis. V, 411 pages. 1983.

Vol. 1021: Probability Theory and Mathematical Statistics. Proceedings, 1982. Edited by K. Itô and J.V. Prokhorov. VIII, 747 pages. 1983.

Vol. 1031: Dynamics and Processes. Proceedings, 1981. Edited by Ph. Blanchard and L. Streit. IX, 213 pages. 1983.

Vol. 1032: Ordinary Differential Equations and Operators. Proceedings, 1982. Edited by W. N. Everitt and R. T. Lewis. XV, 521 pages. 1983.

Vol. 1035: The Mathematics and Physics of Disordered Media. Proceedings, 1983. Edited by B.D. Hughes and B.W. Ninham. VII, 432 pages. 1983.

Vol. 1037: Non-linear Partial Differential Operators and Quantization Procedures. Proceedings, 1981. Edited by S.I. Andersson and H.-D. Doebner. VII, 334 pages. 1983.

Vol. 1041: Lie Group Representations II. Proceedings 1982-1983. Edited by R. Herb, S. Kudla, R. Lipsman and J. Rosenberg. IX, 340 pages. 1984.

Vol. 1045: Differential Geometry. Proceedings, 1982. Edited by A.M. Naveira. VIII, 194 pages. 1984.

Vol. 1047: Fluid Dynamics. Seminar, 1982. Edited by H. Beirão da Veiga. VII, 193 pages. 1984.

Vol. 1048: Kinetic Theories and the Boltzmann Equation. Seminar, 1981. Edited by C. Cercignani. VII, 248 pages. 1984.

Vol. 1049: B. Iochum, Cônes autopolaires et algèbres de Jordan. VI, 247 pages. 1984.

Vol. 1054: V. Thomée, Galerkin Finite Element Methods for Parabolic Problems. VII, 237 pages. 1984.

Vol. 1055: Quantum Probability and Applications to the Quantum Theory of Irreversible Processes. Proceedings, 1982. Edited by L. Accardi, A. Frigerio and V. Gorini. VI, 411 pages. 1984.